ADVANCES IN ORGANOMETALLIC CHEMISTRY RESEARCH

ADVANCES IN ORGANOMETALLIC CHEMISTRY RESEARCH

KENJI YAMAMOTO
EDITOR

Nova Science Publishers, Inc.
New York

For permission to use material from this book please contact us:
Telephone 631-231-7269; Fax 631-231-8175
Web Site: http://www.novapublishers.com

LIBRARY OF CONGRESS CATALOGING-IN-PUBLICATION DATA

Advances in organometallic chemistry research / Kenji Yamamoto, editor.
p. cm.
Includes index.
ISBN-13: 978-1-60021-779-1 (hardcover)
ISBN-10: 1-60021-779-6 (hardcover)
1. Organometallic chemistry--Research. I. Yamamoto, Kenji.
QD411.A355 2006
547'.05--dc22

2007021191

Published by Nova Science Publishers, Inc. ✦ New York

CONTENTS

PREFACE

This new and important book presents recent advances in organometallic chemistry which is based on the reactions and use of a class of compounds (R-M) that contain a covalent bond between carbon and metal. They are prepared either by direct reaction of the metal with an organic compound or by replacement of a metal from another organometallic substance. Research in organometallic chemistry is also conducted in the areas of cluster synthesis, main-group derivatives in unusual oxidation states, organometallic polymers, unstable organometallic compounds and intermediates in matrices, structure determination of organometallic compounds in the solid state [X-ray diffraction] and gaseous states [electron diffraction], and mechanisms of reactions of transient silylenes and related species. In addition to the traditional metals and semimetals, elements such as selenium, lithium and magnesium are considered to form organometallic compounds, e.g. organomagnesium compounds MeMgI, iodo(methyl)magnesium and diethylmagnesium which are Grignard reagents an organo-lithium compound BuLi butyllithium; Organometallic compounds often find practical use as catalysts, the processing of petroleum products and the production of organic polymers.

Chapter 1 - It has been recently reported that, starting from butadiene, the system $CoBr_2(PPh_3)_2$ / MAO (MAO = methylaluminoxane) gives a polymer with mixed *cis*-1,4/1,2 structure, while it is known that the system $CoCl_2$ / MAO gives a highly *cis*-1,4-polybutadiene, meaning that the ligand may influence the polymerization stereospecificity.

To check this indication, we have synthesized and characterized (by IR, RX and elemental analyses) several cobalt complexes by reacting $CoCl_2$ with various types of phosphines (*e.g.* PR_3, $PRPh_2$, PR_2Ph, $R_2P(X)PR_2$; R = alkyl, cycloalkyl or phenyl group; X = (CH_2), $(CH_2)_2$, $(CH_2)_3$). All the obtained complexes have been used in combination with methylaluminoxane (MAO) for the polymerization of 1,3-butadiene. Polymers having different structure (*cis*-1,4, linear or branched; mixed structure *cis*-1,4/1,2; 1,2-syndiotactic; 1,2-isotactic; 1,2-atactic) were obtained, depending on the phosphine ligand, thus confirming the above indication. This is reminiscent of the revolution that occurred in the '80s in the synthesis of stereoregular polyolefins with the introduction of metallocene based catalysts which opened new routes to both traditional and novel polymers.

The results were interpreted on the basis of the previously reported diene polymerization mechanism, by recognizing that the type of phosphorus ligand (*i.e.* its steric hindrance) may influence the reciprocal orientation of the butadienyl group and of the incoming monomer and thus may determine 1) the formation of 1,4 or 1,2 units (regioselectivity), and 2) the formation of iso- or syndiotactic sequences (stereoselectivity).

Currently, the industrial cobalt catalysts used for the production of *cis*-1,4 polybutadiene (*e.g.* $Co(acac)_3$ / $AlEt_2Cl$ / H_2O) and 1,2 syndiotactic polybutadiene (*e.g.* $Co(acac)_3$ / $AlEt_3$ / CS_2) only work in aromatic solvents, while the phosphine cobalt catalysts in this research may be used also in aliphatic solvents (*e.g.* heptane), giving "green" opportunities for novel, aromatic-free polymerization processes. The phosphine cobalt catalysts in our research were also able to polymerize substituted butadienes such as isoprene, 1,3-pentadienes, and 1,3-hexadienes, allowing us to show that the monomer structure also plays a relevant role in determining polymerization regio- and stereoselectivity.

Chapter 2 - Over the last two decades the interest in ruthenium-catalyzed reactions directed to organic synthesis has spectacularly increased and a large number of new highly efficient synthetic approaches are nowadays well documented. On the other hand, the development of organic transformations in aqueous media has become one of the major cornerstones in modern chemistry. Water as a solvent is not only inexpensive and environmentally benign, but also gives, in some cases, a completely new reactivity. In this review, our focus is on ruthenium-catalyzed organic synthesis in aqueous media. In particular, we report on the synthesis, reactivity and catalytic applications of half-sandwich ruthenium(II) complexes (*i.e.* arene- and cyclopentadienyl-type derivatives) containing water-soluble phosphine ligands.

Chapter 3 - An exciting new mechanism of hydrosilylation had been found in experiments with the ruthenium-silylene complex $[Cp^*(i\text{-}Pr_3P)Ru(H)_2Si(H)Ph\cdot OEt_2][B(C_6F_5)_4]$ by Glaser and Tilley. The proposed catalytic active complex shows surprising bond features. The central unit "$Ru(H)_2Si$" should rather be considered as a ruthenium-η^3-silane and not as a ruthenium-silylene. The ruthenium atom in the catalytic active complex has enough coordinative flexibility to accept the hydrogen atoms in order to activate the silicon atom and to stabilize adducts and transition states. The applicability of ruthenium-η^3-element hydride complexes $[Cp(R_3P)Ru\text{-}\eta^3\text{-}EH_4]^{0/+1}$ (E = B, Al, C, Si, Ge) for the insertion of alkenes into E-H bonds is discussed. These insertions should be possible with η^3-complexes of Si, Ge, B, and Al. With B and Al complexes it will probably not work as catalytic reaction, but stoichiometric transformations should be possible.

Chapter 4 - The search for catalysts that promote or accelerate chemical processes is of utmost relevance for the syntheses of organic species otherwise difficult or even impossible to obtain. Higher activities, lower costs and cleaner processes are the actual driving forces for the development of new catalysts that wider the scope of compounds or improve methods to obtain the existing ones.

Within this purpose in the last decades several transition metals were studied as catalysts for promotion of ring closure due to the relevance of those processes for the syntheses of natural products or biological active species.

The formation of aromatic or heterocyclic systems from linear precursors such as alkynes[1,2] continues to be a challenge to fulfil more economical and environmental friendly synthetic processes.

Chapter 5 - The chemistry of organoimido derivatives of polyoxometalates (POMs), an innovative emerging organic chemistry of polyoxometalates, has received increasing interest in the supramolecular chemistry and chemistry of materials recently. Organoimido derivatives of POMs are valuable building blocks for constructing novel nanostructured organic-inorganic hybrid molecular materials with the so-called "value-adding properties" and possible synergistic effects, which could result in unique catalytic, photo and electronic properties dramatically different from the corresponding parent materials, since the π electrons in the organic component may extend their conjugation to the inorganic framework and dramatically modify the electronic structure and redox properties of the corresponding parent POMs. The surface modification of POMs, that is, the substitution of one or more terminal or bridged oxo groups with organoimido ligand, is the fundamental method to synthesize organoimido derivatives of POMs via common organic chemistry reactions. In this chapter of the book, the synthetic chemistry of organoimido derivatives of POMs using different organic reagents will be reviewed, of which the novel DCC-dehydrating protocol to prepare organoimido derivatives of POMs recently developed by us and co-authors will be concentrated. As it can be expected, the reaction chemistry of organoimido derivatives of POMs stands for the fascinating future of the chemistry of organoimido derivatives of POMs since it opens not only a new road to the chemical modification of POMs, but also an exciting research arena where a variety of hybrid materials containing covalently bonded POM clusters and organic conjugated segments can be prepared in a more controllable and rational manner. Hence, in the rest of this chapter, Pd-catalyzed coupling reactions of the functionalized organoimido derivatives of POMs will be introduced in detail for this purpose. Especially, different rationally assembled hybrid molecules and polymers where the organic molecules and inorganic clusters are bonded together by an imido nitrogen atom will then be highlighted.

Chapter 6 - The amazing structural diversity in organotin compounds is discussed in the systems containing Sn-O and Sn-S donor ligands. It is demonstrated that there exist a fascinating range of structural diversity for organotin(IV) complexes, including difference in coordination number and molecular geometry. Difference in the structure is correlated with the nature of tin and ligand bonded R groups. Despite the large number of different structures found in organotin(IV) carboxylates, there is limited range of coordination geometries about the Sn atom. Thus, the four coordinated Sn atom in triorganotin(IV) complexes is invariably distorted tetrahedral, five coordinate Sn is distorted trigonal bipyramidal. A large range has been observed for diorganotin carboxylates structures, where five-, six- and seven- coordinate geometries have been reported. The Sn atom in mono-organotin has only been demonstrated to exist in distorted octahedral geometries (the single exception being a pentagonal bipyramidal geometry).

From the discussion of the structural chemistry of organotin(IV) complexes of S donor ligands, it is clear that there exists a rich diversity in Sn atom geometries and coordination modes of the sulfur donor ligands themselves. As in related carboxylates systems, the assignment of coordination numbers to the Sn centers in some compounds is controversial.

Nevertheless, some general trends are discussed with the most obvious being that as the number of organic substituents at the Sn atom increases the overall coordination number at the Sn atom decreases. This phenomenon is usually achieved by increased asymmetry in the mode of coordination of the sulfur donor ligands.

A theoretical explanation has been proposed, based on valence bond theory, which suggests, in effect that as the number of tin-bound organic substituents increases dative resonance forms of the sulfur donor ligands predominate which leads to an increased asymmetry in the mode of coordination of the sulfur donor ligand in accordance with the experimental observations.

Knowledge of the biological chemistry of the tin is needed for the rational development of more effective compounds and newer uses, for example as antitumour agents. This chapter is particularly concerned with the molecular interactions of tin compounds with enzymes and biological membranes and with the metabolic processing of tin compounds. It deals with biochemical activities of organotin compounds leading from complex formation, through effects on individual enzymes and biological membranes to their inhibition of ATP synthesis and finally to their actions on whole cell systems in the nervous system, in, immune suppression and as antitumour agents.

Chapter 7 - Green chemistry is a strategy of changes in chemical processes aimed at ensuring a safer, cleaner environment in the 21st century. In designing reactions following the green chemistry principles, chemists pay special attention to selecting substrates and chemicals that are as harmless to the environment and to human health as possible. One way to achieve this goal is by replacing traditional toxic solvents with ionic liquids. The use of ionic liquids in C–C bond forming reactions like Heck, Sonogashira, Suzuki, and carbonylation offers many spectacular advantages including facilitation of catalyst separation from the organic products. It is important and of great topical interest because many products of C–C coupling reactions are used as pharmaceuticals or agricultural chemicals, which can contain only very limited amounts of metals. This article presents catalytic systems containing palladium catalyst precursors, both soluble and heterogenized complexes as well as palladium nanocolloids, applied in C–C bond forming processes performed in ionic liquids. The applicability of ionic liquids and the effect of their molecular structure on the reaction course are discussed. Special attention is given to the reactions of ionic liquids with palladium precursors leading to the formation of new species and modification of the catalytic properties of the catalytic system. As an example, reactions of imidazolium halides with palladium precursors are discussed in detail. The article, based on literature data and results obtained in our lab, presents the state of knowledge in the field and formulates some expectations for future design of catalytically active and environmentally friendly palladium-based systems for C–C bond forming reactions.

Chapter 8 - The dismutation (degradation) via alkyl group removal of triorganotins has been detected in the environment and has been attributed to the action of UV light, chemical cleavage and biological degradation by bacteria, although its chemistry has never been elucidated. Interestingly, in contrast to thermally induced solid-state methyl migrations in organic molecules, which have been the subject of extensive investigations, the solid-state methyl transfer phenomenon in triorganotin(IV) compounds have been hardly exploited. Recently, we have shown that the single crystals of Me_3Sn(O,O-donor) complexes with bidentate-chelating O,O-donor ligands such as *N*-nitroso-*N*-phenyl-hydroxylaminato and tropolonato, demethylated into single crystals of $Me_2Sn(O,O\text{-donor})_2$ and volatile Me_4Sn, as a result of a crystal-to-crystal supramolecular structural transformation. In solution, the solvent-induced changes in the tin coordination number followed by specific conformational changes were held responsible for the second order dismutation kinetics, in which the role of the monomeric triorganotin(IV)-solvates could be verified by various temperature- and concentration dependent NMR experiments. We have concluded that in most of the cases, this methyl-migrational dismutation process can be a potential source of the instability of trimethyltin(IV)-complexes, which prevents the preparation of devised structures via crystallization from coordinating solvents.

During the last few decades a large number of triorganotin(IV)-carboxylates have been prepared and a multitude of structural types have been discovered. These complexes have a tendency to self-assemble by dative Sn–O bonds (2.2–2.3 Å) and/or Sn···O secondary interactions (> 2.5 Å) into supramolecular structures. Moreover, the triorganotin(IV) carboxylates have been one of the most extensively studied class of anticancer compounds since it was observed that they significantly reduced the growth rates of tumors. Although, a series of triorganotin(IV)-benzoates has been reported, the trimethyltin(IV)-benzoate, $Me_3Sn(bz)$, remained uncharacterized. This discrepancy between the extremely studied nature of triorganotin(IV)-benzoates and the absence of trimethyltin(IV)-derivative suggests that this might be due to its dismutational reactivity which prevents its isolation. Herein, we report the successful synthesis of $Me_3Sn(bz)$ complex, as well as, its X-ray structural determination and NMR analyses both in solution- and solid-state. The data obtained for $Me_3Sn(bz)$ revealed that the essentially planar triorganotin moieties are bridged by anisobidentate carboxylato groups forming one short Sn–O bond and one relatively short Sn···O secondary interaction. In the resulting polymeric chain the difference between these Sn–O and Sn···O bond lengths is quite large (0.429 Å), therefore the helix is stretched in *S*-configuration and the Sn atoms are no longer collinear along the chain. We also report the first thermally accelerated crystal-to-crystal supramolecular structural transformation of a $Me_3Sn(bz)$ polymer into self-assembled dimeric dibenzoato-tetraorganodistannoxane, $\{[Me_2Sn(bz)]_2O\}_2$. The parallels and deviations in the mechanism of the various dismutation processes of the Me_3Sn(O,O-donor) complexes will be highlighted.

Chapter 9 - The incorporation of the metals into the structure of synthetic polymers offers considerable potential for the preparation of processable materials with properties that significantly differ from those of conventional organic polymers. The presence of metal centers (especially those of transition elements) that, as a consequence of their electronic structure, generally exist in a variety of oxidation states, has been shown to facilitate access to a diverse range of interesting and useful properties. Therefore, the polymer-metal complexes are now a growing research field.

Concurrent with the development in this area, there has also been increased interest in the chemistry of metal complexes containing Si-O bonds. Polysiloxanes are involved either directly with siloxane bond (i.e. metal siloxides containing Si-O-M linkages) or through the attached organic chelating functions in 2D or 3D metal-containing structures. Depending on the used ligands or synthesis pathway, polymers with metals either in the main chain or in the side position can be obtained. Precomplexed metals (e.g., metallocenes) as functionalized structures can be used as monomers in polycondensation reaction with siloxane macromers. Also, polysiloxanes in crosslinking state can serve as matrix for encapsulating of metals nanoparticles or metal ions complexed by a proper chelating groups immobilized on the porous solid siloxane polymers. As a result of the incorporation of the metal atoms, such materials could exhibit attractive electrical, electrochemical, magnetic, optical and catalytic properties. In addition, these materials would also be of interest as pyrolytic precursors to transition metal containing ceramic products.

This chapter is an overview of the recent reports on the silicones involving in some complex structures containing metals (siloxides, coordination polymers, crosslinked structures). Own contributions are also enclosed.

In: Advances in Organometallic Chemistry Research
Editor: Kenji Yamamoto, pp. 1-36
ISBN: 978-1-60021-779-1

Chapter 1

COBALT PHOSPHINE COMPLEXES FOR THE POLYMERIZATION OF 1, 3-DIENES. INFLUENCE OF THE PHOSPHINE LIGAND ON THE POLYMERIZATION CHEMO- AND STEREOSELECTIVITY

Giovanni Ricci [1], Francesco Masi [2], Aldo Boglia [1], Anna Sommazzi [3] and Marco Ricci [3]

[1] CNR - Istituto per lo Studio delle Macromolecole,
via E. Bassini 15, 20133 Milano, Italy

[2] Polimeri Europa
Piazza Boldrini 1, 20097 San Donato Milanese, Italy

[3] Eni S.p.A., Divisione Refining and Marketing, Centro Ricerche Novara
"Istituto Guido Donegani", via G. Fauser 4, 28100 Novara, Italy

ABSTRACT

It has been recently reported that, starting from butadiene, the system $CoBr_2(PPh_3)_2$ / MAO (MAO = methylaluminoxane) gives a polymer with mixed *cis*-1,4/1,2 structure, while it is known that the system $CoCl_2$ / MAO gives a highly *cis*-1,4-polybutadiene, meaning that the ligand may influence the polymerization stereospecificity.

To check this indication, we have synthesized and characterized (by IR, RX and elemental analyses) several cobalt complexes by reacting $CoCl_2$ with various types of phosphines (*e.g.* PR_3, $PRPh_2$, PR_2Ph, $R_2P(X)PR_2$; R = alkyl, cycloalkyl or phenyl group; X = (CH_2), $(CH_2)_2$, $(CH_2)_3$). All the obtained complexes have been used in combination with methylaluminoxane (MAO) for the polymerization of 1,3-butadiene. Polymers having different structure (*cis*-1,4, linear or branched; mixed structure *cis*-1,4/1,2; 1,2-syndiotactic; 1,2-isotactic; 1,2-atactic) were obtained, depending on the phosphine ligand, thus confirming the above indication. This is reminiscent of the revolution that occurred in the '80s in the synthesis of stereoregular polyolefins with the introduction of metallocene based catalysts which opened new routes to both traditional and novel polymers.

The results were interpreted on the basis of the previously reported diene polymerization mechanism, by recognizing that the type of phosphorus ligand (*i.e.* its steric hindrance) may influence the reciprocal orientation of the butadienyl group and of the incoming monomer and thus may determine 1) the formation of 1,4 or 1,2 units (regioselectivity), and 2) the formation of iso- or syndiotactic sequences (stereoselectivity).

Currently, the industrial cobalt catalysts used for the production of *cis*-1,4 polybutadiene (*e.g.* $Co(acac)_3$ / $AlEt_2Cl$ / H_2O) and 1,2 syndiotactic polybutadiene (*e.g.* $Co(acac)_3$ / $AlEt_3$ / CS_2) only work in aromatic solvents, while the phosphine cobalt catalysts in this research may be used also in aliphatic solvents (*e.g.* heptane), giving "green" opportunities for novel, aromatic-free polymerization processes. The phosphine cobalt catalysts in our research were also able to polymerize substituted butadienes such as isoprene, 1,3-pentadienes, and 1,3-hexadienes, allowing us to show that the monomer structure also plays a relevant role in determining polymerization regio- and stereoselectivity.

INTRODUCTION

Cobalt catalysts play a very important role in the field of butadiene polymerization since, depending on the reaction conditions and catalyst formulation, they can produce polymers with different microstructure, including *cis*-1,4-polybutadiene and 1,2-syndiotactic polybutadiene (figure 1), which are the only two polybutadienes produced on an industrial scale [1-5].

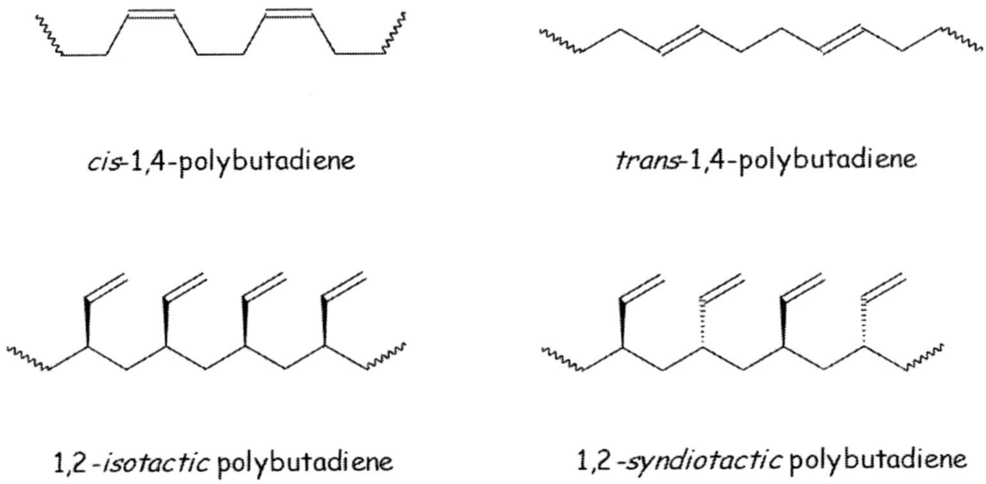

Figure 1. Possible stereoregular polybutadienes.

Thus, cobalt is commonly used, along with other transition metals (including titanium, nickel, and the rare-earth element neodymium) [6-10] to produce *cis*-1,4-polybutadiene, an elastomer with melting point (T_m) of 2 °C and a glass transition temperature (T_g) of -106 °C

[6], which is widely used in tires and as an ingredient in the formulation of high impact polystyrene, HIPS.

Several cobalt compounds can be used as catalytic precursors to polymerize butadiene to *cis*-1,4-polybutadiene with content of *cis* units up to 97-98%, provided that the right co-catalyst (*e.g.* $AlEt_2Cl$ / H_2O) is used [9]. Equibinary *cis*-1,4 / 1,2 polybutadiene is instead obtained with the system $Co(acac)_3$ / $AlEt_3$ / H_2O [10].

Either soluble or insoluble cobalt compounds can be chosen, without dramatically affecting the stereoregularity of the resulting polymer. With soluble cobalt precursors, however, much lower catalyst amounts are required. Commonly employed cobalt compounds are: *i)* acetylacetonate complexes of Co(II) or Co(III) [7,8,11-18]; *ii)* $CoCl_2$ in combination with ethers, pyridine, or other amines [7,19-21]; *iii)* poorly characterized cobalt salts of long chain organic acids, *e.g.* octanoates or versatates [7, 22-29]. The latter precursors have the advantage of significant solubility even in aliphatic solvents.

As co-catalysts, $AlEt_2Cl$, $Al_2Et_3Cl_3$, or $AlEtCl_2$, are usually employed, as well, as their mixtures [30-37]. When $AlEt_2Cl$ is used, some water is needed in order to obtain high catalytic activity (*i.e.* high molecular weights) and high stereoregularity [38,39]. It is generally believed that water reacts with $AlEt_2Cl$ giving chloroaluminoxanes such as $[Et(Cl)Al]_2O$ [9]. If the aluminum derivative has a Cl/alkyl ratio ≥ 1 (*e.g.*, $AlEtCl_2$ or $Al_2Et_3Cl_3$), water is not required [30, 40-42].

The synthesis of *cis*-1,4-polybutadiene with cobalt catalysts is usually performed in benzene or toluene as solvents, at 5-50 °C. Even highly insoluble cobalt precursors (*e.g.* $CoCl_2$) dissolve upon reaction with aluminum alkylchlorides to form clear solutions of active catalysts [43-45].

Cobalt catalysts are commonly used to prepare *cis*-1,4-polybutadiene by several companies, including: Bayer (Germany), UBE Industries (Japan), and BF Goodrich (USA) [46-52].

Cobalt catalysts, however, are also used to prepare 1,2 syndiotactic polybutadiene which, contrary to amorphous *cis*-1,4-polybutadiene, is a crystalline polymer with T_m up to 200-220 °C and a T_g in the range -15/-30°C (melting temperature and glass transition temperature actually depend on the grade of crystallinity and stereoregularity) [6]. 1,2 Syndiotactic polybutadiene is a peculiar polymer material which exhibits various physical properties similar to both plastics and rubber and, therefore, has many uses: for example, it can be blended with various elastomers in the manufacture of tires, shoe soles, sport equipment, etc. [4-6]. The catalysts generally used are obtained by combining a cobalt precursor, such as Co(III) acetylacetonate and a trialkylaluminum (*e.g.* $AlEt_3$) in the presence of carbon disulfide (CS_2) as third component [53-62]. This process is performed by UBE in benzene as solvent, at low temperature, typically 5 °C. Both the regio- (*i.e.* the amount of 1,2 units) and the stereo-regularity of the product are very high (>99%). In the absence of CS_2, only oligomers are obtained [63]. An extremely active and stereospecific catalyst for the preparation of 1,2 syndiotactic polybutadiene is also obtained by combining the allyl cobalt complex $(C_4H_6)(\eta^3$-$C_8H_{13})Co$ [64], prepared several years ago by Giannini, with CS_2 [53-58]. The complex alone is able to dimerize butadiene at room temperature [65], but it gives 1,2 syndiotactic polymers from butadiene at low temperature (-30°C) [57].

Interestingly, a commercially valuable 1,2 syndiotactic polybutadiene is also obtained by using as a catalyst system $CoCl_2$ (or $CoBr_2$) in combination with trialkylaluminum, water and a phosphine (*i.e.* triphenylphosphine) [66-74]. The regioregularity and the crystallinity of the polymers obtained with this system are much lower (80-90% of 1,2 units, T_m 75–90°C) with respect to those of the polymers obtained with the Co(III) acetylacetonate / trialkylaluminum / CS_2 catalyst. Nevertheless, these products find successful application in the shoe soles manufacture and are commercially exploited by JSR Corporation (formerly, Japan Synthetic Rubber) [1,70-74].

An interesting aspect of this catalyst system is that the phosphine seems to have some influence on the catalyst stereospecificity [75-77].

This fact prompted us to systematically explore the output of the butadiene polymerization when different phosphines are used as ligands for cobalt; it is a well known fact that steric and electronic properties of phosphines strongly depend on the type of substituents on the phosphorus atom [75-78].

It turns out that the use of defined cobalt complexes with commercial phosphines in combination with methylaluminoxane (MAO), allows us to manage polybutadiene microstructure, in a similar manner to the polymerization of α-olefins catalyzed by titanium or zirconium metallocenes. Moreover, these complexes work very well at room temperature, in aliphatic solvents, and are very active with rather low Al/Co molar ratio.

Furthermore, these catalysts, based on phosphine cobalt complexes, were able to polymerize not only butadiene, but also several others substituted butadienes such as isoprene, pentadienes and hexadienes, giving homo- and copolymers which have different structures depending on the different monomer used. This fact permitted us to better elucidate some aspects of the diene polymerization mechanism previously proposed [7,79-82].

Synthesis and Characterization of Cobalt Phosphine Complexes

We synthesized several cobalt complexes by reacting $CoCl_2$ with various phosphines: monodentate alkylphosphines (triethylphosphine, tri-*normal*-propylphosphine, dicyclohexylphosphine, di-*tert*-butylphosphine, di-*tert*-butylmethylphosphine, tricyclopentylphosphine, tricyclohexylphosphine, tri-*iso*-propylphosphine, tri-*tert*-butylphosphine), monodentate aromatic phosphines (methyldiphenylphosphine, dimethylphenylphosphine, *normal*-propyldiphenylphosphine, ethyldiphenylphosphine, diethylphenylphosphine, allyldiphenylphosphine, diallylphenylphosphine, *iso*-propyldiphenylphosphine, cyclohexyldiphenylphosphine, dicyclohexylphenyphosphine), bidentate phosphines (bis(dimethylphosphino)methane, 1,2-bis(dimethylphosphino)ethane, 1,2-bis(diethylphosphino)ethane, 1,2-bis(diphenylphosphino) ethane, 1,3-bis(diphenylphosphino)propane). We decided to examine such a large number of phosphines since, as cited above, the phosphine electronic and steric properties can be strongly influenced by varying the type of substituents on the phosphorus atom [75-78].

Table 1. Phosphines used for the cobalt complex syntheses

triethylphosphine	PEt_3	ethyldiphenylphosphine	$PEtPh_2$
tri-*normal*-propylphosphine	P^nPr_3	diethylphenylphosphine	PEt_2Ph
dicyclohexylphosphine	PCy_2H	allyldiphenylphosphine	$P(C_3H_5)Ph_2$
di-*tert*-butylphosphine	P^tBu_2H	diallylphenylphosphine	$P(C_3H_5)_2Ph$
di-*tert*-butylmethylphosphine	P^tBu_2Me	*iso*-propyldiphenylphosphine	P^iPrPh_2
tricyclopentylphosphine	$PCyp_3$	cyclohexyldiphenylphosphine	$PCyPh_2$
tricyclohexylphosphine	PCy_3	dicyclohexylphenyphosphine	PCy_2Ph
tri-*iso*-propylphosphine	P^iPr_3	bis(dimethylphosphino)methane (dmpm)	$Me_2P(CH_2)PMe_2$
tri-*tert*-butylphosphine	P^tBu_3	1,2-bis(dimethylphosphino)ethane (dmpe)	$Me_2P(CH_2)_2PMe_2$
methyldiphenylphosphine	$PMePh_2$	1,2-bis(diethylphosphino)ethane (depe)	$Et_2P(CH_2)_2PEt_2$
dimethylphenylphosphine	PMe_2Ph	1,2-bis(diphenylphosphino)ethane (dppe)	$Ph_2P(CH_2)_2PPh_2$
n-propyldiphenylphosphine	P^nPrPh_2	1,3-bis(diphenylphosphino)propane (dppp)	$Ph_2P(CH_2)_3PPh_2$

Synthesis and Characterization of Aliphatic Phosphine Cobalt Complexes

The cobalt phosphine complexes were prepared and isolated by improving a general experimental procedure already described in the literature for this type of compounds [83]. The following phosphines were used: triethylphosphine (PEt_3), tri-*normal*-propylphosphine (P^nPr_3), dicyclohexylphosphine (PCy_2H), di-*tert*-butylphosphine (P^tBu_2H), di-*tert*-butylmethylphosphine (P^tBu_2Me), tricyclopentylphosphine ($PCyp_3$), tricyclohexylphosphine (PCy_3), tri-*iso*-propylphosphine (P^iPr_3) and tri-*tert*-butylphosphine (P^tBu_3).

Anhydrous cobalt dichloride was dissolved in ethanol, then the phosphine (P/Co molar ratio = 2.5÷3) was added; completely homogeneous blue solutions or heterogeneous blue suspensions were obtained depending on the type of phosphine.

In case of homogeneous solution, the solvent was removed under vacuum and a crystalline product was obtained by continuous extraction of the residue with boiling pentane; in case of heterogeneous suspension, the precipitate was filtered off and then again a crystalline product was obtained by continuous extraction of the residue with boiling pentane. A more detailed description of the cobalt complexes synthesis is reported in our previous papers [84,85] and in the Experimental Section of the present paper.

Single crystals suitable for X-ray structure determination were obtained only for the complexes $CoCl_2(PCy_2H)_2$ and $CoCl_2(PCy_3)_2$ by recrystallization from pentane solution at low temperature (-30 °C). In the case of $CoCl_2(PCyp_3)_2$, several attempts of recrystallization (from pentane, toluene or pentane/toluene mixtures) yielded seemingly good prismatic crystals, which however X-ray analysis proved to be twinned. The large disorder probably affecting the cyclopentyl groups prevented in all cases to solve structure for the different twin components. For the other complexes, only microcrystalline compounds have been obtained.

The molecular structures of $CoCl_2(PCy_2H)_2$ and $CoCl_2(PCy_3)_2$ are shown in figure 2, and some selected bond lengths and angles are reported in table 2. More detailed information on the two crystalline structures are available in our previous papers [84,85].

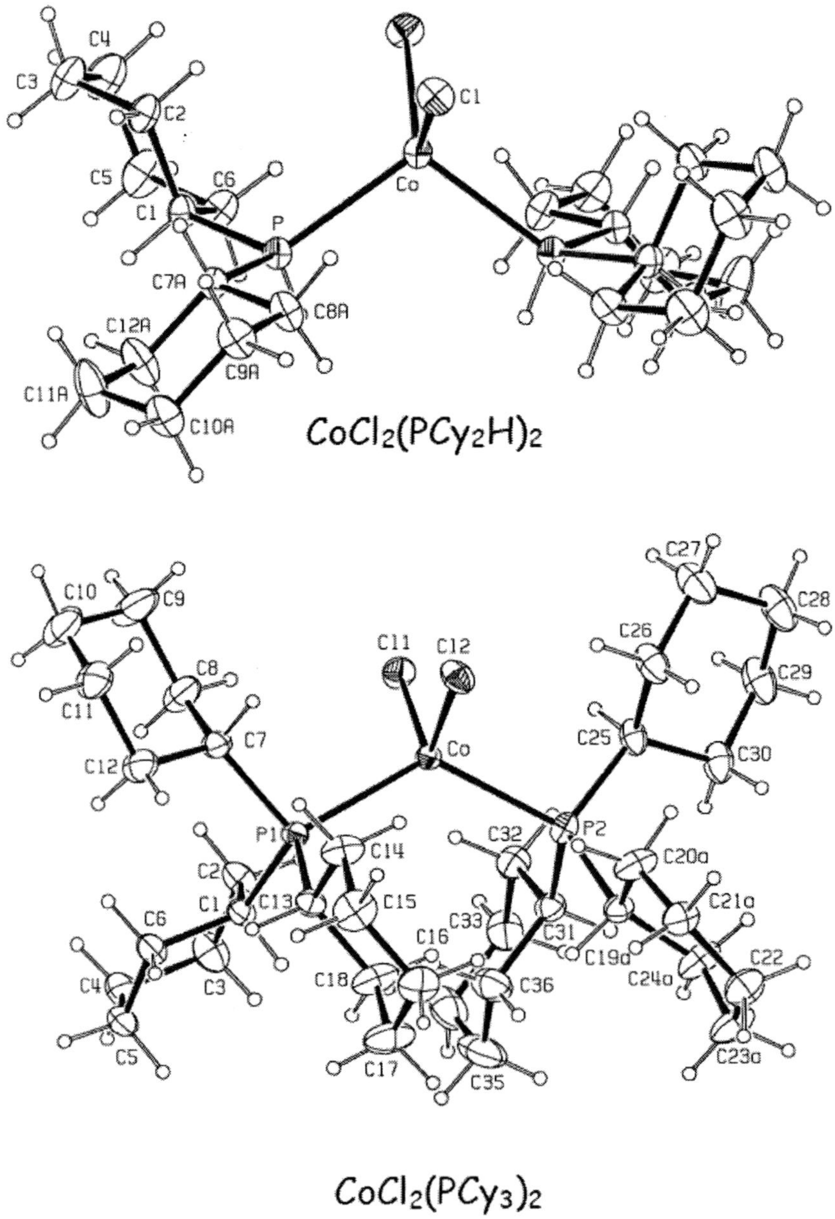

Figure 2. X-Ray crystal structures of $CoCl_2(PCy_3)_2$ and $CoCl_2(PCy_2H)_2$.

Table 2. Selected Bond Lengths (Å) and Angles (deg) for $CoCl_2(PCy_3)_2$ and $CoCl_2(PCy_2H)_2$

	$CoCl_2(PCy_3)_2$	$CoCl_2(PCy_2H)_2$
Co–Cl(1)	2.2319(6)	2.2182(6)
Co–Cl(2)	2.2389(6)	
Co–P(1)	2.4375(6)	2.3942(6)
Co–P(2)	2.4345(6)	
Cl(1)–Co–Cl(2)	112.69(3)	109.81(3)
P(1)–Co–P(2)	125.61(2)	105.87(3)

The effect of the increased dimensions of the phosphine groups on going from $CoCl_2(PCy_2H)_2$ to $CoCl_2(PCy_3)_2$ is clearly visible by comparing corresponding Co–Cl and especially Co–P bond lengths. In both cases they undergo an increase, by 0.0172(6) and 0.0418(6) Å on average, respectively, on going from the former to the latter complex, indicating an electronic rearrangement within the core of the two complexes. The increased sterical demands are also responsible of an enlargement by 19.74(3)° in the P–Co–P bond angle.

Synthesis and Characterization of Aromatic Phosphine Cobalt Complexes

The complexes $CoCl_2(PR_nPh_{3-n})_2$ (R = methyl, ethyl, allyl, *normal*-propyl, *iso*-propyl, cyclohexyl; n = 1,2) were prepared, as occurred for the above aliphatic phosphine complexes, by improving an experimental procedure already reported in the literature [83]. $CoCl_2$ was reacted with an excess of phosphine using ethanol as solvent. After an opportune work up, described in our previous paper [86-90], we succeeded in isolating crystalline products by continuous extraction of the reaction products with boiling pentane or toluene. Single crystals were obtained in the case of $CoCl_2(PEtPh_2)_2$, $CoCl_2(P^nPrPh_2)_2$, $CoCl_2(P^iPrPh_2)_2$ and $CoCl_2[P(allyl)_2Ph]_2$, and their structures solved by X-ray diffraction [86,87,89]. No monocrystals, but microcrystalline powder have been obtained for all the other complexes.

The crystalline molecular structures of $CoCl_2(PEtPh_2)_2$, $CoCl_2(P^nPrPh_2)_2$, $CoCl_2(P^iPrPh_2)_2$ and $CoCl_2[P(allyl)_2Ph]_2$ are shown in figure 3, and some selected bond lengths and angles are reported in table 3. Detailed information on the X-ray structures of the complexes are available in our previous papers [86,87,89].

Table 3. Selected Bond Lengths (Å) and Angles (deg) for $CoCl_2(PEtPh_2)_2$, $CoCl_2(P^nPrPh_2)_2$, and $CoCl_2(P^iPrPh_2)_2$

	$CoCl_2(PEtPh_2)_2$	$CoCl_2(P^nPrPh_2)_2$	$CoCl_2(P^iPrPh_2)_2$
Co(1)–Cl(1)	2.2276(7)	2.2118(9)	2.2276(7)
Co(1)–Cl(2)	2.2201(7)	2.1960(9)	2.2023(8)
Co(1)–P(1)	2.3747(6)	2.3691(8)	2.3708(7)
Co(1)–P(2)	2.3760(6)	2.3640(10)	2.3549(7)
Cl(1)–Co(1)–Cl(2)	110.57(2)	112.60(3)	114.25(3)
P(1)–Co(1)–P(2)	100.63(2)	103.73(3)	110.64(3)

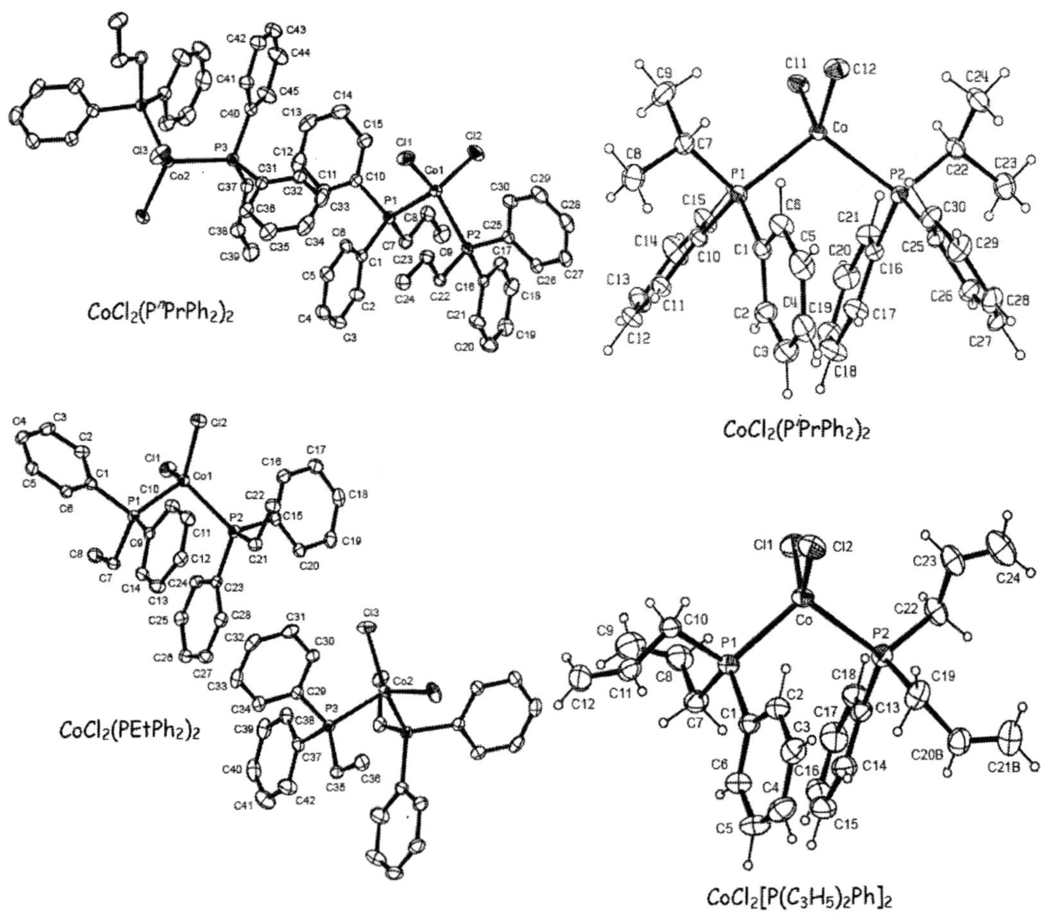

Figure 3. Crystalline X-Ray molecular structures of $CoCl_2(P(C_3H_5)_2Ph)_2$, $CoCl_2(PEtPh_2)_2$, $CoCl_2(P^nPrPh_2)_2$ and $CoCl_2(P^iPrPh_2)_2$.

The complexes show a distorted tetrahedral coordination around the Co(II) centre; the deviations from the tetrahedral coordination are to be ascribed to the mutual repulsion between the chlorine atoms. In all cases, the Cl–Co–Cl angles are found systematically larger than the tetrahedral value, while the opposite trend is observed for the P–Co–P angles.

The crystal structures of $CoCl_2(PEtPh_2)_2$ and $CoCl_2(P^nPrPh_2)_2$ are very similar, as it appears by comparing not only the geometrical parameters, but also the structural data [87]. When comparing these two systems with the $CoCl_2(P^iPrPh_2)_2$ complex, we observe only slight differences, the larger one being a longer P–$C_{alyphatic}$ average distance (1.850 Å) in $CoCl_2(P^iPrPh_2)_2$, which can be ascribed to the larger bulkiness of the isopropyl group with respect to the ethyl and the *n*-propyl groups.

Synthesis and Characterization of Bidentate Phosphine Cobalt Complexes

The following bidentate phosphine were used: bis(dimethylphosphino)methane (dmpm), 1,2-bis(dimethylphosphino)ethane (dmpe), 1,2-bis(diethylphosphino)ethane (depe), 1,2-bis(diphenylphosphino)ethane (dppe) [92], 1,3-bis(diphenylphosphino)propane (dppp). The bidentate phosphine cobalt complexes were prepared according to the following procedure. $CoCl_2$ was suspended in ethanol and then the bidentate phosphine was added. The colour of the suspension turned rapidly from deep blue to green. The mixture was kept under stirring at room temperature for one night, then the solvent was removed under vacuum, and the residue washed with ethanol and pentane and then dried under vacuum. The green products obtained were characterized by elemental analysis and IR spectroscopy; only one phosphine bidentate ligand was found to be coordinated to the cobalt atom, as shown in figure 4.

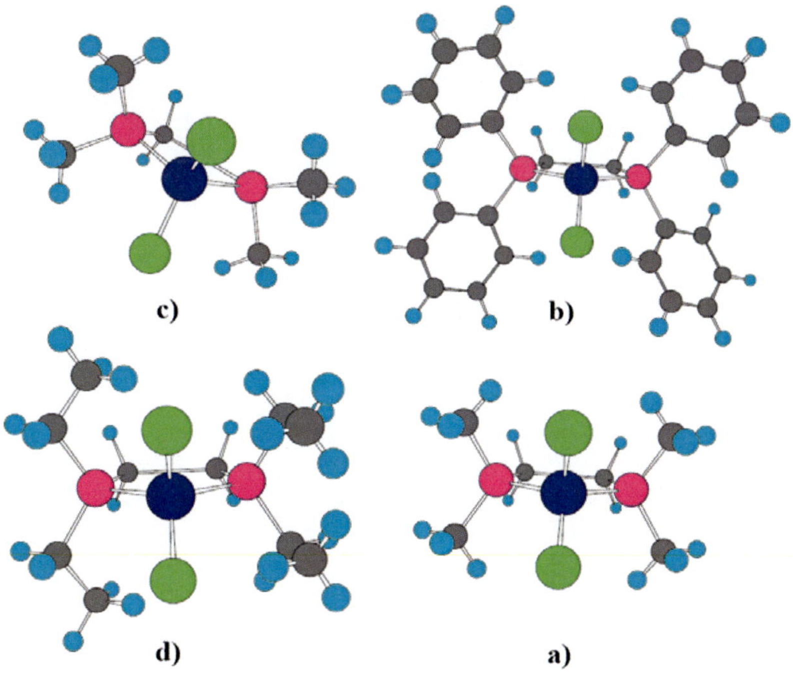

Figure 4. Chem-draw structures of a) $CoCl_2$(dmpe); b) $CoCl_2$(dppe); c) $CoCl_2$(dmpm); d) $CoCl_2$(depe). (cobalt atoms in blue; chlorine atoms in green; carbon atoms in black; hydrogen atoms in azure; phosphorus atoms in magenta)

Detailed information on the synthesis and characterization of the complexes are reported in the Experimental section.

POLYMERIZATION RESULTS

Polymerization of Butadiene with $CoCl_2(PR_3)_2$ / MAO

The results obtained in the polymerization of butadiene with the systems prepared by combining MAO with the various cobalt aliphatic phosphine complexes [84,85] are reported in table 4 and can be summarized as follows.

Table 4. Polymerization of butadiene with $CoCl_2(PR_3)_2$ / MAO[a)]

run	catalyst		polymerization			polymer				
	Co-complex	θ[b)] (deg)	time (min)	conv (%)	N[c)] (min^{-1})	*cis*-1,4[d)] (%)	1,2[d)] (%)	Mw[e)] ($g\times mol^{-1}$)	Mw/Mn[e)]	g[f)]
1	$CoCl_2(PEt_3)_2$	132	30	74.8	129	39.9	60.1			
2	$CoCl_2(P^nPr_3)_2$	132	35	85.5	127	38.2	61.8			
3	$CoCl_2(PCy_2H)_2$	143	30	70.8	122	51.0	49.0	280000	1.9	
4	$CoCl_2(P^tBu_2H)_2$	150	35	60.4	89	59.0	41.0	275000	2.0	
5	$CoCl_2(P^tBu_2Me)_2$	161	40	59	76	94.0	6.0	260000	1.8	0.80
6	$CoCl_2(PCyp_3)_2$	-	39	56.4	85	73.6	26.4	245000	2.0	
7	$CoCl_2(PCy_3)_2$	170	32	44.4	72	77.4	22.6	250000	2.1	
8	$CoCl_2(P^iPr_3)_2$	160	20	57.4	149	94.5	5.5	340000	2.3	0.83
9	$CoCl_2(P^iPr_3)_2$	160	25	50.2	104	92.8	7.2	330000	2.9	0.76
10	$CoCl_2(P^tBu)_3$	182	22	73.0	172	96.0	4.0	360000	2.0	0.82
11	$CoCl_2(P^tBu)_3$	182	25	67.5	140	94.0	6.0	320000	2.6	0.77
12	$CoCl_2$	—	52	68.6	68	95.4	4.6	815000	2.2	1.00

[a)] Polymerization conditions: butadiene, 2 mL; toluene (heptane in runs 9 and 11), 16 mL; MAO, Al/Co = 1000; moles of Co, 5×10^{-6}; temperature, +20°C. [b)] Phosphine cone angle, as reported by Tolman [78]. [c)] N = moles of butadiene polymerized per mol of Co per minute. [d)] Determined by ^1H NMR, as shown in figure 6. *Trans*-1,4 units are almost negligible, as indicated by the fact that in the IR spectra of the polymers no band at 967 cm^{-1} was detected. [e)] Molecular weight and molecular weight distribution determined by GPC distribution. [f)] g is a branching index, calculated by GPC/MALLS analysis (see text).

Predominantly *cis*-1,4 polymers are obtained with catalysts using more hindered ligands (θ ≥ 160°; θ is the phosphine cone angle as reported by Tolman [78]) (table 4, runs 5-11), while polybutadienes with a predominantly 1,2 structure or mixed *cis*-1,4/1,2 structure are obtained in the case of less hindered phosphines (table 4, runs 1-4). It seems indeed that the *cis* content increases with increasing the ligand hindrance; catalysts using cyclic phosphines, PCy_3 and $PCyp_3$, make an exception to this rule, even if they also give a predominantly *cis*-1,4 polybutadiene.

The importance of the ligand in determining the formation of a 1,4 polymer rather than a 1,2 polymer, clearly turns out if we consider the structure of the active site in the polymerization of butadiene with catalysts based on cobalt phosphine complexes (figure 5).

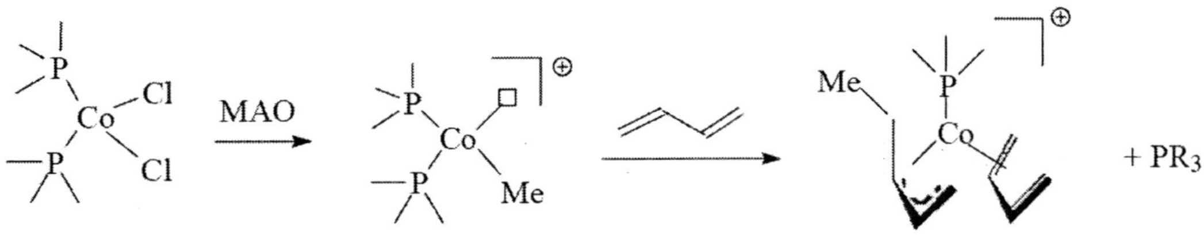

Figure 5. Hypothesis on active site formation, butadiene coordination and first insertion.

It is now universally accepted that in the active site the growing chain is η^3-allyl bonded to the cobalt atom while the incoming butadiene monomer is *cis*-η^4 coordinated [7,79-82]; it follows that only one phosphine ligand can remain coordinated to the cobalt atom because with two phosphine ligands coordinated to the cobalt atom we would have an electron in excess with respect to the rare gas (18e) rule. The phosphine ligand depending on its structure (*i.e.* its steric hindrance) can affect the reciprocal orientation of allyl group and butadiene monomer; with some phosphine the insertion of the monomer at C1 of the allyl group (*i.e.* formation of 1,4 units) may be favoured, with other phosphines the insertion at C3 may be favoured (*i.e.* formation of 1,2 units). A deeper insight on the diene polymerization mechanism is however reported below.

We told above that high *cis* polymers are obtained with catalysts using hindered phosphines (*e.g.* P^tBu_3, P^iPr_3, P^tBu_2Me), but, as cited in the introduction part, also the system $CoCl_2$ / MAO gives such a polymer. However, the *cis*-1,4 polybutadienes obtained with $CoCl_2$ / MAO are linear as indicated by the g value of 1.0 (g is a branching index, calculated by GPC/MALLS analysis. g is the ratio between the radius of gyration of the branched macromolecule and that of the linear macromolecule, with the same molecular weight. A polybutadiene is considered as being branched when $g \leq 0.90$), while the *cis* polybutadienes obtained with $CoCl_2(PR_3)_2$ / MAO ($PR_3 = P^tBu_3, P^iPr_3, P^tBuMe$) are branched (g value around 0.8), as can be deduced from the data of table 4. Moreover, the branching character is higher if the polymerization is carried out in heptane as solvent (table 4; runs 9, 11).

This different behavior seems to indicate that the two systems, although giving polymers having the same microstructure, are in some way different.

When polybutadienes with a mixed structure are obtained, the *cis*-1,4 and 1,2 units are randomly distributed along the polymer chain, as indicated by the fact that polymers fractionation with different solvents always give polymeric fractions having almost the same structure as the crude polymer; it was never possible to isolate polymer fractions having a predominantly *cis*-1,4 or 1,2 structure [93]. This result confirms what already reported in the literature: a random distribution of the different butadiene units (*cis*-1,4 and 1,2) was also observed by Soga in the polymerization of butadiene with the system $CoBr_2(PPh_3)_2$ / MAO [94,95]. The ^1H NMR spectra of some of the polymers of table 4, from which the *cis*-1,4/1,2 content (*trans* units are almost negligible, as indicated by the IR spectra of the polymers in which the band at 967 cm^{-1} is extremely low [96]) can be determined as reported [84,85], are shown in figure 6.

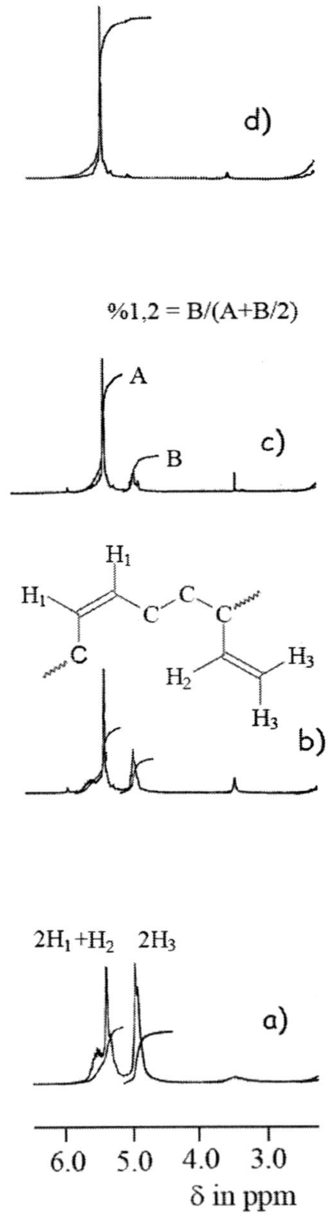

Figure 6. ^{1}H NMR spectra (olefinic protons) of polybutadienes obtained with a) $CoCl_2(P^nPr_3)_2$ / MAO (table 4, run 2); b) $CoCl_2(P^tBu_2H)_2$ / MAO (table 4, run 4); $CoCl_2(PCyp_3)_2$ / MAO (table 4, run 6); d) $CoCl_2(P^tBu_2Me)_2$ / MAO (table 4, run 5).

Polymerization of Butadiene with $CoCl_2(PR_nPh_{3-n})_2$ / MAO (n =1,2)

The results obtained in the polymerization of butadiene with $CoCl_2(PR_nPh_{3-n})_2$-MAO (n = 1,2) [86-88,90] are shown in table 5 and can be summarized as follows.

Table 5. Polymerization of butadiene with $CoCl_2(PR_nPh_{3-n})_2$ / MAO ($R = Me, Et, {}^nPr, allyl, {}^iPr, Cy; n = 1,2$) [a)]

Run	Catalyst	Polymerization				Polymer Microstructure [b)]			T_m [c)]
	Co-compound	Al/Co (molar ratio)	Time (min)	conv (%)	N [d)] (min^{-1})	cis %	1,2 %	rr::mr::mm (molar fraction)	(°C)
1	$CoCl_2(PMePh_2)_2$	1000	5	58.0	602	29.9	70.1	23/47/30	—
2		500	5	59.2	614	27	73.0	23/51/26	—
3		100	5	64.7	671	23.9	76.1	24/52/24	—
4		50	60	26.0	11	27.0	73.0	24/52/24	—
5		20	60	15.2	7	26.0	74.0	25/52/23	—
6	$CoCl_2(PEtPh_2)_2$	100	5	93.7	972	22.3	77.7	42/45/13	—
7	$CoCl_2(P^nPrPh_2)_2$	100	5	100		21.3	78.7	44/43/13	—
8	$CoCl_2[P(allyl)Ph_2]_2$	100	5	76	788	19.8	80.2	41/44/15	—
9	$CoCl_2(P^iPrPh_2)_2$	100	5	100		14.6	85.4	74/26/0	126
10		1000	5	98	1016	13.0	87.0	75/25/0	127
11 [e)]		1000	2	100		14.2	85.8	75/25/0	123
12 [e,f)]		1000	180	100		12.1	87.9	87/13/0	157
13	$CoCl_2(PCyPh_2)_2$	1000	5	72	747	17.3	82.7	71/25/4	117
14 [e)]		1000	2	100		12	88.0	76/22/2	132
15		500	5	70.8	734	20	80.0	68/28/4	106
16		100	5	75.0	778	15.5	84.5	69/27/4	109
17 [e)]		100	3	100			88.0	78/20/2	139
18		20	60	33.4	15	16	84.0	70/27/3	115
19	$CoCl_2(PMe_2Ph)_2$	100	5	39.8	413	27.0	73.0	15/51/34	—
20	$CoCl_2(PEt_2Ph)_2$	100	5	55.4	575	24.7	75.3	19/49/32	—
21	$CoCl_2[P(allyl)_2Ph]_2$	100	5	40.1	520	26.0	74.0	22/49/29	—
22	$CoCl_2(PCy_2Ph)_2$	100	5	55.1	572	20.6	79.4	44/41/15	—

[a)] Polymerization conditions: butadiene, 2 ml; toluene, total volume 16 ml; MAO; 5×10^{-6} moles of Co (1×10^{-5} moles in runs 4,5 and 18); temperature, +20°C. The molecular weight (MW) and molecular weight distribution (MWD) of the amorphous 1,2 polybutadienes are in the range 250÷300000 g/mol and 2÷2.5, respectively. The predominantly syndiotactic 1,2 polybutadienes (runs 9-15) were not soluble in the GPC conditions used. [b)] Determined by ^1H and ^{13}C NMR analysis; *trans*-1,4 units were found to be negligible, as indicated by the extremely low intensity of the band at 967 cm^{-1} in the IR spectra of the polymers. [c)] Determined by DSC. [d)] Moles of butadiene polymerized per mol of cobalt per min. [e)] Heptane as solvent was used instead of toluene. [f)] Carried out at -30°C.

i. The systems $CoCl_2(PRPh_2)_2$ / MAO (R = methyl, *normal*-propyl, ethyl, allyl, *iso*-propyl, cyclohexyl) were found to be extremely active in the polymerization of butadiene; in general complete conversions were reached in a few minutes. The polymers obtained are essentially 1,2, in the range 70÷88 %, and the 1,2 content varies with varying the type of complex and the polymerization conditions. The remaining units are essentially *cis*-1,4 according to the fact that the band at 967 cm^{-1} in the IR spectra of the polymers, indicative of the presence of *trans*-1,4 units, is almost negligible [96].
ii. The polymer tacticity strongly depends on the type of complex, *i.e.* the type of phosphine bonded to the cobalt atom. The syndiotactic index (expressed as percentage of syndiotactic triads *rr*) of the polymers, determined by their ^{13}C NMR spectra (figure 7), increases with increasing the hindrance of the phosphorus ligand,

i.e. the steric demand of the alkyl group bonded to the phosphorus atom (compare, for instance, runs 3,7 and 16 in table 5).

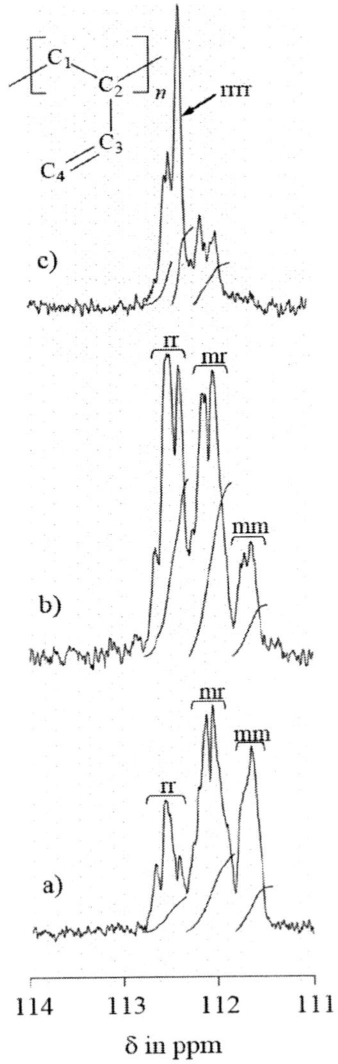

Figure 7. ^{13}C NMR spectra (olefinic region, C4 signal; $C_2D_2Cl_4$, HMDS as internal standard, 103°C) of 1,2 polybutadienes obtained with a) $CoCl_2(PMePh_2)_2$ / MAO (table 5, run 3); b) $CoCl_2(P^nPrPh_2)_2$ / MAO (table 5, run 7); c) $CoCl_2(PCyPh_2)_2$ / MAO (table 5, run 16).

iii. The 1,2 polybutadienes obtained with the cobalt systems using less hindered ligands (*e.g.* $PMePh_2$; $PEtPh_2$; P^nPrPh_2) (table 5, runs 1-7) are amorphous and no melting point could be detected; the T_g values of these polymers were found to be in the range -10 ÷ -15°C. The polymers obtained with systems using more hindered ligands (*e.g.* P^iPrPh_2 and $PCyPh_2$) (table 5, runs 9-18) are crystalline by X-ray with a melting point ranging from 109 to 157°C, depending on the polymerization conditions (figure 8). They exhibited T_g values in the range 3÷5°C and crystallization enthalpies ranging from –10 to –20 J/g (DSC data).

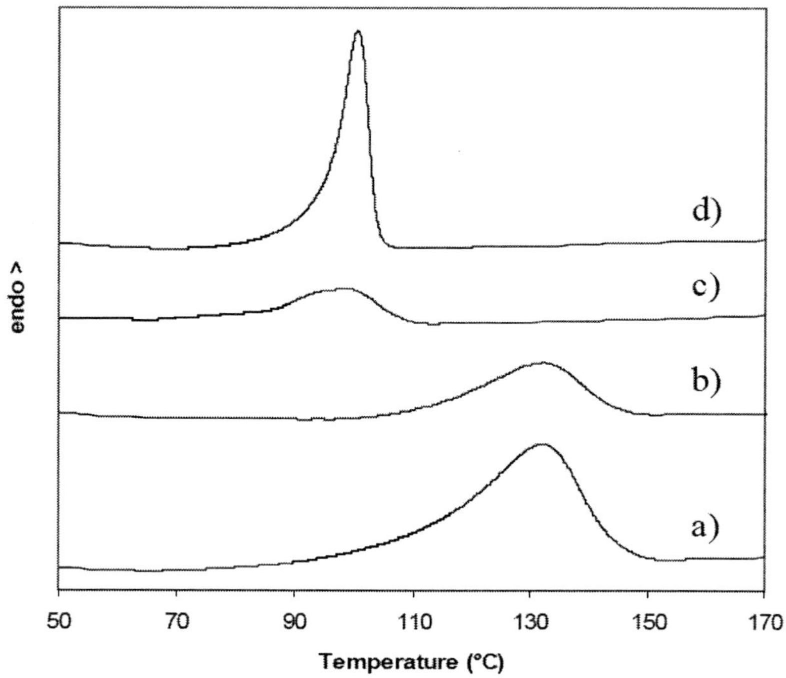

Figure 8. DSC curves of a) 1,2 syndiotactic polybutadiene (table 5, run 14); b) *trans*-1,2 syndiotactic poly(E-1,3-pentadiene) (table 7, run 1); c) *trans*-1,2 syndiotactic poly(1,3-hexadiene) (table 7, run 3); *d) trans*-1,2 isotactic poly(3-methyl-1,3-pentadiene) (table 7, run 6).

iv. In general the Al/Co molar ratio does not affect the stereospecificity, but it has some influence on the catalyst activity. When the Al/Co ratio is in the range 100÷1000 (that is the usual molar ratio used in the olefin polymerization with zirconium metallocenes), the activity is extremely high and complete conversions can be reached in a few minutes (table 5, runs 1-3 and 9-10, 13-17), nevertheless some activity is also observed at very low Al/Co molar ratio (20÷50) (table 5, runs 4,5,18), meaning that the alkylation of the cobalt compound is rather fast and that a certain number of active sites are formed also in these polymerization conditions.

v. The polymerization rate is higher and more stereoregular polymers (higher 1,2 content and higher syndiotacticity) are obtained when heptane instead of toluene is used as polymerization solvent (table 5, runs 11,14,17), indicating that the aromatic solvent can in some way coordinate to the metal atom affecting the catalyst geometry and competing with the incoming monomer. This phenomenon is not however surprising because a similar solvent effect was already observed for instance in the polymerization of 1,3-dienes with neodymium catalysts [91].

vi. The polymerization temperature has some influence on the polymerization stereospecificity too; the 1,2 content and the syndiotacticity seem in fact to increase with decreasing the polymerization temperature (table 5, runs 11,12).

vii. We have also examined the polymerizaton of butadiene with the systems $CoCl_2(PR_2Ph)_2$ / MAO (table 5, runs 19-22); essentially 1,2 polymers were obtained also in this case but the syndiotactic index of the polymers, at the same polymerization conditions, was in general slightly lower with respect to that of the 1,2 polybutadienes obtained with $CoCl_2(PRPh_2)_2$ / MAO.

Polymerization of Butadiene with $CoCl_2[R_2P(CH_2)_nPR_2]$ / MAO (R = Me, Et, Ph; n = 1,2)

The results obtained by polymerizing butadiene with catalysts based on bidentate phosphine cobalt complexes are shown in table 6.

Table 6. Polymerization of 1,3-butadiene with catalyst based on bidentate phosphine cobalt complexes [a)]

Co-complex	polymerization[a)]		polymer			
	time (min)	conv (%)	*cis*-1,4[b)] (%)	Mw[c)] ($g\times mol^{-1}$)	Mw/Mn[c)]	g[d)]
$CoCl_2$	47	74	95.4	615000	2.2	1.00
$CoCl_2$(dmpm)	86	32	95.0	463000	2.7	0.80
$CoCl_2$(dmpe)	91	55	95.2	237000	2.6	0.82
$CoCl_2$(depe)	76	63	94.1	249000	2.0	0.79
$CoCl_2$(dppe)	125	48	96.1	147000	2.3	0.86
$CoCl_2$(dppp)	100	50	96.3	155000	2.4	0.87

[a)] Polymerization conditions : butadiene, 5 mL ; toluene, total volume, 40 mL ; MAO, Al/Co = 1000 ; moles of Co, 1×10^{-5}; temperature, +20°C. [b)] Determined by ^1H NMR, as shown in figure 6. The remaining units are 1,2. [c)] Molecular weight and molecular weight distribution determined by GPC analyses. [d)] g is a branching index, calculated by GPC/MALLS analysis (see text).

High *cis*-1,4 polymers (around 95%) are always obtained, independently on the cobalt complex used; since *cis*-1,4 polymers are also obtained with the system $CoCl_2$ / MAO, one could think that the bidentate phosphine is displaced from the cobalt atom during the polymerization and that the effective catalyst is that originating from the reaction between $CoCl_2$ and MAO. However, the *cis* polymers obtained with $CoCl_2$ / MAO and $CoCl_2[R_2P(CH_2)_nPR_2]$ / MAO are different, the former being linear and the latter branched; a similar behavior was already observed in the polymerization of butadiene with $CoCl_2(PR_3)_2$ / MAO in which PR_3 is a hindered phosphine.

If we admit that in the active site structure the allyl group of the growing chain is η^3-bonded, and the incoming butadiene monomer is cis-η^4-bonded [7,79-82], only one phosphorus atom can coordinate to the cobalt atom because otherwise, with two phosphorus atoms coordinated, we would have an electron in excess with respect to the rare gas (18e) rule. It follows that in the active site formed in the polymerization of 1,3-butadiene with catalysts based on bidentate phosphine cobalt complexes, the bidentate phosphine remains coordinated to the cobalt atom with only one phosphorus atom (figure 9). Such a coordination could be in some way responsible for the formation of branched polymers.

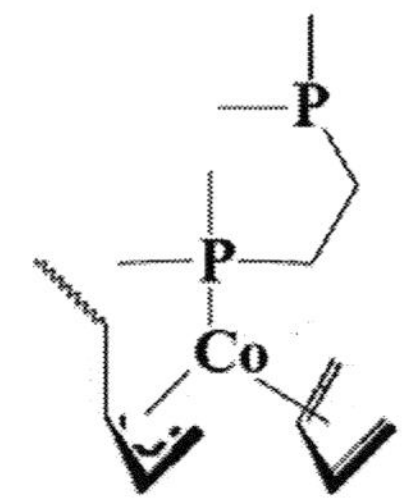

Figure 9. Possible active site in the polymerization of butadiene with catalysts based on $CoCl_2(R_2P(CH_2)_nPR_2)$ (n = 1, 2, 3)

Polymerization of Substituted 1,3-Dienes with $CoCl_2(PRPh_2)_2$ / MAO (R = normal-Propyl; iso-Propyl)

The results obtained in the polymerization of various 1,3-dienes (e.g. 1,3-pentadiene, 1,3-hexadiene, and 3-methyl-1,3-pentadiene) with $CoCl_2(PRPh_2)_2$ / MAO (R = normal-propyl, iso-propyl) are shown in table 7.

Table 7. Polymerization of 1,3-dienes with the systems $CoCl_2(PRPh_2)_2$ / MAO (R= *normal*-propyl, *iso*-propyl)[a)]

run	monomer[b)]	Co_complex	time (min)	conv. (%)	1,2[c)] (%)	rr[d)]	mm[d)]	MW[e)] ($g \times mol^{-1}$)	MWD[e)]	T_m[f)] (°C)
1	EP	$CoCl_2(P^iPrPh_2)_2$	180	78.0	~99	72		220000	1.7	132
2	EP	$CoCl_2(P^nPrPh_2)_2$	30	97.2	~99	45		101000	1.4	n.d.
3	H	$CoCl_2(P^iPrPh_2)_2$	300	49.9	~99	63		74000	1.9	95
4	H	$CoCl_2(P^nPrPh_2)_2$	300	50.4	~99	38		65000	1.7	n.d.
5	3MP	$CoCl_2(P^iPrPh_2)_2$	7200	79.1	~99		≥60	62000	1.6	79
6	3MP	$CoCl_2(P^nPrPh_2)_2$	7200	78.0	~99		≥90	81000	1.2	101

[a)] Polymerization conditions: monomer, 2 ml; toluene, total volume 16 ml; Co, 1×10^{-5} moles; MAO, Al/Co = 100; +20°C. [b)] EP = E-1,3-pentadiene; H=1,3-hexadiene; 3MP=3-methyl-1,3-pentadiene. c) percentage of 1,2 units, determined by NMR analysis (the remaining units are essentially *cis*-1,4, being *trans*-1,4 units almost negligible). [d)] Triads molar ratio, determined by NMR analysis; complementary percentages are mainly due to mr triads. [e)] Molecular weight and molecular weight distribution, determined by GPC analysis. [f)] Determined by DSC analysis.

The systems $CoCl_2(PRPh_2)_2$-MAO gave 1,2 polymers from all the above monomers. The polymers from 1,3-pentadiene and 1,3-hexadiene were predominantly syndiotactic when R = iso-propyl (table 7, runs 1,3), while they had an atactic structure (with predominance of syndiotactic sequences) when R= normal-propyl (table 7, runs 2,4); a similar behavior was observed in the polymerization of 1,3-butadiene with the same catalysts (see table 5). The ^{13}C NMR spectra of the polymers from 1,3-pentadiene and 1,3-hexadiene are shown in figures 10 and 11, respectively, and were attributed as indicated.

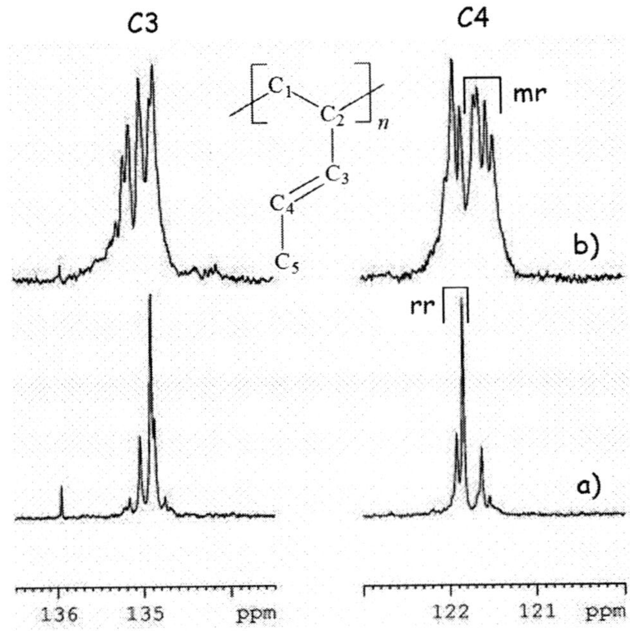

Figure 10. ^{13}C NMR spectra (olefinic region) of trans-1,2 polypentadienes obtained with a) $CoCl_2(P^iPrPh_2)_2$ / MAO (table 7, run 1); b) $CoCl_2(P^nPrPh_2)_2$ / MAO (table 7, run 2) ($C_2D_2Cl_4$, HMDS as internal standard, 103°C).

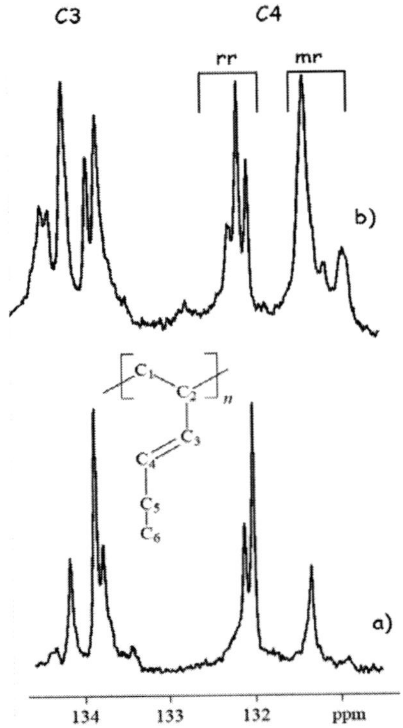

Figure 11. ^{13}C NMR spectra (olefinic region) of *trans*-1,2 polyhexadienes obtained with a) $CoCl_2(P^iPrPh_2)_2$ / MAO (table 7, run 3); b) $CoCl_2(P^nPrPh_2)_2$ / MAO (table 7, run 4) ($C_2D_2Cl_4$, HMDS, 103°C).

The syndiotacticity degree (expressed as percentage of syndiotactic triads rr) was found to be dependent on the type of phosphorus ligand bonded to the metal, being around 60-70% in polymers obtained with the catalytic system $CoCl_2(P^iPrPh_2)_2$ / MAO (table 7, runs 1,3), and much lower (around 40%) with $CoCl_2(P^nPrPh_2)_2$ / MAO (table 7, runs 2,4).

It is worthwhile to note that trans-1,2 syndiotactic polypentadiene and trans-1,2 syndiotactic polyhexadiene obtained with $CoCl_2(P^iPrPh_2)$ / MAO are both novel polymers; all the other systems known to date failed in giving such stereoregular polymers from the above substituted butadienes. On the structural characterization of these two crystalline polymers (figure 12) polymers it has been recently reported [86,97]. Work is in progress on the elucidation of their physical and mechanical properties.

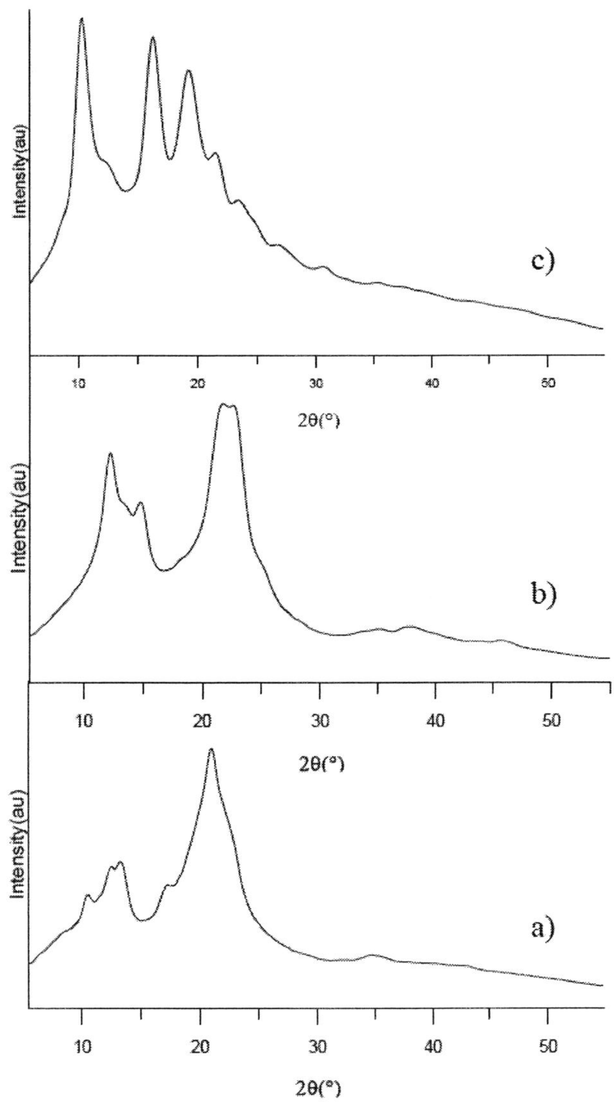

Figure 12. X-Ray powder spectra of a) *trans*-1,2 syndiotactic poly(1,3-pentadiene); b) *trans*-1,2 syndiotactic poly1,3-hexadiene); c) *trans*-1,2-isotactic poly(3-methyl-1,3-pentadiene).

The polymers from 3-methyl-1,3-pentadiene are also crystalline (figure 12), but exhibited an opposite tacticity, isotactic instead of syndiotactic. In particular $CoCl_2(P^nPrPh_2)_2$ / MAO (table 7, run 6), giving the "less syndiotactic" polybutadienes, polypentadienes, and polyhexadienes, gave the "most isotactic" poly(3-methylpentadiene)s and, vice versa, $CoCl_2(P^iPrPh_2)_2$ / MAO (table 7, run 5), giving the "most syndiotactic" polybutadiene, polypentadiene, and polyhexadiene, gave the "less isotactic" poly(3-methylpentadiene). The ^{13}C NMR spectra of the polymers are shown in figure 13, and were attributed as indicated. This polymer is a novel polymer, never reported before; in particular it is one of the few examples of 1,2 isotactic polydiene, being 1,2 isotactic polybutadiene [98] and 1,2 isotactic poly (4-methyl-1,3-pentadiene) [99,100] the other ones. On the characterization and crystalline structure of 1,2 isotactic poly(3-methyl-1,3-pentadiene) a paper has been recently submitted [101].

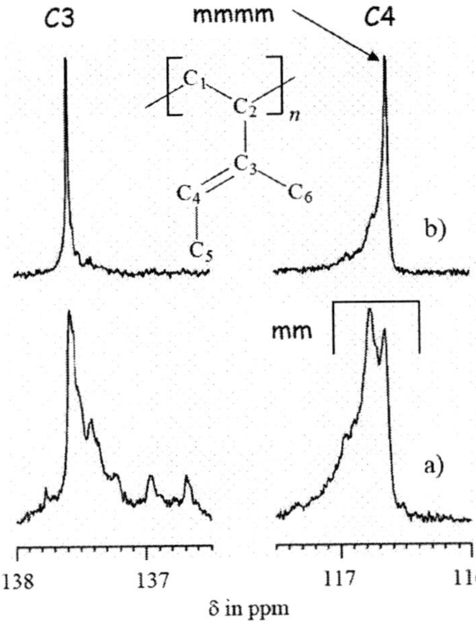

Figure 13. ^{13}C NMR spectra (olefinic region) of 1,2 isotactic poly(3-methyl-1,3-pentadiene) obtained with a) $CoCl_2(P^iPrPh_2)_2$ / MAO (table 7, run 5); b) $CoCl_2(P^nPrPh_2)_2$ / MAO (table 7, run 6) ($C_2D_2Cl_4$, HMDS as internal standard, 103°C).

The different behavior exhibited by 3-methyl-1,3-pentadiene with respect to the other monomers had some mechanistic implications, and allowed us, as illustrated in the paragraph reported below, to highlight for the first time, in comparison with prior art, the influence of monomer structure on polymerization stereoselectivity.

NEW INSIGHTS ON THE DIENE POLYMERIZATION MECHANISM

The results, obtained in the polymerization of 1,3-dienes with the various catalyst systems based on the different cobalt phosphine complexes, turned out to be extremely interesting and useful because they confirmed, and in some cases improved, knowledge on diene polymerization mechanism, as illustrated below.

Before examining the results obtained, it would be useful to sum up what it is known about the diene polymerization mechanism.

It is well known that in the polymerization of 1,3-dienes the growing polymer chain is bonded to the metal atom through a η^3-allyl bond formed by insertion of the monomer on the carbon-metal bond of the catalyst system [7]. According to the polymerisation mechanism proposed by Porri [79-82], the allyl unit can exist in two forms (*anti* and *syn*) and each of the two forms exhibits two reactive positions: C1 and C3 (figure 14).

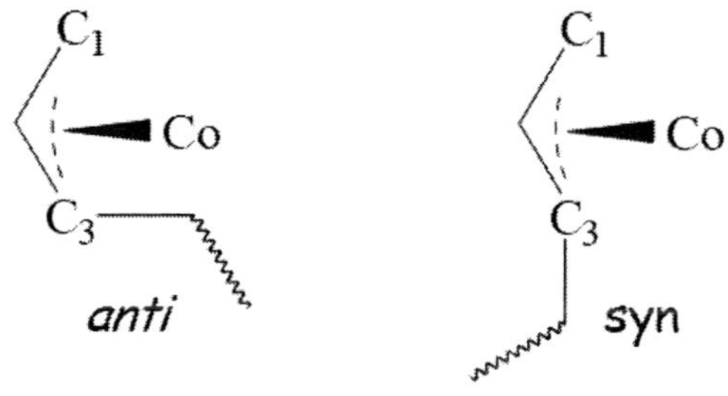

Figure 14. Possible allylic units in the polymerization of 1,3 dienes.

When the new entering diene monomer reacts at C1, we have the formation of a 1,4 unit (*cis* if the allylic unit is in the *anti* form, *trans* if the allylic unit is in the *syn* form); when the incoming monomer reacts at C3 we obtain a 1,2 unit (figure 15).

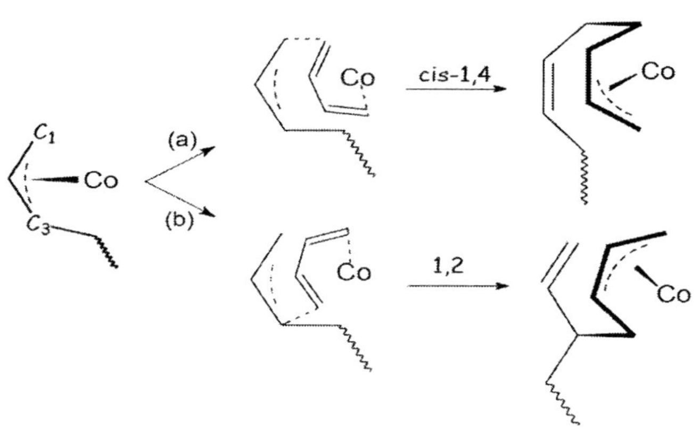

Figure 15. Scheme of formation of 1,4 units *vs.* 1,2 units.

Moreover, the new incoming diene monomer can orient in two different ways with respect to the allylic unit of the polymer growing chain, as in figure 16a (*exo-exo*) or as in figure 16b (*exo-endo*) (for simplicity we have reported only the case of the *anti* unit, but the same is valid in the case of a *syn* unit) [7,79-82,102-105].

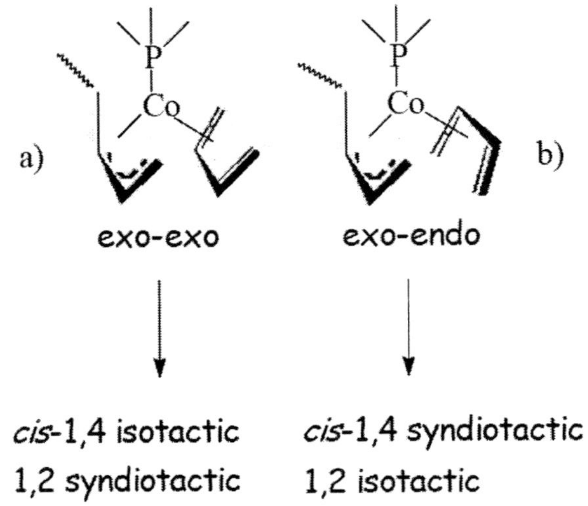

Figure 16. Possible reciprocal orientation of allylic unit and new entering monomer.

From an exo-exo orientation we have the formation of *cis*-1,4 isotactic units if the monomer reacts at C1 or of 1,2 syndiotactic units if the monomer reacts at C3; from an exo-endo orientation we have instead the formation of *cis*-1,4 syndiotactic units when the monomer reacts at C1 or of 1,2 isotactic units when the monomer reacts at C3 [7,79-82].

Some hypotheses have been formulated about the factors determining the formation of 1,4 polymers rather than 1,2 polymers (*i.e.* chemoselectivity), or the formation of isotactic polymers rather than syndiotactic polymers (*i.e.* stereoselectivity); in particular two factors, catalyst structure and monomer structure, were supposed to play a fundamental role in determining the polymerization chemo- and stereoselectivity. The results obtained in the polymerization of 1,3-dienes with the cobalt complexes of our research were found to be extremely useful to better comprehend the diene polymerization mechanism, permitting us to confirm the validity of the mechanism previously proposed [7,79-82] and, in some cases, to put in evidence some new aspects of the diene polymerization mechanism (*e.g.* influence of the monomer structure on the polymerization stereoselectivity). They can be summarized as follows.

Chemoselectivity: Influence of the Cobalt Complex Structure

The results obtained in the polymerization of butadiene with $CoCl_2(PR_3)_2$ / MAO systems (R = alkyl or cycloalkyl group) allowed us to point out the fundamental role played by the ligand nature in determining polymerization chemoselectivity. We obtained polymers ranging from a predominantly 1,2 structure to an essentially *cis*-1,4 structure, depending on the bulkiness of the phosphine ligand [84,85]. If we consider an active site structure (shown in figure 5), with only one phosphine bonded to the cobalt atom, it appears evident that small variations in the reciprocal orientation of incoming monomer and butadienyl group, due to the nature of the ligand on the cobalt atom, may favour the insertion of the monomer at C1 or C3 of the allylic group (figure 15). Thus, hindered ligands seem to favour the insertion of the monomer at C1, while less hindered ligands favour insertion at C3 of the allylic unit.

Chemoselectivity: Influence of the Monomer Structure

It has been shown in previous papers that chemoselectivity may depend on the monomer structure (*i.e.* presence of substituents on the different carbons) [106-108]. For example, in the polymerization of 1,3-dienes with $CpTiCl_3$ / MAO [106,107] we obtained polymers with different microstructure as a product of monomers with different structure and a clear relationship was found between the polymer microstructure and the presence of alkyl substituents on the C1 and C3 position of the allylic units. The insertion of the new incoming monomer preferentially occurs at the less substituted reactive carbons C1 and C3. This indication is now supported and validated by the results obtained in the polymerization of 1,3-dienes with the systems $CoCl_2(PRPh_2)_2$ / MAO. Exclusively 1,2 units were observed in the polymers obtained from 1,3-pentadiene, 1,3-hexadiene and 3-methyl-1,3-pentadiene, while polymers containing a certain amount of *cis*-1,4 units were obtained from butadiene (15-20%) and isoprene (43%) [86]. If we consider the allylic units formed during the polymerization of the above monomers (figure 17), it appears clear that in the case of pentadiene, hexadiene and 3-methylpentadiene, the C1 position, due to the presence of an alkyl substituent, is more hindered and consequently less reactive with respect to the same position in the allylic unit formed in the case of butadiene or isoprene.

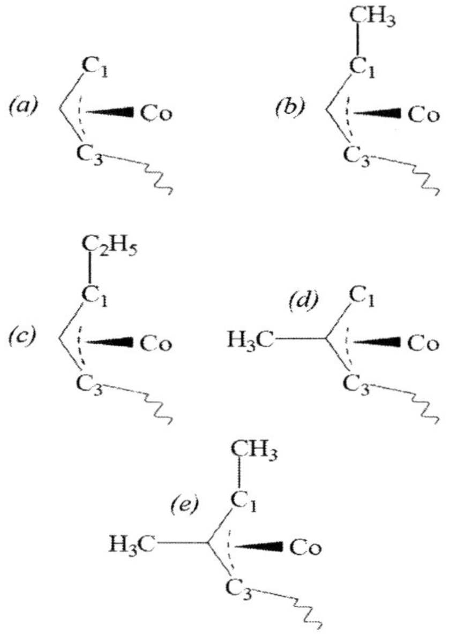

Figure 17. Possible allylic units in the polymerization of a) butadiene; b) 1,3-pentadiene; c) 1,3-hexadiene; d) isoprene; e) 3-methyl-1,3-pentadiene.

It follows that the formation of 1,4 units (insertion of the new monomer at C1) is more probable in the polymerization of butadiene and isoprene than in the polymerization of 1,3-pentadiene, 1,3-hexadiene and 3-methyl-1,3-pentadiene.

Stereoselectivity: Influence of the Cobalt Complex Structure

The type of ligand, however, has been reported to affect the stereoselectivity (formation of iso-or syndiotactic polymers) too; in the polymerization of butadiene with the catalyst systems $CrCl_2(R_2P(CH_2)_nPR_2)_2$ / MAO (R = Me, Et; n = 1,2,3) 1,2 isotactic or syndiotactic polymers were obtained depending on the type of ligand [109,110].

Similar results have been obtained in the polymerization of butadiene with the systems $CoCl_2(PRPh_2)_2$ / MAO (R = alkyl or cycloalkyl group) (see table 5), giving 1,2 polymers [87] with different tacticity depending on the type of ligand (*i.e.* type of alkyl group R bonded to the phosphorus atom and consequently steric hindrance of the ligand). Highly syndiotactic polybutadienes were obtained with $CoCl_2(P^iPrPh_2)_2$ / MAO and $CoCl_2(PCyPh_2)_2$ / MAO in which hindered alkyl groups are present on the P atoms. Atactic polymers with predominance of isotactic sequences were obtained with $CoCl_2(PRPh_2)_2$ / MAO in which the steric demand of the alkyl group R (methyl, ethyl or *n*-propyl) is lower. The formation of polybutadienes having different type and/or degree of stereoregularity was interpreted on the basis of the previously reported diene polymerization mechanism [7,79-82]. The type of phosphorus ligand (*i.e.* its steric hindrance) may influence the reciprocal orientation (*exo-exo* or *exo-endo*) of the butadienyl group and of the incoming monomer thus determining the formation of iso- or syndiotactic polymers (*i.e.* the polymerization stereoselectivity) (figure 16).

Stereoselectivity: Influence of the Monomer Structure

According to the diene polymerization mechanism previously proposed [7,79-82], the reciprocal orientation (*exo-exo* or *exo-endo*) of the butadienyl group and of the incoming monomer should be in principle affected by the monomer structure (*i.e.* presence of substituents on the different carbons of the monomer and consequently of the butadienyl group).

The results obtained in the polymerization of butadiene, 1,3-pentadiene, 1,3-hexadiene, and 3-methyl-1,3-pentadiene with $CoCl_2(P^nPrPh_2)_2$ / MAO and $CoCl_2(P^iPrPh_2)_2$ / MAO (see tables 5 and 7) permitted us to confirm the validity of this hypothesis.

The analysis of the results reported above suggests the following remark. Using the same catalyst, polymers obtained from 3-methyl-1,3-pentadiene have an essentially isotactic structure (the isotactic degree depending on the type of cobalt complex, specifically increasing with decreasing the steric hindrance of the phosphine ligand bonded to the cobalt atom); polymers from butadiene, E-1,3-pentadiene and 1,3-hexadiene exhibit an essentially syndiotactic structure (again the syndiotactic degree depending on the type of catalyst, specifically increasing with increasing the steric hindrance of the phosphine ligand bonded to the cobalt atom). These results clearly indicate that the monomer structure may strongly influence the polymerization stereoselectivity.

The formation of 1,2 syndiotactic polymers from butadiene, E-1,3 pentadiene and 1,3-hexadiene, and of isotactic polymers from 3-methyl-1,3-pentadiene, by using the same polymerization catalyst, can be explained, according to the diene polymerization mechanism [7,79-82], as follows: in case of butadiene (figure 18a), E-1,3-pentadiene (figure 18b), and 1,3-hexadiene (not shown) the favorite incoming monomer-butadienyl group orientation is the *exo-exo* one while, in the case of 3-methyl-1,3-pentadiene (figure 18c), the preferred orientation is the *exo-endo* one.

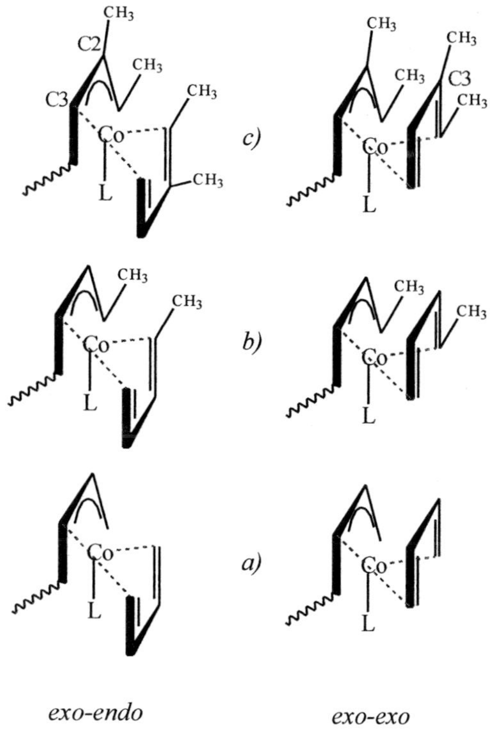

Figure 18. Possible orientations of incoming diene monomer and butadienyl group in the homopolymerization of (a) butadiene; (b) E-1,3-pentadiene; (c) 3-methyl-1,3-pentadiene with $CoCl_2(PRPh_2)_2$ / MAO.

By insertion of the incoming monomer at C3 of the butadienyl group, a 1,2 isotactic diad is obtained in the case of the *exo-endo* orientation, while a 1,2 syndiotactic diad is obtained in the case of the *exo-exo* orientation. Most probably in the polymerization of internal substituted dienes, such as 3-methyl-1,3-pentadiene, the presence of a methyl on the C2 carbon of the butadienyl group and on the C3 carbon of the incoming monomer, forces the two groups to assume an *exo-endo* orientation, even if this particular orientation is not the favourite one when a ligand (less or more hindered) is bonded to the metal. In other words, the incoming monomer prefers suffering the hindrance of the ligand rather than that originating from the butadienyl group.

It has been reported by other authors [102-105], on the basis of conformational and modeling calculations, that actually the energetically favoured reciprocal orientation of the incoming monomer and of the butadienyl group should be the *endo-endo* one and not the *exo-exo*. It should be said that from both the situations (*endo-endo* and *exo-exo*) 1,2 syndiotactic diads are formed by insertion of the incoming monomer at the C3 of the butadienyl group; nevertheless, in this particular case, we prefer to represent the catalytic site with incoming monomer and butadienyl group *exo-exo* oriented because this is the situation found, for example, in the crystal structure of the but-2-enyl-buta-1,3-diene cobalt complex $[(PPh_3)Co(C_4H_6)(C_4H_7)]$ [111,112].

The results reported above for the polymerization of 1,3-dienes with catalysts based on cobalt phosphine complexes, were also observed, even if in less evident measure, in the polymerization of butadiene and isoprene with $CrCl_2(dmpe)_2$ / MAO [113]. A 1,2 syndiotactic polymer was obtained from butadiene, while an atactic polymer, with predominance of isotactic sequences, from isoprene. This result seems to confirm the fundamental role played by the monomer structure in determining the polymerization stereoselectivity.

Copolymerization of 1,3-Dienes

The effective role of monomer structure in determining the polymerization stereoselectivity has been further confirmed through the copolymerization of 1,3-dienes.

Our research on this topic is still in progress. However, some preliminary data have been collected, mainly using $CoCl_2(PRPh_2)$ / MAO (R = *normal*-propyl or *iso*-propyl) to catalyze the copolymerization of E-1,3-pentadiene (EP) and 3-methyl-1,3-pentadiene (3MP) (table 8). These two monomers behave in an opposite way when homopolymerized (table 7), giving syndiotactic and isotactic polymers, respectively.

Table 8. (E)-1,3-pentadiene/3-methyl-1,3-pentadiene copolymerization with $CoCl_2(PRPh_2)_2$ / MAO (R = ^{n}Pr, ^{i}Pr)[a]

run	Co-complex	EP (mmol)	3MP (mmol)	Feed EP/3MP (molar ratio)	Time (h)	Conv. (%)	Composition[b] EP/3MP (molar ratio)	rr[c] (%)	MW (g/mol)	MWD
1		10	—	100/0	2	67	100/0	69	185000	2.0
2		7.5	2.2	77/23	5	33	93/7	65	156000	1.6
3	$CoCl_2(P^iPrPh_2)_2$	5	4.4	53/47	5	35	68/32	44	110000	2.1
4		2.5	6.6	27/73	10	38	64/36	37	100000	1.5
5		—	8.8	0/100	48	67	0/100	—	90000	1.6
6		10	—	100/0	0.5	100	100/0	45	103000	1.7
7		7.5	2.2	77/23	1	39.6	79/21	38	98000	2.0
8	$CoCl_2(P^nPrPh_2)_2$	5	4.4	53/47	1	63.7	67/33	31	105000	1.4
9		2.5	6.6	27/73	2	18.0	51/49	25	92000	1.4
10		—	8.8	0/100	48	77.0	0/100	—	61000	1.3

[a] Polymerization conditions : toluene, total volume 9 ml; MAO/Co = 100; $CoCl_2(PRPh_2)_2$, 1×10^{-5} moles; +20°C. EP = (E)-1,3-pentadiene, 3MP = 3-methyl-1,3-pentadiene. [b] Determined by ^{13}C NMR. [c] Pentadiene syndiotactic triads percentages, determined by ^{13}C NMR

The ^{13}C NMR spectra of the homo- and copolymers of (E)-1,3-pentadiene and 3-methyl-1,3-pentadiene are shown in figures 19 and 20

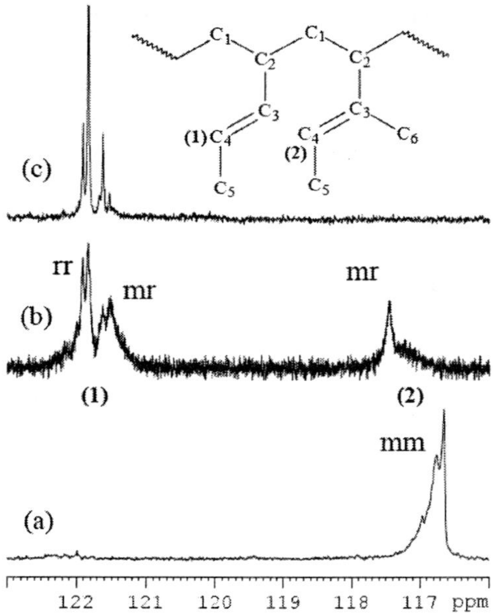

Figure 19. ^{13}C NMR spectra (olefinic region) of (a) poly(3-methyl-1,3-pentadiene) (table 8, run 5); (b) 3-methyl-1,3-pentadieen/1,3-pentadiene copolymer (table 8, run 3); (c) poly(1,3-pentadiene) (table 8, run 1) obtained with $CoCl_2(P^iPrPh_2)_2$ / MAO.

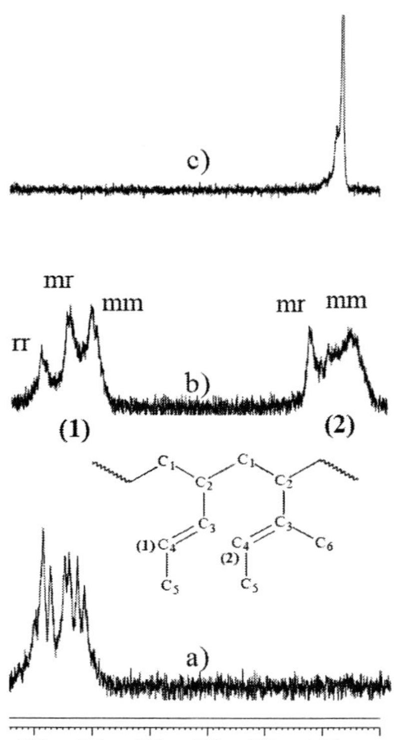

Figure 20. ^{13}C NMR spectra of a) poly(1,3-pentadiene) (table 8, run 6); b) 1,3-pentadiene/3-methyl-1,3-pentadiene copolymer (table 8, run 9); c) poly(3-methyl-1,3-pentadiene) (table 8, run 10) obtained with $CoCl_2(P^nPrPh_2)_2$ / MAO.

It can be clearly observed that 1) the syndiotacticity of the pentads centered on a pentadiene unit (MM(EP)MM; M = EP or 3MP) decreases with increasing the 3MP content in the copolymer (see signals around 124 ppm in the ^{13}C NMR spectra of figures 19 and 20); 2) the isotacticity of the pentads centered on a 3MP unit (MM(3MP)MM; M = EP or 3MP) decreases with increasing the EP content in the copolymers (see signals around 120 ppm in the ^{13}C NMR spectra of figures 19 and 20).

In other words, the occurrence of EP units in syndiotactic pentads is lower in the copolymers than that in its homopolymers; similarly the occurrence of 3MP units in isotactic pentads is lower in the copolymers than in its homopolymers.

These preliminary results seem indeed to confirm the effective role of monomer structure in determining the polymerization stereoselectivity and can be explained in the following way. As shown in figure 21, when the last polymerized unit is a EP unit, a syndiotactic diad is preferentially formed when the incoming monomer is EP (preferred orientation with respect to butenyl group: *exo*, figure 21a); an isotactic diad instead is preferentially formed when the incoming monomer is 3MP (preferred orientation with respect to butenyl group: *endo*, figure 21b). On the contrary, when the last polymerized unit is 3MP (figure 22), an isotactic diad is preferentially formed when the incoming monomer is 3MP (preferred orientation with respect to butenyl group: *endo*, figure 22a); a syndiotactic diad is instead preferentially formed when the incoming monomer is EP (preferred orientation with respect to butenyl group: *exo*, figure 22b). This would explain why the stereoregularity of the copolymers is always lower with respect to that of both homopolymers.

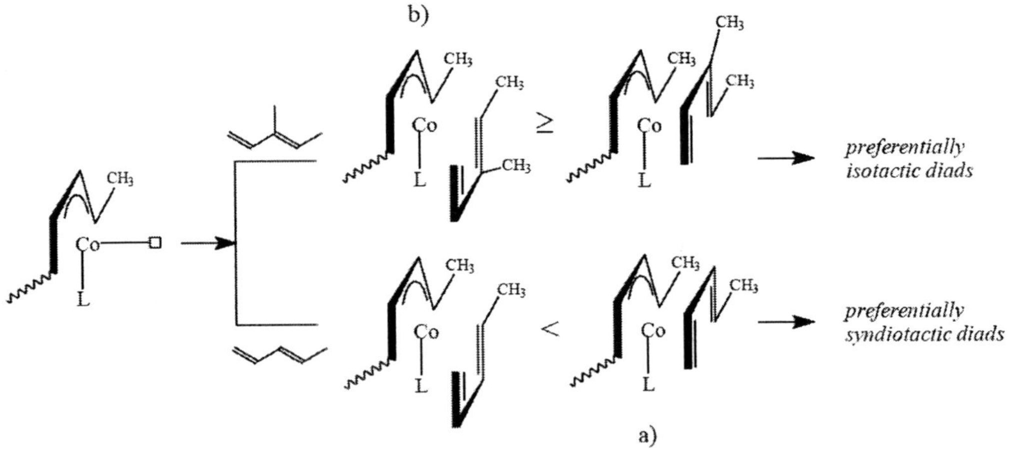

Figure 21. Possible monomer (EP or 3MP) insertions on a pentadienyl unit.

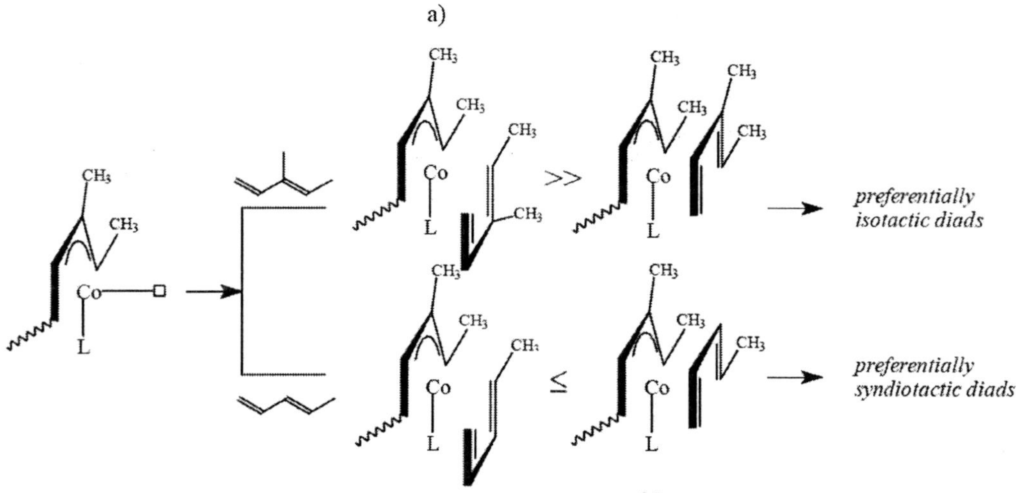

Figure 22. Possible monomer (3MP or EP) insertions on a 3-methyl-pentadienyl unit.

EXPERIMENTAL

Materials

Most of the cobalt phosphine complexes used have been prepared by improving an experimental procedure already described in the literature [83,84,87,89,92]. The syntheses of the complexes not yet reported, essentially bidentate phosphine cobalt complexes, are described below. Infrared spectra of the cobalt complexes were recorded as KBr disks with a Bruker IFS 48 instrument.

Methylaluminoxane (MAO) (Crompton, 10 wt._% solution in toluene) was used as received. 1,3-Butadiene (Air Liquide, >99.5% pure) was evaporated from the container prior to each run, dried by passing through a column packed with molecular sieves and condensed into the reactor which had been pre-cooled to –20 °C. (E)-1,3-pentadiene (Fluka, ≥96% pure), 1,3-hexadiene (Aldrich, 95% pure, mixture of (Z) and (E) isomers) and 3-methyl-1,3-pentadiene (Aldrich, 98% pure, mixture of (Z) and (E) isomers) were refluxed over calcium hydride for ca. 3 hours, then distilled trap-to-trap and stored under dry nitrogen.

Synthesis of $CoCl_2(P^iPr_3)_2$

A solution of tri-*iso*-propylphosphine (0.96 g, 6×10^{-3} moles) in ethanol (20 ml) is added dropwise, under stirring, to a solution of $CoCl_2$ (0.31 g, 2.4×10^{-3} moles) in ethanol (30 ml). A light, pale blue suspension is formed, which is kept under stirring for about 20 h. It is subsequently dried under vacuum obtaining a pitchy dark blue-coloured residue. Various washings are effected with pentane (3×20 ml) and with ethanol at a low temperature, in order to remove traces of excess phosphine and non-reacted $CoCl_2$.

The residue is dried under vacuum at room temperature for a whole night. 0.55 g of a pale blue microcrystalline powder are recovered, with a yield equal to 50.6%. Anal. Calcd. for $C_{18}H_{42}CoCl_2P_2$: Co, 13.09 ; Cl, 15.75; P, 13.76. Found: Co, 12.97; Cl, 15.81; P, 14.0.

Spectroscopic data: IR (KBr) ν (cm^{-1}) 2955 (m); 2926 (s); 2854 (m); 1463 (s); 1377 (m).

Synthesis of $CoCl_2(P^tBu_3)_2$

A solution of tri-*tert*-butylphosphine (1.14 g, 5.64×10^{-3} moles) in ethanol (20 ml) is added dropwise, under stirring, to a solution of $CoCl_2$ (0.30 g, 2.26×10^{-3} moles) in ethanol (30 ml). A pale blue precipitate is formed.

The suspension is left under magnetic stirring for about 20 h, and is then filtered. The residue is dried under vacuum, washed with pentane (2×15 ml) and with ethanol (2×10 ml) at a low temperature in order to remove traces of excess phosphine and non-reacted $CoCl_2$.

The residue is dried under vacuum at room temperature for a whole night: 0.75 of pale blue product are recovered, with a yield of 62.1%. Anal. Calcd. for $C_{24}H_{54}CoCl_2P_2$: Co, 11.03; Cl, 13.27; P, 11.59. Found: Co, 11.09; Cl, 13.01; P, 11.41.

Spectroscopic data: IR (KBr) ν (cm^{-1}) 2954 (m); 2924 (s); 2854 (m); 1465 (s); 1378(m).

Synthesis of $CoCl_2$(dmpm)

Bis(dimethylphosphino)methane (1.04 g, 7.5 mmol) was added to a solution of cobalt(II) chloride (0.836 g, 6.3 mmol) in 40 mL of dry ethanol under nitrogen. The colour of the suspension turned immediately from deep blue to brilliant green. The mixture was kept under stirring for about 24 h, and then the solvent was removed under vacuum. The green residue was washed with pentane (2×10 ml) in order to remove excess phosphine, and successively dried under vacuum. Yield 1.59 g, 95%, brilliant green powder.

Anal. Calcd. for $C_5H_{14}CoCl_2P_2$: Co, 22.16 ; Cl, 26.66; P, 23.29. Found: Co, 21.97; Cl, 26.401; P, 22.88.

Spectroscopic data: IR (KBr) ν (cm^{-1}) 2960 (s); 1461 (s); 1377 (ms); 1300 (m); 1150 (ms); 936 (ms); 898 (m); 759 (m); 723 (m).

Synthesis of $CoCl_2$(dmpe)

1,2-bis(dimethylphosphino)ethane (12 g, 6.7 mmol) was added to a solution of cobalt(II) chloride (0.71 g, 5.51 mmol) in 40 mL of dry ethanol under nitrogen. The colour of the suspension turned immediately from deep blue to green. The mixture was kept under stirring for about 24 h, and then the solvent was removed under vacuum. The green residue was washed with pentane (2×10 ml) in order to remove excess phosphine, and successively dried under vacuum. Yield 1.30 g, 88%, green powder.

Anal. Calcd. for $C_6H_{16}CoCl_2P_2$: Co, 21.05 ; Cl, 25.33; P, 22.13. Found: Co, 20.87; Cl, 25.20; P, 22.39.

Spectroscopic data: IR (KBr) ν (cm^{-1}) 2909 (m); 1420 (m); 1303 (m); 1132 (s); 951 (s); 897 (s); 743 (m); 714 (m); 654 (m).

Synthesis of $CoCl_2$(depe)

1,2-bis(diethylphosphino)ethane (2 g, 9.7 mmol) was added to a solution of cobalt(II) chloride (1.06 g, 8.1 mmol) in 40 mL of dry ethanol under nitrogen. The colour of the suspension turned immediately from deep blue to green. The mixture was kept under stirring for about 24 h, and then the solvent was removed under vacuum. The green residue was washed with pentane (2×10 ml) in order to remove excess phosphine, and successively dried under vacuum. Yield 2.5 g, 92%, green powder.

Anal. Calcd. for $C_{10}H_{24}CoCl_2P_2$: Co, 17.54 ; Cl, 21.10; P, 18.43. Found: Co, 17.28; Cl, 20.90; P, 18.20.

Spectroscopic data: IR (KBr) ν (cm^{-1}) 2961 (s); 1460 (s); 1410 (m); 1416 (s); 1372 (s); 1110 (m); 1028 (s); 872 (m); 820 (m); 758 (s).

Synthesis of $CoCl_2$(dppp)

A suspension of 1,3-bis(diphenylphosphino)propane (2.1 g, 5 mmol) in 50 mL of dry ethanol was added to a solution of cobalt(II) chloride (0.55 g, 4.2 mmol) in 25 mL of dry ethanol under nitrogen. The colour of the suspension turned from deep blue to blue turquoise. The mixture was heated to reflux for 3 h, cooled to room temperature and the precipitate was filtered off through a Schlenk frit under nitrogen, washed with dry ethanol and pentane and dried in vacuum. Yield 1.96 g, 86%, blue turquoise powder.

Anal. Calcd. for $C_{27}H_{26}CoCl_2P_2$: Co, 10.87 ; Cl, 13.08; P, 11.42. Found: Co, 11.20; Cl, 13.25; P, 11.30.

Spectroscopic data: IR (KBr) ν (cm^{-1}) 3053 (m); 1485 (m); 1435 (s); 1159 (m); 1104 (s); 998 (m); 966 (m); 744 (s); 694 (s); 546(m); 509 (sm).

Polymerization

All operations were carried out under an atmosphere of dry dinitrogen. A standard procedure is reported. 1,3-Butadiene was condensed into a dried glass reactor kept at –20°C, then toluene was added and the solution so obtained was brought to the desired polymerization temperature. Methylaluminoxane (MAO) and the cobalt complex were then added, as toluene solutions, in the order. The polymerization was terminated with methanol containing a small amount of hydrochloric acid, the polymer was coagulated and repeatedly washed with methanol, then dried in vacuum at room temperature.

Polymer Characterization

^{13}C NMR and ^{1}H NMR measurements were performed with a Bruker Avance 400 instrument. The spectra were obtained in $C_2D_2Cl_4$ at 103°C (hexamethyldisiloxane, HMDS, as internal standard) or in $CDCl_3$ (tetramethylsilane, TMS, as internal standard). The concentration of polymer solutions was about 10 wt._%. Differential scanning calorimetry (DSC) scans were carried out on a Perkin Elmer Pyris 1 instrument. Typically, ca. 10 mg of polymer were analyzed in each run, while scan speed was ca. 20K/min under a dinitrogen atmosphere. X-ray diffraction powder spectra of the polymers were recorded on an Italstructure θ/θ diffractometer. The infrared spectra were performed with a Bruker IFS 48 instrument, using polymer films on KBr disk. The films were obtained by deposition from solutions in benzene or hot (ca.100°C) solutions in 1,2,4-trichlorobenzene.

CONCLUSION

Several new cobalt complexes with mono- and bidentate phosphines have been prepared and fully characterized; for some of them single crystals were obtained and their crystalline molecular structures were identified. These complexes are easy to synthesize and rather stable; by combining them with MAO, a new family of catalysts for the polymerization of 1,3 dienes was obtained.

These catalysts allow us to obtain butadiene polymers with controlled microstructure (*cis*-1,4; 1,2, iso- and syndiotactic; mixed *cis*-1,4/1,2), giving the possibility of tailoring the polymers for different applications. These new catalyst systems are ready to be applied in an industrial context, aromatic free, which promises "greener" possibilities for the future.

In view of the industrial interest of *cis*-1,4 and 1,2 polybutadienes, the use of these novel systems is particularly attractive, since they *i)* are highly active and stereospecific; *ii)* work very well in aliphatic solvents, contrary to most of the currently used systems requiring aromatic solvents; *iii)* do not need any further harmful component (*e.g.* CS_2).

Furthermore, they were able to polymerize substituted butadienes such as 1,3-pentadienes and 1,3-hexadiene, giving novel polymers never obtained before. Polymers with different structure were obtained from the different monomers. This fact, together with the observed dependence of the polymer microstructure on the type of phosphine ligand, has permitted us to confirm, and in some cases to improve, the knowledge on diene polymerization mechanism. In particular, the influence of ligand and monomer structures on polymerization chemo- and stereoselectivity, have been clearly elucidated.

These systems were also able to copolymerize various 1,3 dienes. The preliminary data so far obtained suggest the possibility of modulating the stereoregularity and melting point of 1,2 polymers by the synergistic effect of monomer and phosphine ligand structures.

In conclusion, what reported above highlights the fundamental role of organic and metallorganic chemistry in the field of macromolecular science. This is reminiscent of the revolution that occurred in the '80s in the synthesis of stereoregular polyolefins with the introduction of metallocene based catalysts which opened new routes to both traditional and novel polymers, *e.g.* syndiotactic polypropylene [115].

REFERENCES

[1] Takeuchi, Y.; Sekimoto, A.; Abe, M. *New Industrial Polymers, Am. Chem. Soc. Symp. Ser.4,* Deanin, R. D., Ed. American Chemical Society, ACS, Washington DC; 1974, *26,* 15.

[2] Halasa, A. F. *US Patent* 4,051,308 (1977) to Firestone Tire and Rubber Co.

[3] PERP (Process Evaluation Research Planning) *Report 02/03S1, Styrene Butadiene Rubber-Butadiene Rubber,* Chem Systems, New York, (23 Dec. 2001).

[4] Halasa, A. F.; Massiein, J.M. in Kirk-Othmer *Encyclopedia of Chemical Technology,* 4th ed., edited by J. I. Kroschwitz. John Wiley and Sons, New York, 1989, vol. 8.

[5] Tate, D. P.; Bethea, T. W. in *Encyclopedia of Polymer Science and Engineering,* 2nd ed., edited by H.F. Mark, et al., John Wiley and Sons, New York, 1989, vol. 2.

[6] Kerns, M.; Henning, S.; Rachita, M. *Butadiene Polymers.* In *Encyclopedia of Polymer Science and Technology*; H. F. Mark, Ed.; Wiley; 2003; Vol. 5, pp 317-356.

[7] Porri, L.; Giarrusso, A. *Conjugated Diene Polymerization.* In *Comprehensive Polymer Science,* Eastmond, G.; Edwith, A.; Russo, S.; Sigwalt, P.; Eds.; Pergamon Press Ltd., Oxford, 1989, vol. 4, pp 53-108.

[8] Thiele, S. K. H.; Wilson, D. R. *J. Macromol. Sci. Part C:Polymer Reviews* 2003, Vol. C43, No. 4, pp 581-628.

[9] Racanelli, P.; Porri, L. *Eur. Polym. J.* 1970, *6*, 751.

[10] Furukawa, J.; Haga, K.; Kobayashi, E.; Iseda, Y.; Yoshimoto T.; Sakamoto, K. *Polym. J.* 1971, *2*, 371.

[11] Winter, H. et al. *GB Patent* 1,042,173 (1966) to Huels.

[12] Winter, H. *GB Patent* 1,042,801 (1966) to Huels.

[13] *DE Patent* 1,930,633 (1970) to Shell.

[14] Natta, G. et al. *GB Patent* 924,427 (1963) to Montecatini.

[15] *GB Patent* 936,061 (1963) to Montecatini.

[16] *GB Patent* 940,396 (1963) to Montecatini.

[17] Natta, G.; Carbonaro, A.; Porri, L.; Lionetti, A. *GB Patent* 1136743 (1968) to Montecatini.

[18] Endo K.; Hatakeyama, N. *J. Polymer Sci., Part A: Polymer Chem.* 2001, *39*, 2793-2798.

[19] *GB Patent* 941,373 (1963) to Montecatini.

[20] *GB Patent* 594,618 (1958) to Montecatini.

[21] Gippin, M. *Ind. Eng. Chem. Prod. Res. Devel.* 1962, *1*, 32.

[22] *GB Patent* 916,384 (1963) to Goodrich.

[23] *FR Patent* 1,441,168 (1966) to Huels.

[24] *US Patent* 1,930,633 (1970) to Shell.

[25] *GB Patent* 890139 (1962) to Shell.

[26] Natta, G. et al. *Makrom. Chem.* 1964, *71*, 207-211.

[27] Costner, K. et al. *US Patent* 5,811,499 (1998) to Goodyear.

[28] Throckmorton, M. *US Patent* 3,528,957 (1970) to Goodrich.

[29] Gippin, M. *Ind. Eng. Chem. Prod. Res. Devel.*, 1965, *4*, 160-167.

[30] Tucker, H. *US Patent* 3,094,514 (1963) to Goodrich.

[31] Van de Kamp, F. P. *Angew. Chem.* 1962, *74*, 661.

[32] Childers, C. W. *J. Am. Chem. Soc.* 1963, *85*, 229.
[33] Cooper, W.; Eaves, D. E.; Vaughan, G. *Makromol. Chem.* 1963, *67*, 229.
[34] Cooper, W. *Rubber Plastics Age* 1963, *44,* 44.
[35] Matsuzaki, K.; Yasukawa, T. *Polym. Lett.* 1965, *3*, 393.
[36] Porri, L.; Carbonaro, A. *Makromol. Chem.* 1963, *60*, 236.
[37] Natta, G.; Porri, L. *Adv. Chem. Ser.* 1966, *52*, 24.
[38] Gavrilova, L.V. et al. *Polym. Sci. USSR* 1968, *10*, 2324.
[39] Volkov, L. A. et al. *Polym. Sci. USSR* 1973, *15*, 2681.
[40] Timosfeyeva, G. V. et al. *Polym. Sci. USSR* 1969,11, 677.
[41] Price, F. C. *Chem. Eng.* 1963, *70*, 84.
[42] Gippin, M. *Macromolecular Syntheses,* John Wiley and Sons, Inc., New York, 1978, Collective Vol. 1, p.129.
[43] Longiave, C.; Castelli, R.; Croce, G.F. *Chim. Ind.* (Milan) 1961, *43*, 625.
[44] Okamoto, N. et al. *JP Patent* 2006054808 (2006) to UBE.
[45] Tsujimoto, N. et al. *US Patent* 5,905,125 (1999) to UBE.
[46] Osman, A. *WO Patent* 0014130 (2000) to Bayer.
[47] Ueno, H. et al. *US Patent* 4,503,202 (1985) to Ube.
[48] Okanoto, N. et al. *WO Patent* 2006054808 (2006) to Ube.
[49] Suzuki, K. *JP Patent* 9052912 (1997) to Ube.
[50] Odar, J. *US Patent* 4303769 (1981) to Goodrich.
[51] Odar, J. *US Patent* 4314045 (1982) to Goodrich.
[52] Odar, J. *US Patent* 4224426 (1980) to Goodrich.
[53] Ashitaka, H.; Isikawa, H.; Veno, H. *J. Polym. Sci. Pol. Chem. Ed.* 1983, *21*, 1853.
[54] Ashitaka, H.; Jinda, K.; Veno, H. *J. Polym. Sci. Pol. Chem. Ed.* 1983, *21*, 1951.
[55] Ashitaka, H.; Inaishi, K.; Veno, H. *J. Polym. Sci. Pol. Chem. Ed.* 1983, *21*, 1973.
[56] Susa, E. *J. Polym. Sci., Part C* 1963, *4*, 399.
[57] Ricci, G., Italia, S.; Porri, L. *Polym. Commun.* 1988, *29*, 305.
[58] Ashitaka, H.; Jinda, K.; Veno, H. *J. Polym. Sci. Pol. Chem. Ed.* 1983, *21*, 1989.
[59] Sugiura, S. et al. *US Patent* 3,778,424 (1970) to UBE.
[60] Fujii, T. et al. *JP Patent* 2006143764 (2006) to UBE.
[61] Tsujimoto, N. et al. *US Patent* 5,677,400 (1997) to UBE.
[62] Wong, T. *US Patent* 5756606 (1998) to Goodyear.
[63] Saito, T. et al. *Kogyo Kagaku Zasshi* 1963, *66*, 1099.
[64] Natta, G.; Giannini, U.; Pino P.; Cassata, A. *Chim. Ind.* (Milan) 1965, *47*, 524.
[65] Tamai, K.; Saito, T.; Uccida, Y.; Risono, A. *Bull. Chem. Soc. Japan* 1965, *38*, 1578.
[66] Brady, C. et al. *US Patent* 5879805 (1999) to Union Carbide.
[67] Hahn, K. et al. *US Patent* 4324939 (1982) to Huels.
[68] Kampf, W. et al. *US Patent* 3966697 (1976) to Huels.
[69] Beck, M. et al. *US Patent* 4285833 (1981) to Bayer.
[70] Ichikawa, M.*US Patent* 3498963 (1970) to JSR.
[71] Ichikawa, M. *US Patent* 3,522,332 (1970) to JSR.
[72] Makino, K. *US Patent* 4,182,813 (1980) to JSR.
[73] Goto, K. et al. *US Patent* 5,548,045 (1996) to JSR.
[74] Shibata, M. et al. *US Patent* 7,009,013 (2006) to JSR.
[75] Dierkes, P.; van Leeuwen, P. W. N. M. *J. Chem. Soc., Dalton Trans.* 1999, 1519.

[76] van Leeuwen, P. W. N. M.; Kamer, P. C. J.; Reek, J. N. H.; Dierkes, P. *Chem. Rev.* 2000, *100*, 2741.
[77] Freixa, Z.; van Leeuwen, P. W. N. M. *Dalton Trans.* 2003, 1890.
[78] Tolman, C. *Chemical Reviews.* 1977, *77*, 313.
[79] Porri, L. in *Structural Order in Polymers*, Ciardelli, F.; Giusti, P. Eds. Pergamon Press Ltd,. Oxford, 1981, p.51.
[80] Porri, L.; Giarrusso, A; Ricci, G. *Progress Polym. Sci.* 1991, *16*, 405.
[81] Porri, L.; Giarrusso, A.; Ricci, G. *Makromol. Chem., Macromol. Symp.* 1991, *48/49*, 239.
[82] Porri, L.; Giarrusso, A.; Ricci, G. *Makromol. Chem., Macromol. Symp.* 2002, *178*, 55.
[83] Chatt J.; Shaw, B.L. *J. Chem. Soc.* 1961, 285.
[84] Ricci, G.; Forni, A.; Boglia, A.; Motta, T. *J. Mol. Cat. A: Chemical* 2005, *226*, 235.
[85] Ricci, G.; Masi, F.; Sommazzi, A.; Santi, R.; Battistella, M.; Righi, S.; Savini, G. *WO Patent* 2003018649 (2003) to Polimeri Europa.
[86] Ricci, G.; Forni, A.; Boglia, A.; Motta, T.; Zannoni, G.; Canetti, M.; Bertini, F. *Macromolecules* 2005, *38*, 1064.
[87] Ricci, G.; Forni, A.; Boglia, A.; Sommazzi A.; Masi, F. *J. Organomet. Chem.* 2005, *690*, 1845.
[88] Ricci, G. ; Boglia, A.; Santi, R.; Sommazzi, A.; Masi, F. *It. Patent* Mi03A001807 (2003) to Polimeri Europa.
[89] Ricci, G. ; Boglia, A.; Forni, A.; Santi, R.; Sommazzi, A.; Masi *It. Patent* Mi03A001808 (2003) to Polimeri Europa.
[90] Ricci, G.; Boglia, A.; Masi, F.; Santi, R.; Sommazzi, A. *It. Patent* Mi03A001809 (2003) to Polimeri Europa.
[91] Ricci, G.; Boffa G.; Porri, L. *Makromol. Chem., Rapid Commun.* 1986, *7*, 355.
[92] Alt, F. P.; Heitz, W. *Macromol. Chem. Phys.* 1998, *199*, 1951–195.
[93] Unpublished results of our laboratory.
[94] Takeushi, M.; Shiono, T.; Soga, K. *Polymer International* 1992, *29*, 209.
[95] Takeushi, M.; Shiono, T.; Soga, K. *Macromol. Rapid Commun.* 1995, *16*, 373.
[96] Morero, D.; Santambrogio, A.; Porri, L.; Ciampelli, F. *Chim. Ind.* (Milan) 1959, *41*, 758.
[97] Ricci, G.; Motta, T.; Boglia, A.; Alberti, E.; Zetta, L.; Bertini, F.; Arosio, P.; Famulari, A.; Meille, S. V. *Macromolecules* 2005, *38*, 8345-8352.
[98] Natta, G.; Corradini, P. *J. Polym. Sci.* 1956, *20*, 251.
[99] Ricci, G.; Porri, L. *Polymer* 1997, *38*, 4499.
[100] Proto, A.; Capacchione, C.; Venditto, V.; Okuda, J. *Macromolecules* 2003, *36*, 9249.
[101] Ricci, G.; Boglia, A; Motta, T.; Bertini, F; Boccia, A. C.; Zetta, L.; Alberti, E.; Arosio, P.; Famulari, A.; Meiile, S. V. *Macromolecules,* submitted.
[102] Guerra, G.; Cavallo, L.; Corradini, P.; Fusco, R. *Macromolecules* 1997; *4*, 677.
[103] Peluso, A.; Improta, R.; Zambelli, A. *Organometallics* 2000, *19*, 411.
[104] Costabile, C.; Milano, G.; Cavallo, L.; Guerra, G. *Macromolecules* 2001, *34*, 7952.
[105] Costabile, C.; Milano, G.; Cavallo, L.; Longo, P.; Guerra, G:; Zambelli, A. *Polymer* 2004, *45*, 467.
[106] Ricci, G.; Italia, S.; Giarrusso A.; Porri, L. *J. Organomet. Chem.* 1993, *451*, 67.
[107] Ricci, G.; Giarrusso, A.; Porri, L. *Macromol. Symp.* 1995, *89*, 383.
[108] Ricci, G.; Bosisio, C.; Porri, L. *Macromol. Rapid Commun.* 1996, *17*, 781.

[109] Ricci, G.; Forni, A.; Boglia, A.; Sonzogni, M. *Organometallics.* 2004, *23*, 3727.
[110] Ricci, G.; Boglia, A.; Motta, T. *J. Mol. Cat A: Chemical* 2007, *267*, 102.
[111] Porri, L.; Vitulli, G.; Zocchi, M.; Allegra, G. *Chem. Commun.* 1969, 276.
[112] Vitulli, G.; Porri, L.; Segre, A. L. *J. Chem. Soc. (A)* 1971, 3246.
[113] Ricci, G.; Battistella, M.; Porri, L. *Macromolecules* 2001, *34*, 5766.
[114] Zhang, Y.; Zhang, H. J.; Ma, H. M.; Wu, Y. *J. Mol. Catal.* 1982, *17*, 65.
[115] Masi, F.; Santi, R.; Sommazzi, A. *Chim. Ind.* (Milan) 2003, *85*, 1.

In: Advances in Organometallic Chemistry Research ISBN: 978-1-60021-779-1
Editor: Kenji Yamamoto, pp. 37-65

Chapter 2

HALF-SANDWICH RUTHENIUM COMPLEXES CONTAINING WATER-SOLUBLE PHOSPHINE LIGANDS: CATALYTIC APPLICATIONS

Victorio Cadierno* and Pascale Crochet**

Departamento de Química Orgánica e Inorgánica, Instituto Universitario de Química Organometálica "Enrique Moles" (Unidad Asociada al CSIC), Facultad de Química, Universidad de Oviedo, Julián Clavería 8, E-33006 Oviedo, Principado de Asturias, Spain

ABSTRACT

Over the last two decades the interest in ruthenium-catalyzed reactions directed to organic synthesis has spectacularly increased and a large number of new highly efficient synthetic approaches are nowadays well documented. On the other hand, the development of organic transformations in aqueous media has become one of the major cornerstones in modern chemistry. Water as a solvent is not only inexpensive and environmentally benign, but also gives, in some cases, a completely new reactivity. In this review, our focus is on ruthenium-catalyzed organic synthesis in aqueous media. In particular, we report on the synthesis, reactivity and catalytic applications of half-sandwich ruthenium(II) complexes (*i.e.* arene- and cyclopentadienyl-type derivatives) containing water-soluble phosphine ligands.

INTRODUCTION

Ruthenium holds a prominent position among the efficient transition-metals involved in catalytic processes directed to organic synthesis. The wide variety of ruthenium complexes actually known, featuring several oxidation states, coordination numbers and geometries,

* V. Cadierno; E-mail: vcm@uniovi.es
** P. Crochet; E-mail: pascale@fq.uniovi.es

provides a great potential for the development of new catalytic reactions with a large panel of applications. This fact has allowed the preferred choice of ruthenium complexes as catalysts (or precatalysts) and thereby a large number of highly efficient synthetic approaches have been recently disclosed. Several books [1] and review articles [2] on ruthenium-catalyzed organic synthesis have appeared during the past decade, and they are recommended to readers seeking a broader introduction to the topic.

On the other hand, new achievements in synthetic organic chemistry have appeared very rapidly as part of the increasing academic and industrial interest in fulfilling the fundamental principles of "Green Chemistry" [3]. Thus, among the most important challenges of modern chemistry, organic reactions in water are of primary interest especially when they are performed under catalytic conditions [4]. As a matter of fact, the development of metal-mediated catalytic transformations in aqueous media represents nowadays one of the major cornerstones of organometallic chemistry [5,6]. Water is undoubtedly the most "green" solvent one can imagine in terms of cost, availability, safety and environmental impact. Moreover, although the hydrophobic character of most organic compounds may be a drawback, the recovery of the organic products from an aqueous medium can be easily achieved by simple decantation or extraction processes. Introduction of hydrophilic ligands, specially phosphines, in the coordination sphere of a transition-metal is probably the most popular method for the preparation of water-soluble catalysts [5,6]. Thus, a wide variety of functionalized phosphine ligands containing highly polar sulfonated, ammonium, phosphonium, phosphonate, carboxylate, carbohydrate or hydroxyalkyl groups are actually known and their effectiveness to induce the solubility of transition-metal complexes in water largely demonstrated [5,6].

Half-sandwich-type complexes, *i.e.* η^5-cyclopentadienyl (**A**), η^5-indenyl (**B**) or η^6-arene (**C**) derivatives (see Figure 1), are probably the most studied organometallic ruthenium species due to their rich and versatile coordination chemistry [7] and catalytic properties [1,2,8]. Beside catalytic transformations in classical organic media, a variety of applications in water have very recently emerged by the aid of water-soluble ligands. Thus, in this contribution, we provide a detailed overview of the chemistry of half-sandwich ruthenium complexes containing water-soluble phosphines, paying special attention to their catalytic applications. This review article, which covers the literature up to January 2007, is roughly divided into sections based on the nature of the coordinated *P*-donor ligand.

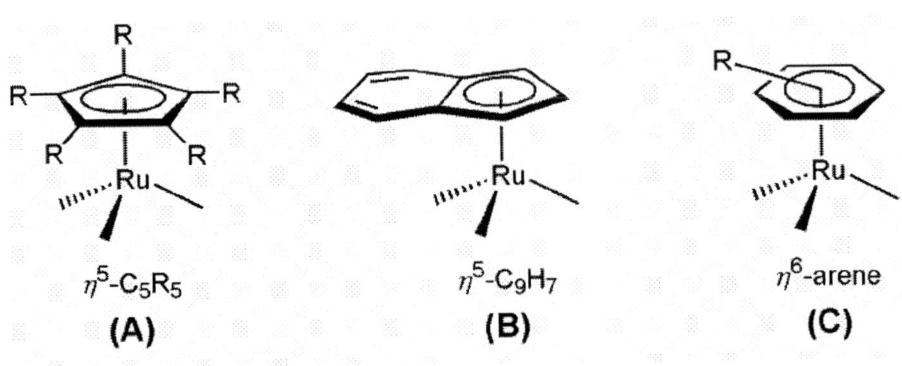

Figure 1. The half-sandwich ruthenium fragments **A**, **B** and **C**.

PHOSPHINE LIGANDS CONTAINING SULFONATED GROUPS

Phosphines containing sulfonated substituents are undoubtedly the most widely used ligands to induce the solubility in water of transition-metal complexes [5,6]. Among them, (3-sulfonatophenyl)diphenylphosphine (TPPMS) and tris(3-sulfonatophenyl)phosphine (TPPTS) (Figure 2) are the most popular due to their easy access by direct sulfonation of inexpensive triphenylphosphine (PPh_3) with oleum [9]. In general, both are used as their corresponding sodium salts, their solubility in water being extremely high (S = 80 $g{\cdot}L^{-1}$ and 1100 $g{\cdot}L^{-1}$, respectively, at room temperature) [9]. In this section we will discuss on the coordination chemistry of this type of ligands in half-sandwich ruthenium(II) fragments and the catalytic applications of the resulting water-soluble complexes.

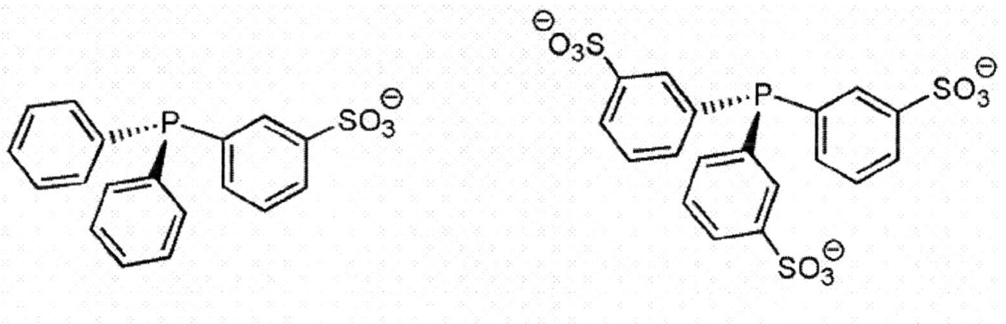

Figure 2. Structure of the sulfonated phosphines TPPMS and TPPTS.

Thus, concerning the monosulfonated phosphine TPPMS, the cyclopentadienyl derivative $[RuCl(\eta^5\text{-}C_5H_5)(TPPMS)_2]$ has been synthesized following two alternative routes: (*i*) by reacting $RuCl_3{\cdot}3H_2O$ with an excess of the TPPMS sodium salt and cyclopentadiene in refluxing ethanol (56% yield) [10] or (*ii*) by displacement of the PPh_3 ligands in $[RuCl(\eta^5\text{-}C_5H_5)(PPh_3)_2]$ upon treatment with two equivalents of TPPMS in refluxing toluene (90% yield) [11]. This ruthenium(II) complex has found to be a good catalytic precursor for the hydrogenation of 1-hexene in water/toluene biphasic medium. Thus, under moderate reaction conditions (T = 100°C, $p(H_2)$ = 500 psi, substrate / catalyst ratio = 500:1), 82% conversion of 1-hexene into a mixture containing mainly *n*-hexane (60%) and minor amounts of *cis*-2-hexene (13%) and *trans*-2-hexene (8%) could be attained. In addition, the active ruthenium species is fairly stable in aqueous phase being recovered and reused four consecutive runs without losing its catalytic activity [10]. Complex $[RuCl(\eta^5\text{-}C_5H_5)(TPPMS)_2]$ cleanly reacts with $SnCl_2$ to generate the heterobimetallic derivative $[Ru(SnCl_3)(\eta^5\text{-}C_5H_5)(TPPMS)_2]$ which is able to catalyze the electrooxidation of methanol to dimethoxymethane. The best results in terms of selectivity (up to 100%) and current efficiency (moles of product formed per mol of charge passed up to 90.1%) were obtained performing the reactions in neat methanol and using a potential range of 1.25-1.40 V (maximum TON = 40) [11]. Remarkably, under similar reaction conditions, its chloride counterparts $[RuCl(\eta^5\text{-}C_5H_5)(TPPMS)_2]$ and $[RuCl(\eta^5\text{-}C_5H_5)(PPh_3)_2]$ were found to be much less active and selective [11].

The reaction of two equivalents of the potassium salt of TPPMS with the dimeric precursors [{RuCl(μ-Cl)(η^6-arene)}$_2$], in methanol at r.t., generates the water-soluble mononuclear derivatives [RuCl$_2$(η^6-arene)(TPPMS)] (arene = benzene, *p*-cymene) [12] *via* the expected chloride-bridges cleavage [7,13]. The disubstituted species [RuCl(η^6-arene)(TPPMS)$_2$][Cl] (arene = benzene, *p*-cymene) could be also prepared by using four equivalents of TPPMS [12]. Remarkably, the solid state X-ray analysis for [RuCl$_2$(η^6-*p*-cymene)(TPPMS)] revealed that complex anions are held together in the crystal lattice by weak electrostatic interactions with potassium cations, leading to a linear chain structure which is retained in non-polar solvents. The nitrile adduct [RuCl(NCMe)(η^6-*p*-cymene)(TPPMS)][SbF$_6$] and the hydride derivative [RuHCl(η^6-*p*-cymene)(TPPMS)] were synthesized from [RuCl$_2$(η^6-*p*-cymene)(TPPMS)] *via* chloride abstraction with AgSbF$_6$ in the presence of acetonitrile or by treatment of a methanolic solution with NaOMe, respectively [12]. Complexes [RuCl$_2$(η^6-arene)(TPPMS)] and [RuCl(η^6-arene)(TPPMS)$_2$][Cl] proved to be excellent catalysts for transfer hydrogenation of cyclohexanone using 2-propanol as solvent and hydrogen source (turnover frequencies TOF up to 2982 h^{-1}; Scheme 1), being much more active than the related indenyl-ruthenium(II) derivative [RuCl(η^5-C$_9$H$_7$)(TPPMS)$_2$] (TOF = 70 h^{-1}). Other aliphatic and aromatic ketones could be also efficiently reduced using [RuCl$_2$(η^6-*p*-cymene)(TPPMS)] as catalyst. Unfortunately, the activity of these arene-Ru(II) complexes decreases drastically when water is used as solvent both under homogeneous or biphasic conditions [12].

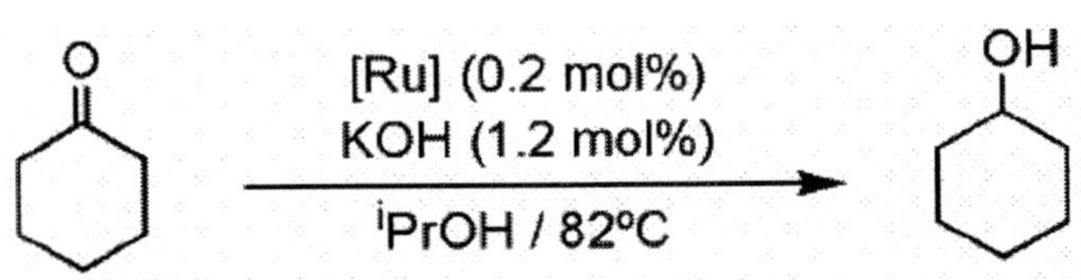

catalyst	time	yield	TOF
[RuCl$_2$(η^6-*p*-cymene)(TPPMS)]	15 min	>99%	1994 h^{-1}
[RuCl$_2$(η^6-benzene)(TPPMS)]	10 min	>99%	2982 h^{-1}
[RuCl(η^6-*p*-cymene)(TPPMS)$_2$][Cl]	45 min	>99%	663 h^{-1}
[RuCl(η^6-benzene)(TPPMS)$_2$][Cl]	30 min	>99%	1000 h^{-1}

Scheme 1. Transfer hydrogenation of cyclohexanone catalyzed by Ru-TPPMS complexes.

The ruthenium-benzocrownether complex **1** (Figure 3), conventionally synthesized from [{RuCl(μ-Cl)(η^6-benzo-15-crown-5)}$_2$] and the sodium salt of TPPMS, has also been described in the literature albeit it has not been applied in catalysis [14]. Remarkably, as inferred by NMR measurements, the complexation of the alkali metal by the crownether unit imposes the sulfonate-substituted phenyl ring to adopt a specific orientation *via* Na$^+$·····SO$_3^-$ electrostatic interactions.

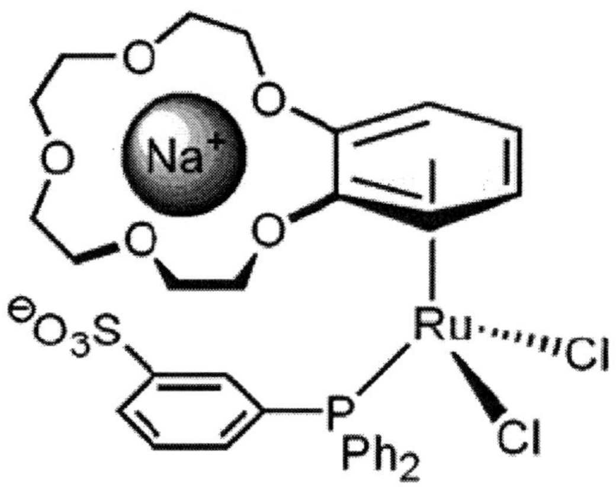

Figure 3. Structure of complex **1**.

Concerning the trisulfonated phosphine ligand TPPTS, the chloride complexes [RuCl$_2$(η^6-*p*-cymene)(TPPTS)] [15] and [RuCl(η^6-benzene)(TPPTS)$_2$][Cl] [16] are known. They have been obtained by refluxing water or methanol solutions of the appropriate dimer [{RuCl(μ-Cl)(η^6-arene)}$_2$] with two or four equivalents of TPPTS (trisodium salt), respectively. The neutral derivative [RuCl$_2$(η^6-*p*-cymene)(TPPTS)] is a highly efficient catalyst for the hydrogenation of benzene and monosubstituted arenes in aqueous media (Scheme 2), being much more active than the related PTA-based complexes [RuCl$_2$(η^6-*p*-cymene)(PTA)] and [RuCl(η^6-*p*-cymene)(PTA)$_2$][BF$_4$] (PTA = 1,3,5-triaza-7-phosphatricyclo[3.3.1.1]decane) [15]. Remarkably, the addition of mercury to the hydrogenations using [RuCl$_2$(η^6-*p*-cymene)(TPPTS)] results in significantly reduced turnovers, suggesting that the active species is a colloid stabilised by free TPPTS which acts as a surfactant. In contrast, the catalytic activity shown by the PTA-based catalysts remains unaffected by the addition of mercury, the formation of a trinuclear ruthenium cluster and a mononuclear complex, respectively, as the active species, being in these cases proposed. In the context of pulp yellowing prevention, related studies directed to the hydrogenation of milled wood lignin aromatics have also been reported [17].

Hydride derivatives containing the trisulfonated phosphine TPPTS (as the corresponding trisodium salt), *i.e.* [RuH(η^6-arene)(TPPTS)$_2$][Cl] (arene = benzene, toluene, *p*-xylene, ethylbenzene, cumene, tetraline), are known. They have been synthesized in high yields (87-92%) by reacting an aqueous solution of [{RuCl(μ-Cl)(TPPTS)$_2$}$_2$] with an excess of the appropriate arene ligand under H_2 pressure (35 bar) [18]. Formation of [RuH(η^6-arene)(TPPTS)$_2$][Cl] (arene = benzene, *p*-cymene) by hydrogenation of aqueous solutions of the dimeric precursors [{RuCl(μ-Cl)(η^6-arene)}$_2$] in the presence of TPPTS has also been reported [19]. Complex [RuH(η^6-benzene)(TPPTS)$_2$][Cl] has been checked as catalyst for the hydrogenation of aqueous bicarbonate to formate, showing only a modest activity [19]. A similar behaviour was also observed in the hydrogenation of CO_2 to formic acid using the catalytic system [{RuCl(μ-Cl)(η^6-benzene)}$_2$] / TPPTS (phosphine / Ru ratio = 3:1) and methanol as solvent [20].

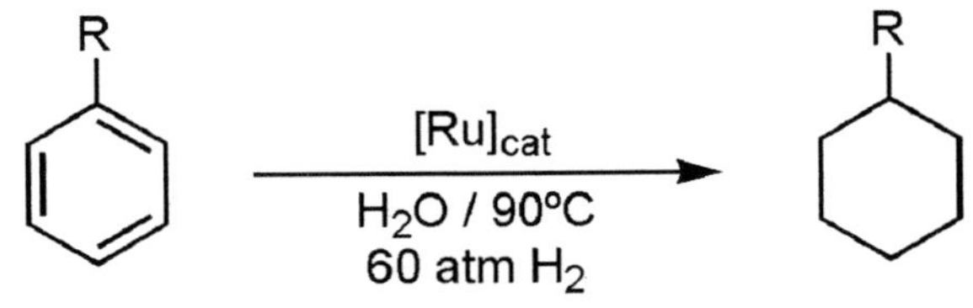

substrate	catalyst	TOF (h^{-1})
C_6H_6	$[RuCl_2(\eta^6\text{-}p\text{-cymene})(TPPTS)]$	488
C_6H_5Me	$[RuCl_2(\eta^6\text{-}p\text{-cymene})(TPPTS)]$	365
C_6H_5Et	$[RuCl_2(\eta^6\text{-}p\text{-cymene})(TPPTS)]$	245
$C_6H_5{}^iPr$	$[RuCl_2(\eta^6\text{-}p\text{-cymene})(TPPTS)]$	98
C_6H_6	$[RuCl_2(\eta^6\text{-}p\text{-cymene})(PTA)]$	170
C_6H_5Me	$[RuCl_2(\eta^6\text{-}p\text{-cymene})(PTA)]$	130
C_6H_5Et	$[RuCl_2(\eta^6\text{-}p\text{-cymene})(PTA)]$	122
C_6H_6	$[RuCl(\eta^6\text{-}p\text{-cymene})(PTA)_2][BF_4]$	150
C_6H_5Me	$[RuCl(\eta^6\text{-}p\text{-cymene})(PTA)_2][BF_4]$	129
C_6H_5Et	$[RuCl(\eta^6\text{-}p\text{-cymene})(PTA)_2][BF_4]$	72

Conditions: catalyst (30 mg), water (10 mL), substrate (1 mL), 90°C, 1 h.

Scheme 2. Ruthenium-catalyzed hydrogenation of arenes in water.

A series of mononuclear arene-ruthenium complexes containing the chelating sulfonated diphosphine DPPBTS (1,2-bis(di-4-sulfonatophenylphosphino)benzene tetrasodium salt) has been described (Figure 4). They have been prepared in moderate to good yields (47-80%) by direct reaction of the appropriate dimer $[\{RuCl(\mu\text{-}Cl)(\eta^6\text{-arene})\}_2]$ with two equivalents of DPPBTS in a mixture water-methanol (1:1), being formulated as anionic species, *i.e.* $[RuCl(\eta^6\text{-arene})(DPPBTS)][Na]_3$ (arene = benzene, *p*-cymene, [2.2]paracyclophane) [21]. These complexes are active catalysts for styrene hydrogenation in water leading to the selective and quantitative formation of ethylbenzene within 2 hours (T = 100°C, $p(H_2)$ = 45 bar, substrate / catalyst ratio = 2000). By the use of high gas pressure NMR techniques and electrospray ionization mass spectrometry (ESI-MS), the active species in the hydrogenation have been indirectly identified as the dihydrogen complexes $[Ru(H_2)(\eta^6\text{-arene})(DPPBTS)][Na]_2$, which also catalyze H/D isotope exchange. In addition, the hydride derivative $[RuH(\eta^6\text{-ethylbenzene})(DPPBTS)][Na]_3$ could be identified in the reaction mixtures by ESI-MS, pointing out that arene ligand exchange takes place during catalysis [21].

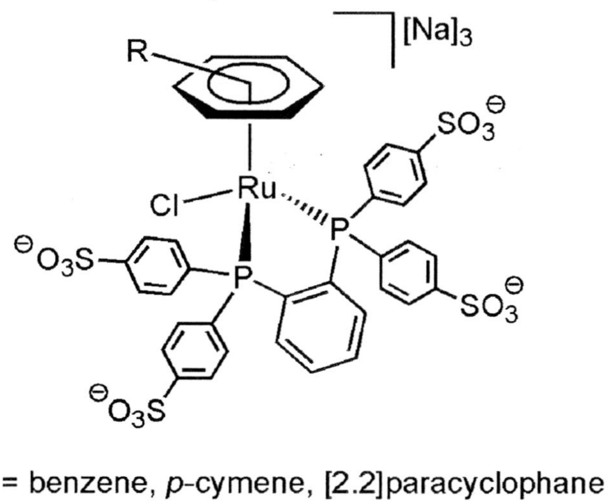

Figure 4. Water-soluble arene-Ru(II) complexes containing the DPPBTS ligand.

The cyclopentadienyl derivative [RuCl(η^5-C_5H_5)(DPPBTS)] is also known, being synthesized by reacting [RuCl(η^5-C_5H_5)(COD)] (COD = 1,5-cyclooctadiene) with DPPBTS [22]. This complex catalyzes the selective *anti*-Markovnikov hydration of 1-hexyne. Thus, using a 5 mol% of Ru, in a mixture 2-propanol / water (3:1 v/v) at 100°C, hexanal is formed in 84% yield after 12 h. Nevertheless, we note that, despite its remarkable activity, the effectiveness shown by this complex is by far lower than that of [RuCl(η^5-C_5H_5)(dppm)] (dppm = bis(diphenylphosphino)methane) which, under similar reaction conditions, leads to a quantitative conversion after 12 h using a catalyst loading of only 1 mol% [22]. At this point, it should be mentioned that the catalytic system [{RuCl(μ-Cl)(η^6-benzene)}$_2$] / TPPTS (phosphine / Ru ratio = 1:4) has also been used to promote the *anti*-Markovnikov hydration of terminal alkynes, resulting much less competitive than the cyclopentadienyl-based catalysts [16].

The application of half-sandwich ruthenium(II) complexes containing water-soluble chiral sulfonated phosphine ligands in asymmetric catalysis has been described. Thus, the η^6-benzene complex **2**, bearing the optically active dibenzofuran-based diphosphine (*R*)-2,2′-bis(diphenylphosphino)-1,1′-dibenzofuranyl-8,8′-disulfonic acid dipotassium salt, has been used as catalyst for the asymmetric hydrogenation of (*Z*)-acetamidocinnamic acid and methyl acetoacetate (Scheme 3) [23]. Thus, hydrogenation of the former resulted in a product with 72% *e.e.* in methanol as well as in a 1:1 water / ethylacetate mixture. Nevertheless, it should be noted that the reaction rate in the aqueous medium was extremely low due to the poor solubility of the substrate in the two-phase system. In contrast, the hydrogenation of methyl acetoacetate in water proceeds at the same rate than in methanol, a slight drop in the enantiomeric excess of the hydrogenated product being in this case observed.

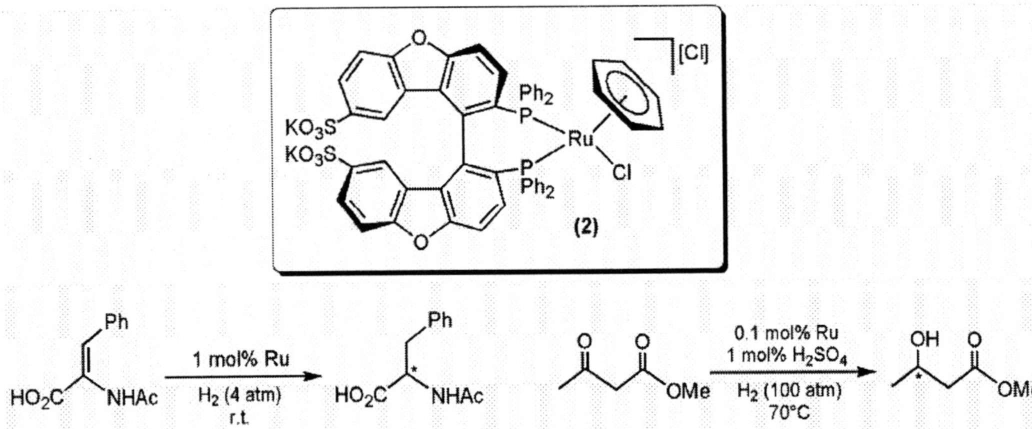

solvent	time	conversion	e.e.
MeOH	48 h	100%	72% (*R*)
H_2O / EtOAc	48 h	18%	72% (*R*)

solvent	time	conversion	e.e.
MeOH	2 h	100%	100% (*R*)
H_2O	2 h	100%	86% (*R*)

Scheme 3. Asymmetric hydrogenation of (*Z*)-acetamidocinnamic acid and methyl acetoacetate using complex **2**.

Finally, it should be also mentioned that (η^6-arene)-ruthenium complexes containing the atropisomeric C_2-symmetric diphosphines **3-6** (Figure 5; for all of them the *R* and *S* enantiomers have been described), and their utility for the asymmetric hydrogenation of olefins, ketones or imines in aqueous media, have been patented [24-26].

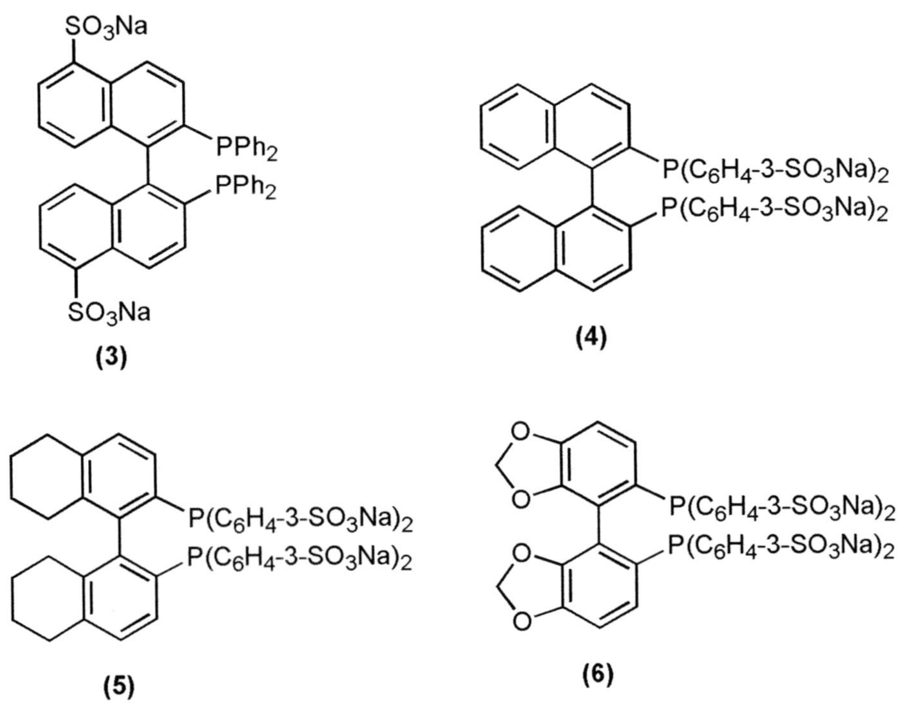

Figure 5. Structure of the optically active diphosphines **3-6**.

PTA AND RELATED *P*-DONOR LIGANDS

The PTA Ligand

Although it is known since 1974 [27], the cage-like 1,3,5-triaza-7-phosphatricyclo[3.3.1.1]decane ligand (usually abbreviated as PTA; see Figure 6) and its transition-metal complexes have received a renewed interest in recent years due to their high water-solubility [28]. This property is generally attributed to the ability of the nitrogen atoms to form N···H hydrogen bonds with water and their easy protonation in protic solvents to generate ammonium species, the latter phenomena being more or less important depending on the pH of the media. In addition to its high solubility in water (S = 235 $g\cdot L^{-1}$ in pure water; S = 350 $g\cdot L^{-1}$ in 0.1M HCl), the PTA ligand presents the advantage of being very stable against oxidation and, therefore, easy to handle. Furthermore, it is readily accessible in multigram-scale from commercially available tetrakis(hydroxymethyl)phosphonium chloride, formaldehyde and hexamethylenetetraamine [29].

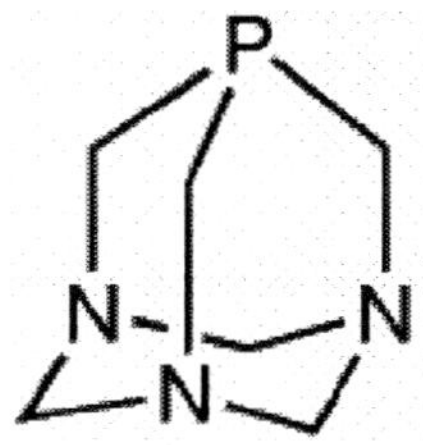

Figure 6. Structure of the PTA ligand.

The syntheses of a wide range of arene-ruthenium(II) complexes containing the coordinated PTA ligand have been carried out through conventional methodologies [7,13]. Thus, the chloride-bridged dimeric precursors [{RuCl(μ-Cl)(η^6-arene)}$_2$] readily react with two equivalents of PTA to give the corresponding water-soluble mononuclear derivatives [$RuCl_2$(η^6-arene)(PTA)] (arene = benzene [19], *p*-cymene [30], toluene, hexamethylbenzene, PhC(=O)OEt, **7** and **8** (Figure 7) [31], $PhCH_2CH_2OH$, $PhCH_2CH_2CH(Me)OH$, $PhCH_2NMe_2$, [$PhCH_2NMe_2H$][Cl], [$PhCH_2CH_2NH_3$][Cl] [32]) in high yields. Analogous iodide, bromide and isothiocyanate compounds [RuX_2(η^6-*p*-cymene)(PTA)] (X = I, Br, SCN) have been similarly prepared starting from the appropriate dinuclear precursors [{RuX(μ-X)(η^6-*p*-cymene)}$_2$] [33].

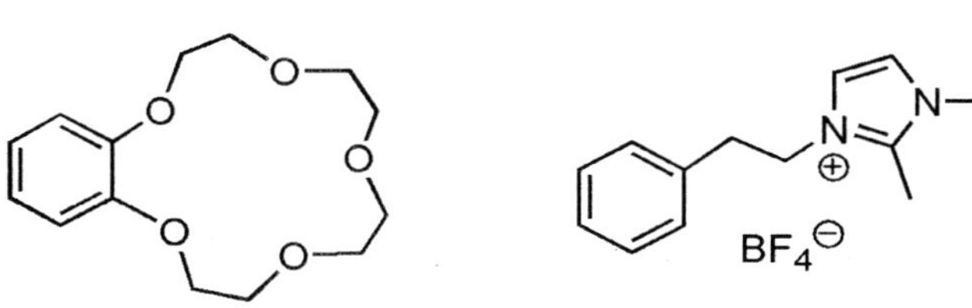

(7) **(8)**

Figure 7. The arene ligands **7** and **8**.

The related complexes [Ru(κ^2-*O,O*-C_2O_4)(η^6-*p*-cymene)(PTA)] (**9**) and [Ru(κ^2-*O,O*-$C_6H_6O_4$)(η^6-*p*-cymene)(PTA)] (**10**), containing the dianionic chelating ligands oxalate (C_2O_4) and 1,1-cyclobutanedicarboxylate ($C_6H_6O_4$), are also known [34]. They have been obtained through a two steps methodology involving the initial treatment of [{RuCl(μ-Cl)(η^6-*p*-cymene)}$_2$] with $Ag_2C_2O_4$ or $Ag_2C_6H_6O_4$ salts, respectively, and subsequent reaction with PTA. Interestingly, the replacement of the chloride ligands by the $C_2O_4^{2-}$ or $C_6H_6O_4^{2-}$ groups induces a significant increase in the water-solubility of the resulting complexes, **9** and **10** being five to ten times more soluble than [$RuCl_2$(η^6-*p*-cymene)(PTA)].

As a general trend, when [$RuCl_2$(η^6-arene)(PTA)] complexes are dissolved in water they are prone to dissociate one or two chloride ligands giving rise to the formation of mono- and/or diaqua-species, *i.e.* [RuCl(OH_2)(η^6-arene)(PTA)][Cl] and [Ru(OH_2)$_2$(η^6-arene)(PTA)][Cl]$_2$, respectively. However, these ligand-substitution processes can be partially or completely suppressed by adding NaCl or KCl to the solution [19,31]. The presence or not of chloride salts in the reaction media plays also a crucial role in the catalytic activity of this type of complexes. Thus, the efficiency of [$RuCl_2$(η^6-*p*-cymene)(PTA)] in the hydrogenation of benzene into cyclohexane is significantly different when the reaction is carried out in "halide-free" or in "chloride-contaminated" ionic liquids (Scheme 4) [35]. The inhibiting effect of the chloride anion has also been observed in the hydrogenation of toluene, ethylbenzene and chlorobenzene into the corresponding saturated cycloalkanes. Alternatively, these arenes can be reduced by [$RuCl_2$(η^6-*p*-cymene)(PTA)] using water as solvent, nevertheless, lower activities are obtained in this case (Scheme 4). Both in aqueous or ionic liquid media, the catalytic systems can be recycled at least 4 times without significant loss of activity.

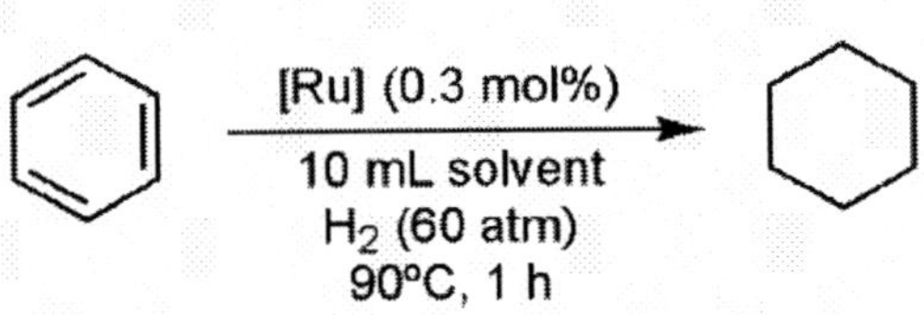

solvent	TOF
[bmim][BF_4][a,b]	140 h^{-1}
[bmim][BF_4][a,c]	206 h^{-1}
water	170 h^{-1}

[a] bmim: 1-butyl-3-methylimidazolium
[b] halide-contaminated, [Cl^-] = 0.2 mol/kg
[c] halide-free

Scheme 4. Hydrogenation of benzene catalyzed by [$RuCl_2$(η^6-*p*-cymene)(PTA)].

Treatment of an aqueous solution of the *N*-heterocyclic carbene complex [$RuCl_2$(η^6-*p*-cymene)(L)] (L = 1-butyl-3-methylimidazol-2-ylidene) with a stoichiometric amount of PTA generates the cationic derivative [RuCl(η^6-*p*-cymene)(PTA)(L)][Cl] (**11**) [36]. In absence of KCl, partial dissociation of the coordinated chloride ligand takes place giving rise to a mixture of [RuCl(η^6-*p*-cymene)(PTA)(L)][Cl] and [Ru(η^6-*p*-cymene)(OH_2)(PTA)(L)][Cl]$_2$ derivatives. This mixture, preformed *in situ*, has shown to be active in the catalytic hydrogenation of carbonylic and olefinic substrates in aqueous media (Figure 8). Under these conditions, both cinnamaldehyde and benzylideneacetone are preferentially hydrogenated at the C=C double bond [36].

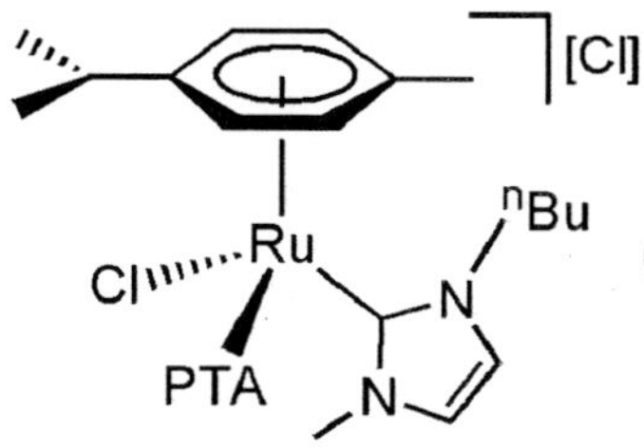

(11)

substrate	conversion (%)	TOF
acetone	98.2%	139 h^{-1}
acetophenone	46.1%	65 h^{-1}
allyl alcohol	95.6%[a]	135 h^{-1}
benzylideneacetone	42.3%	60 h^{-1}
cinnamaldehyde	42.0%	59 h^{-1}
propanal	86.2%	122 h^{-1}
4-styrenesulfonic acid Na salt	27.4%	39 h^{-1}

Conditions: $p(H_2)$ = 10 bar, T = 80°C, [Ru] = 4.73 mM in water, [substrate] = 667mM, 1 h, pH = 6.90 (phosphate buffer). [a] Isomerization into propanal included in the total conversion.

Figure 8. Hydrogenation reactions catalyzed by the water-soluble complex **11**.

The hydride bis-PTA derivatives $[RuH(\eta^6\text{-arene})(PTA)_2][Cl]$ (arene = benzene (**12**), *p*-cymene (**13**)) can be selectively generated by reacting dimers $[\{RuCl(\mu\text{-}Cl)(\eta^6\text{-arene})\}_2]$ with PTA in water under hydrogen pressure (100 bar) [19]. Nevertheless, it should be noted that, depending on the reaction conditions (*i.e.* temperature, PTA concentration), traces of the monophosphine complexes $[RuClH(\eta^6\text{-arene})(PTA)]$ and $[RuH(\eta^6\text{-arene})(OH_2)(PTA)][Cl]$, as well as the arene-free derivatives $[RuClH(PTA)_4]$, $[RuH(OH_2)(PTA)_4][Cl]$, $[RuH_2(PTA)_4]$ and $[RuH(PTA)_5][Cl]$, can also be formed. Complexes **12** and **13** have been proposed as the catalytically active species in the hydrogenation of sodium bicarbonate into sodium formate using as catalyst a mixture of $[\{RuCl(\mu\text{-}Cl)(\eta^6\text{-arene})\}_2]$ (arene = benzene, *p*-cymene) and PTA in aqueous media (0.2 mol% Ru; PTA / Ru ratio = 2; T = 60°C, $p(H_2)$ = 100 bar). Under these conditions, almost quantitative conversions in sodium formate are attained after 24 hours [19].

The water-soluble tethered arene-ruthenium(II) derivative $[RuCl(\eta^6{:}\kappa^1\text{-}N\text{-}C_6H_5CH_2CH_2NH_2)(PTA)][BF_4]$ is also known, being easily prepared by deprotonation of the organometallic ammonium salt $[RuCl_2(\eta^6\text{-}C_6H_5CH_2CH_2NH_3)(PTA)][Cl]$ and subsequent treatment with $AgBF_4$ [32]. The biological activities of this derivative, as well as those of most of the PTA-complexes described above, have been investigated [30-34,37,38], some of them showing interesting anticancer, antimicrobial and antiviral properties.

The PTA ligand has also been coordinated to indenyl and cyclopentadienyl-type ruthenium(II) fragments. Thus, treatment of the appropriate bis-triphenylphosphine precursor $[RuCl(\eta^5\text{-}C_nH_m)(PPh_3)_2]$ with a stoichiometric amount of PTA affords the corresponding monosubstituted derivatives $[RuCl(\eta^5\text{-}C_nH_m)(PPh_3)(PTA)]$ (C_nH_m = C_5H_5, C_9H_7, 1,2-

dihydropentalenyl) in moderate to good yields (50-89%) [39,40]. When a two-fold excess of PTA is employed, the disubstituted species [RuCl(η^5-C_nH_m)(PTA)$_2$] (C_nH_m = C_5H_5, 1,2-dihydropentalenyl) are selectively formed [40,41,42]. The related complex [RuCl(η^5-$C_5H_4CH_2CH_2NEt_2$)(PTA)$_2$], containing an amino-functionalized cyclopentadienyl ring, has been similarly prepared starting from the bis-triphenylphosphine derivative [RuCl(η^5-$C_5H_4CH_2CH_2NEt_2$)(PPh_3)$_2$] [43]. In contrast to these results, the treatment of the indenyl precursor [RuCl(η^5-C_9H_7)(PPh_3)$_2$] with two equivalents of PTA generates the cationic species [Ru(η^5-C_9H_7)(PPh_3)(PTA)$_2$][Cl] selectively [40]. The pentamethylcyclopentadienyl complex [RuCl(η^5-C_5Me_5)(PTA)$_2$] is also known, being accessible through two alternative synthetic routes: (*i*) by displacement of both PPh_3 ligands in [RuCl(η^5-C_5Me_5)(PPh_3)$_2$], upon treatment with two equivalents of PTA in refluxing toluene, or *(ii)* by reduction of the ruthenium(III) precursor [{RuCl(μ-Cl)(η^5-C_5Me_5)}$_2$] with zinc dust in the presence of PTA [41]. The preparation of [RuCl(η^5-C_5H_5)(PTA)$_2$] starting from [RuCl(η^5-C_5H_5)(COD)] (COD = 1,5-cyclooctadiene) has also been described [42].

A series of cationic PTA-containing cyclopentadienyl-ruthenium(II) compounds are known. Thus, the acetonitrile derivatives [Ru(η^5-C_5H_5)(NCMe)$_2$(PTA)][PF_6] and [Ru(η^5-C_5H_5)(NCMe)(PTA)$_2$][PF_6] have been easily prepared, under smooth conditions, by addition of one or two equivalents of PTA, respectively, to dichloromethane solutions of [Ru(η^5-C_5H_5)(NCMe)$_3$][PF_6] [43]. In contrast, the related species [Ru(η^5-C_5Me_5)(NCMe)(PTA)$_2$][PF_6] and [Ru(η^5-$C_5H_4CH_2CH_2NEt_2$)(NCMe)(PTA)$_2$][PF_6] were obtained starting from the corresponding chloride precursors, *i.e.* [RuCl(η^5-C_5Me_5)(PTA)$_2$] and [RuCl(η^5-$C_5H_4CH_2CH_2NEt_2$)(PTA)$_2$], *via* halide abstraction with $TlPF_6$ or NH_4PF_6 in the presence of acetonitrile [43]. Complex [Ru(η^5-$C_5H_4CH_2CH_2NEt_2$)(NCMe)(PTA)$_2$][PF_6] can be alternatively synthesized by treatment of a methanolic solution of [RuCl(η^5-$C_5H_4CH_2CH_2NEt_2$)(PTA)$_2$] with NH_4PF_6 and subsequent reaction of the isolable tethered intermediate [Ru(η^5:κ^1-*N*-$C_5H_4CH_2CH_2NEt_2$)(PTA)$_2$][PF_6] with acetonitrile [43].

Remarkably, besides the expected product [Ru(η^5-C_5H_5)(DMSO)(PTA)$_2$][OTf], the treatment of a dimethylsulfoxide (DMSO) solution of [RuCl(η^5-C_5H_5)(PTA)$_2$] with AgOTf generates the water-soluble polymer [{Ru(η^5-C_5H_5)(DMSO)(μ^2-*P,N*-PTA)$_2$}{$AgCl_2$}]$_n$ **(14)** (Scheme 5). This organometallic polymer, which is constituted by ruthenium cations [Ru(η^5-C_5H_5)(DMSO)(PTA)$_2$]$^+$ and bridging anionic [$AgCl_2$]$^-$ units coordinated through the nitrogen atoms of two different PTA ligands, can be selectively formed if a chloride source (NH_4Cl) is present in the reaction medium. Moreover, when the same reaction is carried out in the absence of DMSO (*i.e.* in chloroform solution) the related polymeric compound {[NH_4][{Ru(η^5-C_5H_5)(OTf)(PTA)$_2$}{$AgCl_2$}]}$_n$ (**15**), in which the coordination sphere of ruthenium is completed by the triflate anion, is formed (Scheme 5). When dissolved in DMSO, **15** is quantitatively converted into **14** *via* NH_4OTf elimination. We note also that, as assessed by dynamic light-scattering experiments, the polymeric structure of **14** is retained in water (only the substitution of the coordinated DMSO ligand by H_2O is observed) [44].

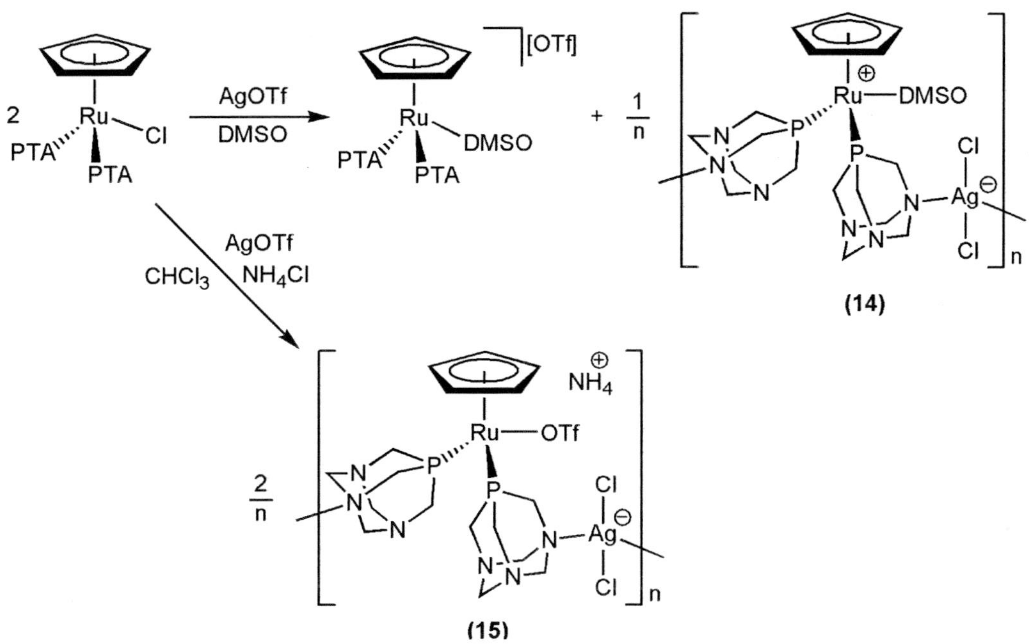

Scheme 5. Synthesis of the PTA-based water-soluble polymers **14** and **15**.

Another polymeric PTA-based compound, namely $[\{Ru(\eta^5\text{-}C_5H_5)(\mu^2\text{-}P,N\text{-}PTA)_2\}\{\mu^2\text{-}B,B\text{-}Ph_3B_3O_3\}]_n$ (**16**), has been obtained through the reaction of $[RuCl(\eta^5\text{-}C_5H_5)(PTA)_2]$ with 3 equivalents of $PhB(OH)_2$ in dichloromethane (Figure 9) [45]. The bridging triphenylboroxine unit, which links the $[RuCl(\eta^5\text{-}C_5H_5)(PTA)_2]$ fragments, results from the cyclocondensation of 3 molecules of phenylboronic acid. Unlike **14**, the structure **16** does not remain polymeric in aqueous solution, presumably due to hydrolysis of the coordinated triphenylboroxine.

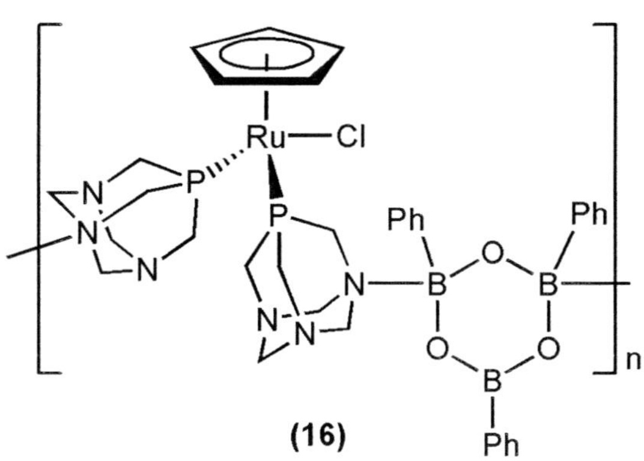

Figure 9. Structure of the organometallic polymer **16**.

Water-soluble cyclopentadienyl-ruthenium hydrides, containing at least one PTA ligand, are also known. Thus, complex $[RuH(\eta^5\text{-}C_5H_5)(PTA)_2]$ has been obtained in moderate yield (43%) by heating a mixture of $[RuCl(\eta^5\text{-}C_5H_5)(PTA)_2]$ and potassium hydroxide in methanol [42]. Similarly, treatment of $[RuCl(\eta^5\text{-}C_nH_m)(PPh_3)(PTA)]$ with sodium formate or sodium methoxide in hot methanol gives rise to the high-yield formation of [RuH(η^5-C_nH_m)(PPh_3)(PTA)] (C_nH_m = C_5H_5, C_9H_7, 1,2-dihydropentalenyl) [40]. In D_2O solution, the bis-PTA hydride derivative [RuH(η^5-C_5H_5)$(PTA)_2$] undergoes H/D exchange to generate [RuD(η^5-C_5H_5)Cp$(PTA)_2$] ($t_{\frac{1}{2}}$ = 127 min), whereas no evolution is observed in deuterated methanol [42]. In contrast, the mixed PTA-triphenylphosphine complexes [RuH(η^5-C_nH_m)(PPh_3)(PTA)] (C_nH_m = C_5H_5, C_9H_7) experiment H/D exchange in CD_3OD, the rate of the process being strongly dependent on the cyclopentadienyl-type ligand ($t_{\frac{1}{2}}$ << 10 min for C_5H_5 *vs.* $t_{\frac{1}{2}} \approx 5.5$ days for C_9H_7) [40].

Treatment of [RuH(η^5-C_5H_5)$(PTA)_2$] with a stoichiometric amount of HBF_4 leads to [RuH(η^5-C_5H_5)(PTA)(PTA·H)][BF_4], in which one PTA ligand has been protonated on a nitrogen atom. In the presence of an excess of HBF_4, double protonation takes place yielding [RuH(η^5-C_5H_5)$(PTA{\cdot}H)_2$][BF_4]$_2$ [42]. Both complexes have been spectroscopically detected but not isolated. In contrast, the related chloride species [RuCl(η^5-C_5H_5)(PTA)(PTA·H)][PF_6] and [RuCl(η^5-C_5H_5)$(PTA{\cdot}H)_2$][PF_6]$_2$, generated from [RuCl(η^5-C_5H_5)$(PTA)_2$] and HPF_6, could be fully characterized by NMR spectroscopy and solid-state X-ray diffraction techniques [46]. A more detailed study on the nature of the organometallic species formed from [RuH(η^5-C_5H_5)$(PTA)_2$] in aqueous solutions as function of the pH has evidenced that this complex is only stable at pH above 7. Lowering the pH to 5.9 affords a mixture of [RuH(η^5-C_5H_5)(PTA)(PTA·H)]$^+$ and the dihydride derivative [Ru$(H)_2$(η^5-C_5H_5)$(PTA)_2$]$^+$ (4:1 ratio), resulting from the protonation of the PTA ligand or the metal center, respectively. Both species are presumably in equilibrium through a dihydrogen intermediate [Ru(η^5-C_5H_5)(η^2-H_2)$(PTA)_2$]$^+$. At pH = 3.6 a mixture of both compounds is still present in the medium (ratio = 1.67:1), but at pH = 2.1 dihydride [Ru$(H)_2$(η^5-C_5H_5)$(PTA)_2$]$^+$ and the dicationic species [RuH(η^5-C_5H_5)$(PTA{\cdot}H)_2$]$^{2+}$ are the only species observed in solution [46].

This latter study evidences the complexity of the cyclopentadienyl-PTA-ruthenium(II) chemistry in aqueous media and, in particular, the crucial role of pH. Therefore, it is not surprising that the catalytic activity, as well as the selectivity, of these species is strongly pH-dependent. As a representative example, the performance shown by [RuH(η^5-C_5H_5)$(PTA)_2$] in the catalytic hydrogenation of benzylideneacetone is completely different in acid or basic media (Scheme 6) [46]. Thus, in the absence of buffer, as well as at pH ≤ 3.6 or pH ≥ 7, low conversions are observed (from 7 to 58% after 60 h, entries 1-4 and 8-9). However, high rate is reached at pH = 4.7, being the reaction almost complete within 24 h (entry 5). Interestingly, at a given pH (*i.e.* pH = 2.1) the nature of the buffer used (HCl/NaH_2PO_4 or HBF_4/NaH_2PO_4) has also a great influence both on the catalytic activity (7% *vs* 46% yield, entries 2-3) and on the selectivity. In the former case, benzylideneacetone is preferently hydrogenated at the C=O double bond affording 4-phenylbut-3-en-2-ol (**F**), while in the latter case 4-phenylbutan-2-one (**D**), resulting from the C=C reduction is formed predominantly. A similar effect of the anions Cl^- and BF_4^- on the reaction rate has been evidenced at pH = 4.7, using KCl and $NaBF_4$ as additives (entries 10-11), the optimal reaction conditions being reached with $NaBF_4$. Thus, by controlling the pH and the salinity of the medium, an almost quantitative hydrogenation of benzylideneacetone can be attained under smooth conditions (T = 25°C, p(H_2) = 60 psi) in only 6 hours (entry 11).

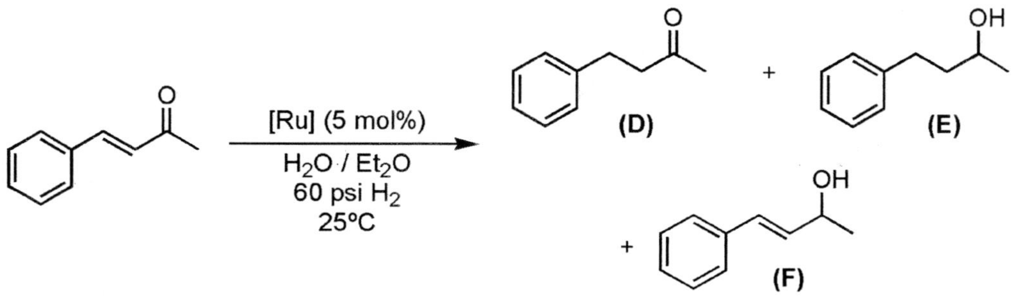

entry	buffer (pH)	conversion	D	E	F
1	-	7%	90%	5%	5%
2	2.1[a]	7%	22.5%	<1%	77.5%
3	2.1[b]	46%	>99%	<1%	<1%
4	3.6	90%	>99%	<1%	<1%
5	4.7	97%[c]	>99%	<1%	<1%
6	5.9	87%	99%	1%	0%
7	6.5	>99%	>99%	<1%	<1%
8	7.0	58%	>99%	<1%	<1%
9	9.8	11%	97%	3%	<1%
10[d]	4.7	5%[e]	ns	ns	ns
11[f]	4.7	>95%[e]	ns	ns	ns

Conditions: 5 mol% of Ru, 14.2 mg of substrate, H_2O / Et_2O = 5 mL / 5 mL, 3 mL of 0.1M phosphate buffer, 60 h, 25°C. [a] Buffer: HCl/NaH_2PO_4. [b] Buffer: HBF_4/NaH_2PO_4. [c] After 24 h. [d] Using 1.6 mmol of KCl. [e] After 6 h. [f] Using 1.6 mmol of $NaBF_4$. ns = not specified.

Scheme 6. Biphasic hydrogenation of benzylideneacetone catalyzed by $[RuH(\eta^5\text{-}C_5H_5)(PTA)_2]$.

More drastic conditions (T = 80-130°C, $p(H_2)$ = 450 psi) are necessary in the hydrogenations of benzylideneacetone promoted by the chloride derivatives $[RuCl(\eta^5\text{-}C_5R_5)(PTA)_2]$ (R = H, Me) (Scheme 7) [41]. In the course of these processes, hydride species $[RuH(\eta^5\text{-}C_5R_5)(PTA)_2]$ are assumed to be formed. As a general trend, the reduction of C=C *vs* C=O double bond is preferred but the selectivity decreases at high temperature. The best performances in term of activity were obtained with the cyclopentadienyl catalyst while the higher selectivity was observed using the pentamethylcyclopentadienyl one. Remarkably, complex $[RuCl(\eta^5\text{-}C_5H_5)(PTA)_2]$ can be recycled up to 3 consecutive runs.

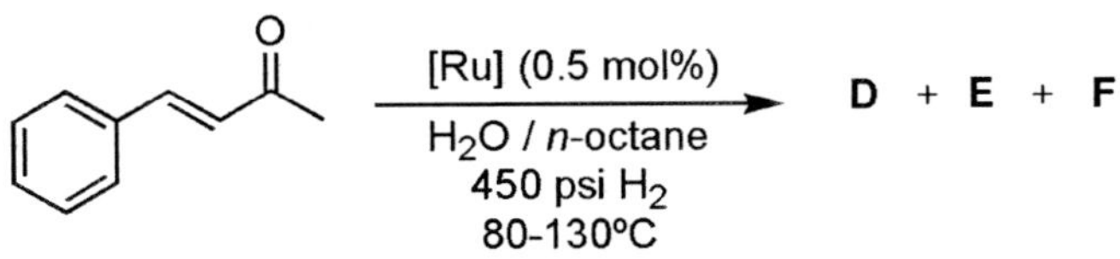

catalyst	T (°C)	conversion (time)	D	E	F
$[RuCl(\eta^5\text{-}C_5H_5)(PTA)_2]$	80	26.3% (13 h)	90%	6%	4%
$[RuCl(\eta^5\text{-}C_5H_5)(PTA)_2]$	130	78.5% (13 h)	76.4%	17.3%	6.3%
$[RuCl(\eta^5\text{-}C_5H_5)(PTA)_2]$	130	99.7% (21 h)	75%	22%	3%
$[RuCl(\eta^5\text{-}C_5Me_5)(PTA)_2]$	80	40.0% (13 h)	93.5%	4.3%	2.2%
$[RuCl(\eta^5\text{-}C_5Me_5)(PTA)_2]$	130	38.7% (13 h)	84%	4%	12%

Conditions: 0.5 mol% of Ru, 1.8 mmol of substrate, H_2O / *n*-octane = 15 mL / 30 mL, p(H_2) = 450 psi.

Scheme 7. Biphasic hydrogenation of benzylideneacetone catalyzed by $[RuCl(\eta^5\text{-}C_5R_5)(PTA)_2]$.

The selectivity in 4-phenylbutan-2-one (**D**) using catalysts $[RuCl(\eta^5\text{-}C_5R_5)(PTA)_2]$ (R = H, Me) can be enhanced up to 99.6% and 100%, respectively, by increasing the quantity of water (H_2O / *n*-octane = 35 mL / 10 mL instead of H_2O / *n*-octane = 15 mL / 30 mL) [43]. Under these new conditions, the catalytic activity of $[Ru(\eta^5{:}\kappa^1\text{-}N\text{-}C_5H_4CH_2CH_2NEt_2)(PTA)_2][PF_6]$, $[Ru(\eta^5\text{-}C_5H_4CH_2CH_2NEt_2)(NCMe)(PTA)_2][PF_6]$, $[Ru(\eta^5\text{-}C_5H_5)(NCMe)_2(PTA)][PF_6]$ and $[Ru(\eta^5\text{-}C_5R_5)(NCMe)(PTA)_2][PF_6]$ (R = H, Me) has also been investigated but their efficiency does not exceed that of $[RuCl(\eta^5\text{-}C_5H_5)(PTA)_2]$. It is also interesting to note that the reaction rate can be increased by lowering the H_2 pressure (Scheme 8). This fact allows to perform efficiently the hydrogenation process at lower temperatures (25-40°C; entries 3-4).

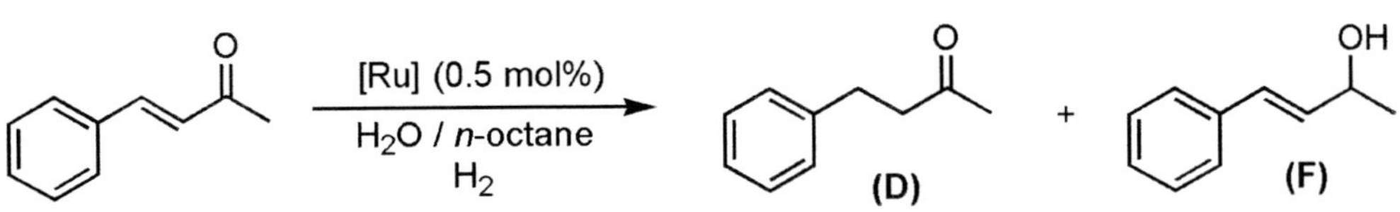

entry	p(H_2)	T (°C)	conversion (time)	D	F
1	450 psi	80	76.3% (16 h)	99.6%	0.4%
2	200 psi	80	99.9% (16 h)	99.6%	0.4%
3	200 psi	40	99.7% (6 h)	99.4%	0.6%
4	200 psi	25	88.5% (16 h)	99.7%	0.3%

Conditions: 0.5 mol% of $[RuCl(\eta^5\text{-}C_5H_5)(PTA)_2]$, 1.8 mmol of substrate, H_2O / *n*-octane = 35 mL / 10 mL. Formation of **E** not observed.

Scheme 8. Influence of the hydrogen pressure on the catalytic activity of $[RuCl(\eta^5\text{-}C_5H_5)(PTA)_2]$.

Complexes $[RuCl(\eta^5\text{-}C_5R_5)(PTA)_2]$ (R = H, Me) are also able to promote transfer hydrogenations of α,β-unsaturated carbonyl compounds in aqueous media, using sodium formate as the hydrogen source (Scheme 9) [43]. Unlike hydrogenation, the best performances for this process were obtained with the pentamethylcyclopentadienyl catalyst

(R = Me). Thus, this complex can reduce quantitatively benzylideneacetone into 4-phenylbutan-2-one with 99.5% of selectivity within 6 h. Transfer hydrogenation of cinnamaldehyde was also more rapid with [RuCl(η^5-C_5Me_5)(PTA)$_2$] (99.4% of conversion *vs* 11.1 after 5 hours). Nevertheless, in this case the selectivity of the catalysts was completely different. Dihydrocinnamaldehyde (**D**) was formed preferentially using [RuCl(η^5-C_5H_5)(PTA)$_2$], while the fully saturated product, namely 3-phenyl-1-propanol (**E**), was rather obtained with [RuCl(η^5-C_5Me_5)(PTA)$_2$]. In the latter case, high proportions of intermediates **D** and **F** were observed at the beginning of the reaction (entry 4), but they further evolve to **E**. The catalytic activity of complexes [Ru(η^5-C_5R_5)(NCMe)(PTA)$_2$][PF_6] (R = H, Me) and [Ru(η^5-C_5H_5)(NCMe)$_2$(PTA)][PF_6] has also been explored in the transfer hydrogenation of α,β-unsaturated carbonyls, giving rise to low conversions and selectivities [43].

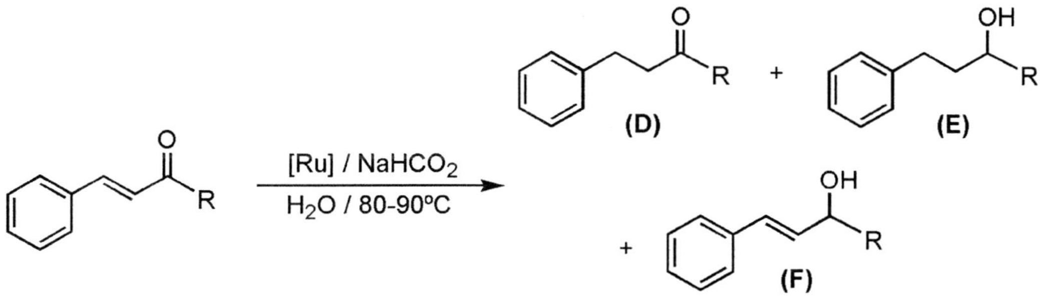

catalyst	R	conversion (time)	D	E	F
[RuCl(η^5-C_5H_5)(PTA)$_2$]	Me[a]	36.1% (6 h)	93.9%	0%	6.1%
[RuCl(η^5-C_5Me_5)(PTA)$_2$]	Me[a]	97.4% (6 h)	99.5%	0.5%	0%
[RuCl(η^5-C_5H_5)(PTA)$_2$]	H[b]	11.1% (5 h)	85.6%	7.2%	7.2%
[RuCl(η^5-C_5Me_5)(PTA)$_2$]	H[b]	76.7% (15 min)	34.6%	30.8%	34.8%
[RuCl(η^5-C_5Me_5)(PTA)$_2$]	H[b]	99.4% (5 h)	5.6%	87.0%	7.4%

[a] 1 mol% of Ru, 0.75 mmol of substrate, 7.5 mmol of $NaHCO_2$, H_2O / MeOH = 3 mL / 3 mL, 90ºC. [b] 1.5 mol% of Ru, 2 mmol of substrate, 22 mmol of $NaHCO_2$, 6 mL of H_2O, 80ºC.

Scheme 9. Transfer hydrogenation of benzylideneacetone and cinnamaldehyde in aqueous media.

The mixed triphenylphosphine-PTA compounds [RuH(η^5-C_nH_m)(PPh_3)(PTA)] (C_nH_m = C_5H_5, 1,2-dihydropentalenyl) are also able to promote transfer hydrogenations of cinnamaldehyde, benzylideneacetone and chalcone using formic acid as reducing agent (Scheme 10) [40]. In the case of the enal, both complexes selectively reduce the C=O bond prior to the reduction of the C=C bond (entries 1-2). However, by increasing the quantity of HCO_2H present in the reaction media hydrocinnamyl alcohol (**E**) is exclusively formed (entry 3). In contrast, transfer hydrogenation of the enones occurs predominantly at the C=C double bond (entries 4-6). This tendency is also observed when formic acid is replaced by sodium formate, although different proportions are in this case obtained (entry 7).

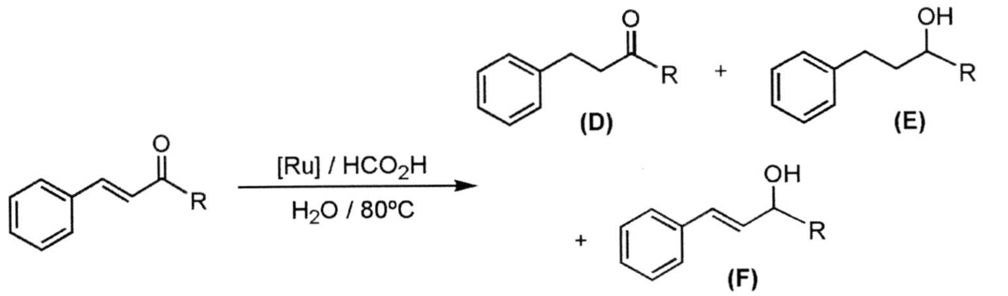

catalyst	R	HCO_2H	conversion	D	E	F
$[RuCl(\eta^5\text{-}C_5H_5)(PPh_3)(PTA)]$	H[a]	40 μL	99%	0%	19.7%	80.3%
$[RuCl(\eta^5\text{-}Dp)(PPh_3)(PTA)]$	H[a]	40 μL	99%	0%	21.8%	78.2%
$[RuCl(\eta^5\text{-}C_5H_5)(PPh_3)(PTA)]$	H[a]	100 μL	89.5%	0%	>99%	<1%
$[RuCl(\eta^5\text{-}C_5H_5)(PPh_3)(PTA)]$	Me[b]	40 μL	79.4%	55.9%	34.7%	0%
$[RuCl(\eta^5\text{-}Dp)(PPh_3)(PTA)]$	Me[b]	40 μL	51.1%[c]	100%	0%	0%
$[RuCl(\eta^5\text{-}C_5H_5)(PPh_3)(PTA)]$	Ph[d]	40 μL	47.9%	100%	0%	0%
$[RuCl(\eta^5\text{-}C_5H_5)(PPh_3)(PTA)]$	Ph[d]	-[e]	39.8%	65.1%	34.9%	0%

Conditions: 5 mol% of Ru, 2 mL of H_2O, indicated volume of 88% HCO_2H, 80°C. [a] 4.5 μL of substrate, 6 h. [b] 4.3 mg of substrate, 24 h. [c] After 6 h. [d] 7.6 mg of substrate, H_2O / MeOH = 2 mL / 3 mL, 13 h. [e] Using 61.2 mg of $NaHCO_2$ instead of HCO_2H. (Dp = 1,2-dihydropentalenyl)

Scheme 10. Transfer hydrogenations promoted by mixed PPh_3-PTA ruthenium complexes.

The *N*-Alkylated PTA Salts

The PTA phosphine ligand selectively reacts with alkyl halides or methyl triflate to give the corresponding *N*-alkylated salts [28]. In particular, the water-soluble *N*-methylated PTA ligand has been prepared as its chloride (**17-Cl**), iodide (**17-I**) and triflate (**17-OTf**) salts (Figure 10).

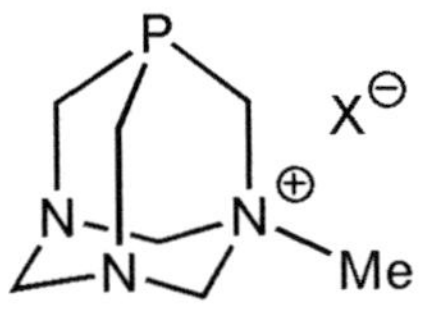

X = I **(17-I)**
X = OTf **(17-OTf)**
X = Cl **(17-Cl)**

Figure 10. Structure of the *N*-methylated PTA salts.

This methylated species presents a reactivity towards arene-ruthenium(II) fragments similar than that observed for PTA. Thus, the reaction of two-fold excess of **17-Cl** with dimers $[\{RuCl(\mu\text{-}Cl)(\eta^6\text{-arene})\}_2]$ (arene = toluene, *p*-cymene) results in the high-yield formation (74-90%) of the corresponding mononuclear derivatives $[RuCl_2(\eta^6\text{-arene})$(**17-Cl**)] [31,37]. Nevertheless, it is much less nucleophile than PTA, as clearly evidenced in the reaction of the cyclopentadienyl complex $[RuCl(\eta^5\text{-}C_5H_5)(PPh_3)_2]$ with **17-I** or **17-OTf**. In contrast to that observed with PTA (see above), the substitution of both triphenylphosphine is not possible neither in the presence of an excess of ligand nor under refluxing conditions.

Thus, the heating of [RuCl(η^5-C_5H_5)(PPh_3)$_2$] with 2 equivalents of **17-OTf** in acetone selectively affords [RuCl(η^5-C_5H_5)(PPh_3)(**17-OTf**)], whereas the treatment of this organometallic precursor with excess of **17-I** in hot isopropanol leads to the formation of [RuI(η^5-C_5H_5)(PPh_3)(**17-Cl**)], as the result of the replacement of one PPh_3 ligand and concomitant chloride/iodide metathesis [39]. An alternative synthetic strategy, based on the reaction of $RuCl_3{\cdot}nH_2O$, cyclopentadiene and **17-OTf** in refluxing ethanol, has been used to obtain selectively the disubstituted derivatives [RuCl(η^5-C_5H_5)(**17-OTf**)$_2$] albeit in very low yield (13%) [39]. No catalytic application of these methylated species has been reported to date.

The DAPTA Ligand

The 3,7-diacetyl-1,3,7-triaza-5-phosphabicyclo[3.3.1]nonane phosphine (Figure 11), usually abbreviated as DAPTA, can be prepared in multigram-scale, albeit with moderate yield (47%), through the reaction of PTA with acetic anhydride in water at 0°C [9c,47]. Surprisingly, despite its extremely high water-solubility (S = 1696 $g{\cdot}L^{-1}$ at room temperature), the coordination chemistry of the DAPTA ligand has been scarcely studied. Concerning half-sandwich ruthenium(II) complexes, compounds [RuCl$_2$(η^6-$C_6H_5CH_2CH_2OH$)(DAPTA)] and [RuCl$_2$(η^6-$C_6H_5CH_2CH_2NH_3$)(DAPTA)][Cl] are the unique DAPTA-derivatives reported up to now in the literature [32]. As their PTA-counterparts, these water-soluble compounds present anticancer activity.

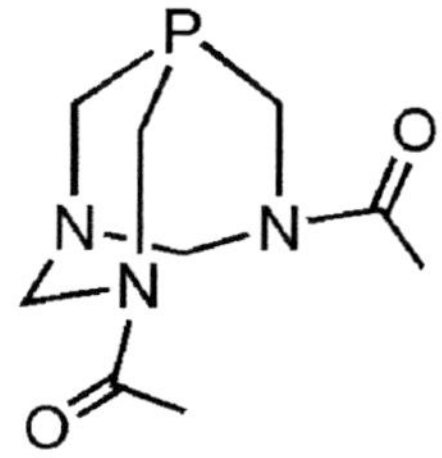

Figure 11. Structure of the DAPTA ligand.

The THPA Ligand and Its *N*-Methylated Form

Although the 2,4,10-trimethyl-1,2,4,5,7,10-hexaaza-3-phosphatricyclo[3.3.1.1^{3,7}]-decane ligand, abbreviated as trihydrazinophosphaadamantane or THPA (Figure 12), is known since 1982 [48], its coordination chemistry remains almost unexplored [49,50]. Like the other cage-like phosphines described above, the THPA ligand is also soluble in water although in a less extend (S = 47 $g{\cdot}L^{-1}$).

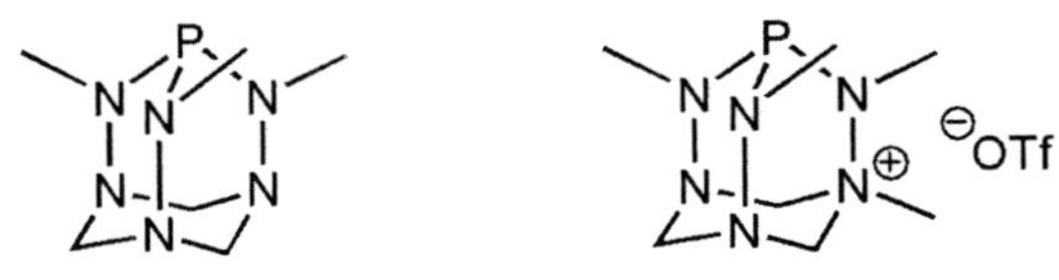

Figure 12. Structure of the THPA ligand and its *N*-methylated salt **18**.

The THPA ligand readily reacts with dimers [{RuCl(μ-Cl)(η^6-arene)}$_2$] in dichloromethane at room temperature to give the corresponding water-soluble complexes [RuCl$_2$(η^6-arene)(THPA)] (arene = benzene, *p*-cymene, 1,3,5-trimethylbenzene, hexamethylbenzene) in good yields (73-85%) [50]. Treatment of [RuCl$_2$(η^6-benzene)(THPA)] and [RuCl$_2$(η^6-*p*-cymene)(THPA)] with methyltriflate results in the *N*-alkylation of the trihydrazinophosphaadamantane unit affording complexes [RuCl$_2$(η^6-benzene)(**18**)] and [RuCl$_2$(η^6-*p*-cymene)(**18**)], respectively [50]. All these ruthenium complexes have shown to be highly active in the catalytic isomerization of allylic alcohols into the corresponding ketones both in organic and in aqueous media. Thus, using THF as solvent, 1-octen-3-ol is quantitatively converted into 3-octanone within only 5 min using complexes [RuCl$_2$(η^6-arene)(THPA)] (arene = benzene, *p*-cymene, 1,3,5-trimethylbenzene, hexamethylbenzene) as pre-catalysts (1 mol% of Ru, 5 mol% of KO^{t}Bu, T = 75°C) [50]. Other allylic alcohols, such as 1-hepten-3-ol, 1-hexen-3-ol, 1-penten-3-ol and α-vinylbenzyl alcohol, can also be efficiently transformed into the corresponding saturated ketones under the same conditions (TOF up to 1200 h^{-1}). More interestingly, complexes [RuCl$_2$(η^6-arene)(THPA)] and [RuCl$_2$(η^6-arene)(**18**)] also isomerize quantitatively allylic alcohols in aqueous medium (representative examples are shown in Scheme 11), the best activities being observed with the benzene and *p*-cymene derivatives. At the end of the reactions, these catalysts, which present a high affinity to water, could be easily separated from the products by a simple extraction process and therefore, they could be recycled up to 4 times [50].

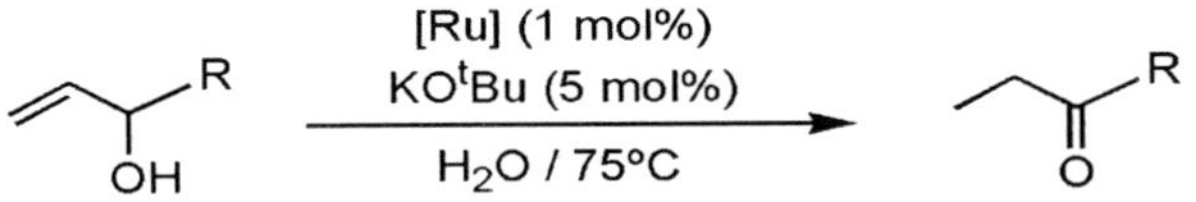

catalyst	R	time	yield	TOF
[RuCl$_2$(η^6-benzene)(THPA)]	^{n}Pent	45 min	99%	132 h^{-1}
[RuCl$_2$(η^6-*p*-cymene)(THPA)]	^{n}Pent	1 h	99%	99 h^{-1}
[RuCl$_2$(η^6-C$_6$H$_3$Me$_3$)(THPA)]	^{n}Pent	2 h	99%	50 h^{-1}
[RuCl$_2$(η^6-C$_6$Me$_6$)(THPA)]	^{n}Pent	3.5 h	100%	29 h^{-1}
[RuCl$_2$(η^6-benzene)(**18**)]	^{n}Pent	1 h	99%	99 h^{-1}
[RuCl$_2$(η^6-*p*-cymene)(**18**)]	^{n}Pent	30 min	99%	200 h^{-1}
[RuCl$_2$(η^6-*p*-cymene)(**18**)]	^{n}Bu	45 min	99%	132 h^{-1}
[RuCl$_2$(η^6-*p*-cymene)(**18**)]	^{n}Pr	1.5 h	96%	64 h^{-1}
[RuCl$_2$(η^6-*p*-cymene)(**18**)]	Et	3.5 h	100%	26 h^{-1}
[RuCl$_2$(η^6-*p*-cymene)(**18**)]	Ph	2 h	99%	50 h^{-1}

Conditions: 4 mmol of substrate, 20 mL of water.

Scheme 11. Aqueous isomerization of allylic alcohols catalyzed by THPA-based Ru complexes.

The THPA-derivatives [$RuCl_2(\eta^6$-arene)(THPA)] and [$RuCl_2(\eta^6$-arene)(**18**)] have also been employed to promote the cycloisomerization of (*Z*)-3-methylpent-2-en-4-yn-1-ol into 2,3-dimethylfuran in water (Scheme 12) [50]. The best performances were reached using the *N*-methylated THPA-complex [$RuCl_2(\eta^6$-benzene)(**18**)] which leads to an almost quantitative transformation in 5 h at 80°C.

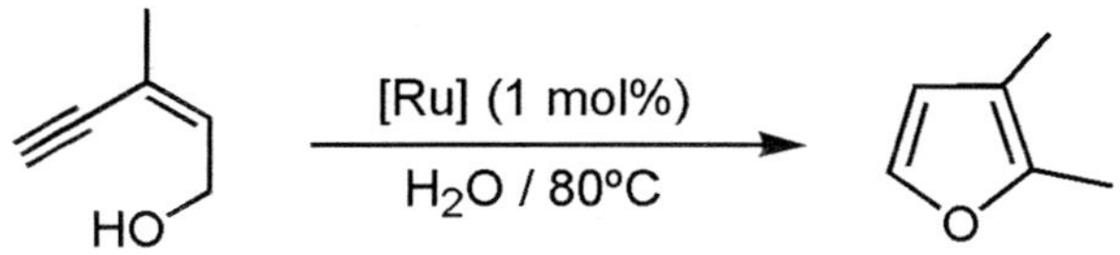

catalyst	time	yield
[$RuCl_2(\eta^6$-benzene)(THPA)]	5 h	78%
[$RuCl_2(\eta^6$-*p*-cymene)(THPA)]	5 h	93%
[$RuCl_2(\eta^6$-$C_6H_3Me_3$)(THPA)]	5 h	92%
[$RuCl_2(\eta^6$-C_6Me_6)(THPA)]	5 h	77%
[$RuCl_2(\eta^6$-benzene)(**18**)]	5 h	98%
[$RuCl_2(\eta^6$-*p*-cymene)(**18**)]	6 h	92%

Conditions: 5 mmol of enynol, 1 mL of water.

Scheme 12. Catalytic synthesis of 2,3-dimethylfuran using THPA-based Ru complexes.

PHOSPHINE LIGANDS CONTAINING HYDROXYALKYL CHAINS

Phosphines containing hydroxyalkyl substituents constitute an important class of water-soluble ligands [5,6]. Nevertheless, it should be noted that the solubility of this type of compounds in water is strongly dependent on the number of hydroxyalkyl groups present in the molecule, being only significant when several CH_2OH units are present. The commercially available tris(hydroxymethyl)phosphine $P(CH_2OH)_3$ is undoubtedly the most popular water-soluble hydroxyalkyl-phosphine, a short account covering the synthesis and catalytic applications in water of platinum metals containing this ligand being available [51].

Treatment of THF solutions of [$RuCl(\eta^5$-$C_5Me_5)(CO)_2$] with a stoichiometric amount of $P(CH_2OH)_3$ and $P\{(CH_2)_3OH\}_3$, under UV-light irradiation, results in the high-yield formation of the water-soluble compounds [$RuCl(\eta^5$-$C_5Me_5)(CO)\{P(CH_2OH)_3\}$] and [$RuCl(\eta^5$-$C_5Me_5)(CO)(P\{(CH_2)_3OH\}_3)$], respectively, *via* a classical CO ligand exchange process [52]. These complexes, as well as the *in situ* formed cationic derivative [$Ru(\eta^5$-$C_5Me_5)(CO)(P\{(CH_2)_3OH\}_3)$][OTf], were found to be active catalysts for the two-phase (water / *n*-heptane) hydrogenation of sorbic acid, the efficiency and selectivity of the process being strongly affected by the nature of the catalyst employed (Scheme 13) [52]. Thus, the best results in terms of activity were obtained using the cationic derivative [$Ru(\eta^5$-$C_5Me_5)(CO)(P\{(CH_2)_3OH\}_3)$][OTf] which is able to reduce sorbic acid in 62% yield after 4 h, leading to a mixture of products with *cis*-3-hexenoic acid (**H**) as the major component. In contrast, when its neutral chloride precursor is used as catalyst hexanoic acid (**J**) is selectively formed, albeit a longer reaction time (15 h) is required to attain only a 52% of conversion.

Although these results are not exceptional, they are much better than those obtained using as catalysts the related ruthenium complexes containing the sulfonated phosphine ligand TPPTS [52].

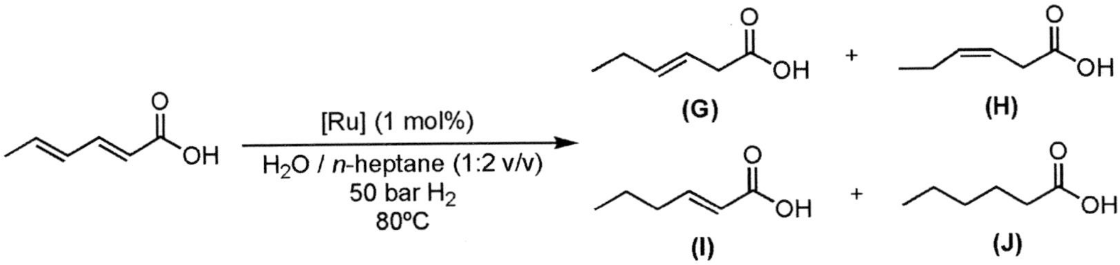

catalyst	time	conversion	G	H	I	J
$[RuCl(\eta^5\text{-}C_5Me_5)(CO)\{P(CH_2OH)_3\}]$	15 h	40%	10%	41%	7%	41%
$[RuCl(\eta^5\text{-}C_5Me_5)(CO)(P\{(CH_2)_3OH\}_3)]$	15 h	52%	0%	3%	5%	92%
$[Ru(\eta^5\text{-}C_5Me_5)(CO)(P\{(CH_2)_3OH\}_3)][OTf]^a$	4 h	62%	7%	66%	5%	22%
$[RuCl(\eta^5\text{-}C_5Me_5)(CO)(TPPTS)]$	15 h	8%	20%	20%	20%	40%
$[Ru(\eta^5\text{-}C_5Me_5)(CO)(TPPTS)][OTf]^a$	4 h	14%	85%	0%	15%	0%

[a] Generated *in situ* by treatment of the corresponding chloride precursor with AgOTf.

Scheme 13. Biphasic hydrogenation of sorbic acid catalyzed by water-soluble ruthenium complexes.

The (η^6-arene)-ruthenium(II) dimers $[\{RuCl(\mu\text{-}Cl)(\eta^6\text{-arene})\}_2]$ (arene = benzene, *p*-cymene, hexamethylbenzene) readily react with two or four equivalents of $P(CH_2OH)_3$, in a mixture dichloromethane/methanol at r.t., to afford the water-soluble complexes $[RuCl_2(\eta^6\text{-arene})\{P(CH_2OH)_3\}]$ and $[RuCl(\eta^6\text{-arene})\{P(CH_2OH)_3\}_2][Cl]$, respectively [53]. These compounds, associated with Cs_2CO_3, are efficient catalysts in the redox isomerization of 1-octen-3-ol into octan-3-one under biphasic conditions (Scheme 14). Although quantitative and selective transformations were in all cases observed, the neutral complexes $[RuCl_2(\eta^6\text{-arene})\{P(CH_2OH)_3\}]$ are much more active than their cationic counterparts $[RuCl(\eta^6\text{-arene})\{P(CH_2OH)_3\}_2][Cl]$ (TOF = 29-67 h^{-1} *vs* 4-6 h^{-1}). In addition, the catalytic efficiency of these complexes was found to be strongly dependent on the nature of the arene ligand. Thus, the rate order observed in both series, *i.e.* C_6H_6 > *p*-cymene > C_6Me_6, indicates that the less sterically demanding and electron-rich arene the higher performances are found. Interestingly, the catalytically active ruthenium species could be in all cases recycled, leading to conversions ≥86% in a second catalytic run, but only the neutral derivative $[RuCl_2(\eta^6\text{-}C_6H_6)\{P(CH_2OH)_3\}]$ showed good performances in a third and fourth cycle. Other allylic alcohols, such as 1-hepten-3-ol, 1-hexen-3-ol, 1-penten-3-ol, 3-buten-2-ol and 2-buten-1-ol, could also be efficiently isomerized into the corresponding carbonyl compounds using $[RuCl_2(\eta^6\text{-arene})\{P(CH_2OH)_3\}]$ (arene = benzene, *p*-cymene) as catalysts under the same reaction conditions. Excellent results in terms of both activity and catalyst recycling (up to 8 consecutive runs) were obtained when 3-buten-2-ol was used as substrate, leading to TOF values up to 600 h^{-1} and TON values up to 782 [53].

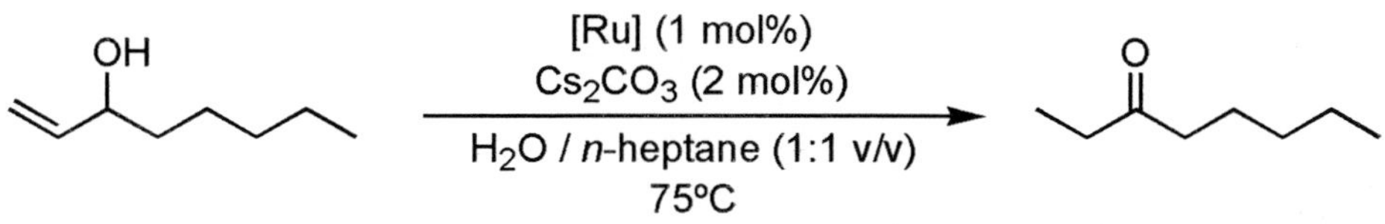

catalyst	time	yield	TOF
$[RuCl_2(\eta^6\text{-benzene})\{P(CH_2OH)_3\}]$	1.5 h	100%	67 h^{-1}
$[RuCl_2(\eta^6\text{-}p\text{-cymene})\{P(CH_2OH)_3\}]$	2.25 h	100%	44 h^{-1}
$[RuCl_2(\eta^6\text{-}C_6Me_6)\{P(CH_2OH)_3\}]$	3.5 h	100%	29 h^{-1}
$[RuCl(\eta^6\text{-benzene})\{P(CH_2OH)_3\}_2][Cl]$	18 h	100%	6 h^{-1}
$[RuCl(\eta^6\text{-}p\text{-cymene})\{P(CH_2OH)_3\}_2][Cl]$	21 h	100%	5 h^{-1}
$[RuCl(\eta^6\text{-}C_6Me_6)\{P(CH_2OH)_3\}_2][Cl]$	24 h	100%	4 h^{-1}

Scheme 14. Isomerization of 1-octen-3-ol catalyzed by Ru-$P(CH_2OH)_3$ complexes.

The neutral complexes $[RuCl_2(\eta^6\text{-arene})\{P(CH_2OH)_3\}]$ (arene = benzene, *p*-cymene, hexamethylbenzene) are also active catalysts for the hydration of terminal alkynes in aqueous medium (Scheme 15). Concerning the regioselectivity of the process, the addition of water across the C≡C bond follows the Markovnikov's rule to produce the corresponding ketone as the major or unique product [53].

The closely related water-soluble complexes $[RuCl_2\{\eta^6\text{-}C_6H_5(CH_2)_3OH\}\{P(CH_2OH)_3\}]$ and $[RuCl\{\eta^6\text{-}C_6H_5(CH_2)_3OH\}\{P(CH_2OH)_3\}_2][Cl]$, containing both an hydroxyalkyl appended arene and $P(CH_2OH)_3$ ligands, are known [54]. Nevertheless, their catalytic applications have not been explored. Other half-sandwich ruthenium(II) complexes containing hydroxyalkyl phosphine ligands, *i.e.* $[RuCl_2(\eta^6\text{-}p\text{-cymene})(Ph_2PCH_2OH)]$ [55], $[RuCl_2(\eta^6\text{-}p\text{-cymene})\{P(CH_2OH)_2(CH_2Fc)\}]$ (Fc = ferrocenyl) [56,57], and (R_C)- and (S_C)-$[RuCl_2(\eta^6\text{-}C_6Me_6)\{Ph_2PCH_2CH(Me)CH_2OH\}]$ [58], have also been described in the literature, but neither their solubility in water nor their potential as catalysts have been determined.

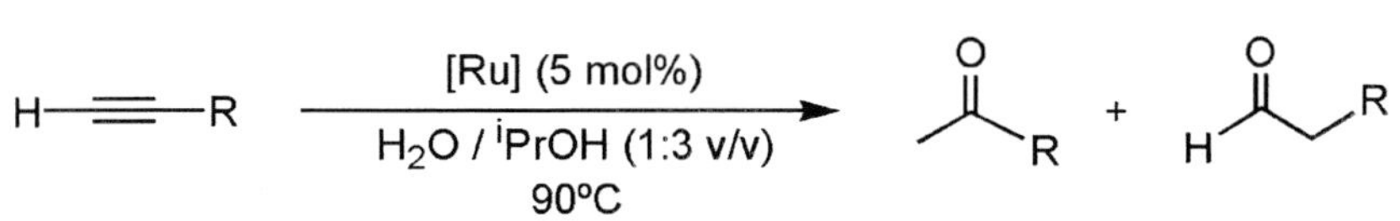

catalyst	R	time	conversion	ketone / aldehyd
$[RuCl_2(\eta^6\text{-benzene})\{P(CH_2OH)_3\}]$	Ph	24 h	88%	100:0
$[RuCl_2(\eta^6\text{-}p\text{-cymene})\{P(CH_2OH)_3\}]$	Ph	24 h	89%	100:0
$[RuCl_2(\eta^6\text{-}C_6Me_6)\{P(CH_2OH)_3\}]$	Ph	24 h	89%	99:1
$[RuCl_2(\eta^6\text{-benzene})\{P(CH_2OH)_3\}]$	^{n}Hex	8 h	100%	84:16
$[RuCl_2(\eta^6\text{-}p\text{-cymene})\{P(CH_2OH)_3\}]$	^{n}Hex	8 h	100%	69:31
$[RuCl_2(\eta^6\text{-}C_6Me_6)\{P(CH_2OH)_3\}]$	^{n}Hex	8 h	100%	72:28

Scheme 15. Hydration of terminal alkynes catalyzed by complexes $[RuCl_2(\eta^6\text{-arene})\{P(CH_2OH)_3\}]$.

Other Water-Soluble Phosphine Ligands

The water-soluble complexes [$RuCl_2(\eta^6$-*p*-cymene){κ^1-*P*-$PPh_{3-n}(OCH_2CH_2NMe_3)_n$}][$SbF_6]_n$ (n = 1 (**19**), 2 (**20**), 3 (**21**)), containing ammonium-functionalized phosphine ligands, have been prepared from the neutral precursors [$RuCl_2(\eta^6$-*p*-cymene){κ^1-*P*-$PPh_{3-n}(OCH_2CH_2NMe_2)_n$}] (n = 1, 2, 3) and MeI *via* quaternization of the nitrogen atom of the terminal aminoalkyl groups, and subsequent counteranion exchange upon treatment with $AgSbF_6$ (Scheme 16) [59].

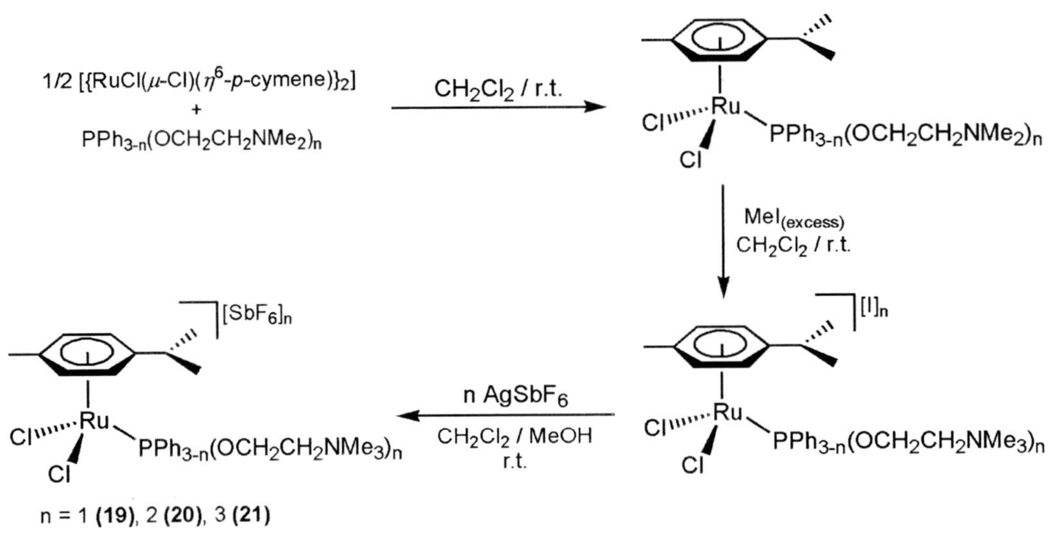

Scheme 16. Water-soluble arene-Ru(II) complexes containing ammonium-functionalized ligands.

Complexes **19-21** have proven to be highly active catalysts for the redox isomerization of allylic alcohols into the corresponding carbonyl compounds in aqueous media (Scheme 17). Moreover, they can be reused after a simple extraction process at the end of the reaction. In particular, [$RuCl_2(\eta^6$-*p*-cymene){κ^1-*P*-$P(OCH_2CH_2NMe_3)_3$}][$SbF_6]_3$ (**21**) presents a remarkable capability to be recycled, being active during 10 successive runs which has no precedent in the literature [59].

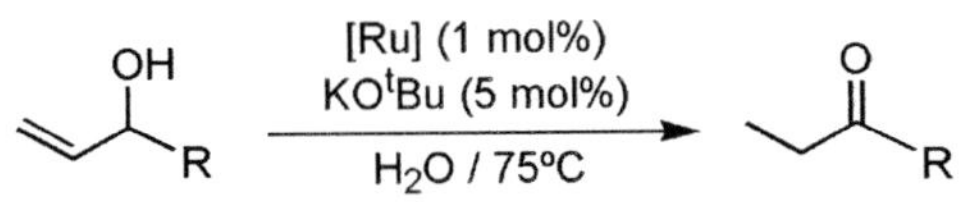

catalyst	R	time	yield	TOF
19	^{n}Pent	10 min	99%	594 h^{-1}
19	Et	15 min	97%	388 h^{-1}
20	^{n}Pent	15 min	99%	396 h^{-1}
20	Et	5 min	99%	1188 h^{-1}
21	^{n}Pent	35 min	99%	170 h^{-1}
21	Et	400 min	99%	15 h^{-1}

Scheme 17. Isomerization of allylic alcohols catalyzed by complexes **19-21**.

It has also been reported that, in the presence of the bisphosphorylated BINAP-type ligand **22**, the dimeric precursor [{RuCl(μ-Cl)(η^6-*p*-cymene)}$_2$] (Ru / ligand = 1:1) is able to catalyze the asymmetric hydrogenation of dimethyl itaconate under aqueous biphasic conditions (Scheme 18). Nevertheless, the enantioselectivity was only moderate (*e.e.* = 79.4%), a slight drop in the *e.e.* value being observed when compared to the monophasic hydrogenation using the non-functionalized BINAP ligand (*e.e* = 91.5%) [60].

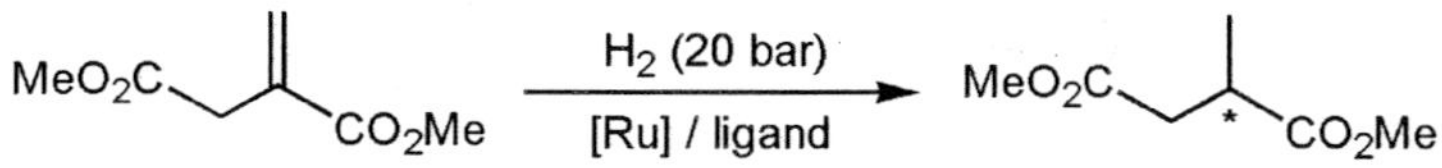

ligand	solvent	yield	*e.e.*
22	H_2O / EtOH / hexane (5:1:5)	99.4%	79.4%
BINAP	EtOH	99.8%	91.5%

Reactions performed at 60°C for 2 h. Substrate / Ru ratio = 300:1.

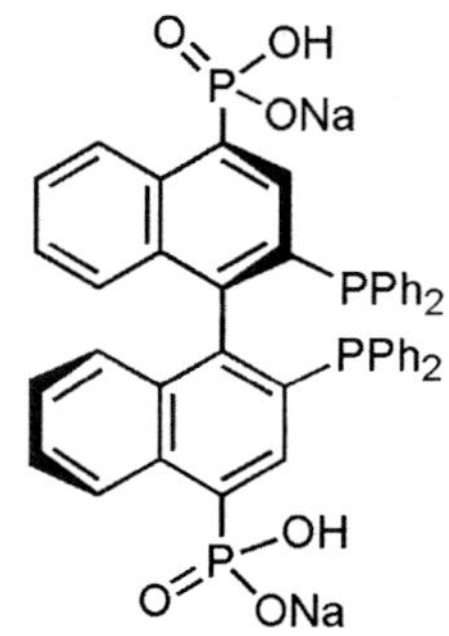

(22)

Scheme 18. Asymmetric hydrogenation of dimethyl itaconate.

Although phosphine ligands containing carboxylic functional groups are some of the earliest water-soluble ligands used in catalysis [5,6], no application of half-sandwich ruthenium complexes containing such type of ligands has been described to date. Only the synthesis and electrochemical behaviour of the cationic arene derivatives [Ru(η^6-*p*-cymene){κ^2-*N,O*-o-$H_2NC_6H_4CO_2$}(L)][BF_4] (L = $Ph_2PCH_2CO_2H$, Ph_2P-2-$C_6H_4CO_2H$, Ph_2P-3-$C_6H_4CO_2H$, Ph_2P-4-$C_6H_4CO_2H$) [61] and [Ru(η^6-*p*-cymene){κ^2-*N,O*-8-quinolinate}(L)][BF_4] (L = $Ph_2PCH_2CO_2H$, Ph_2P-2-$C_6H_4CO_2H$, Ph_2P-3-$C_6H_4CO_2H$, Ph_2P-4-$C_6H_4CO_2H$) [62] has been reported.

Conclusion

Extraordinary attention has been paid to synthetic organic chemistry in water in the past decade and the research in this field is still increasing exponentially, specially when the reactions are performed under catalytic conditions. This review is a summary of recent progress in aqueous catalysis using half-sandwich ruthenium complexes containing water-soluble phosphine ligands. In comparison with the well-studied catalytic chemistry of ruthenium in classical organic solvents, catalytic applications in water have been largely unexplored. In fact, we are still at the beginning to learn what the potential of this type of compounds in organic synthesis is. Three of the most desirable characteristics of a water-soluble ruthenium catalyst are: (*i*) high solubility to ensure fast reactions, (*ii*) aqueous stability, and (*iii*) the presence of suitable chiral elements to promote asymmetric transformations with high enantioselectivities. Although there is already a body of work in this area, it is clear that further advances in this field can be expected in the near future as more active catalysts are discovered and new water-soluble ligands are designed.

ACKNOWLEDGEMENTS

Financial support by the Spanish Ministerio de Educación y Ciencia (Project CTQ2006-08485/BQU and "Ramón y Cajal" program) and FICYT of Asturias (Project IB05-035) is gratefully acknowledged.

REFERENCES

[1] (a) *Ruthenium in Organic Synthesis*; Murahashi, S.-I., Ed.; Wiley-VCH: Weinheim, 2004; (b) *Ruthenium Catalysts and Fine Chemistry*; Bruneau, C., Dixneuf, P. H., Eds.; Springer: Berlin, 2004.

[2] (a) Bruneau, C.; Dixneuf, P. H. *Chem. Commun.* 1997, 507-512; (b) Naota, T.; Takaya, H.; Murahashi, S.-I. *Chem. Rev.* 1998, *98*, 2599-2660; (c) Trost, B. M.; Toste, F. D.; Pinkerton, A. B. *Chem. Rev.* 2001, *101*, 2067-2096; (d) Trnka, T. M.; Grubbs, R. H. *Acc. Chem. Res.* 2001, *34*, 18-29; (e) Kondo, T.; Mitsudo, T. *Curr. Org. Chem.* 2002, *6*, 1163-1179; (f) Ritleng, V.; Sirlin, C.; Pfeffer, M. *Chem. Rev.* 2002, *102*, 1731-1769; (g) Trost, B. M. *Acc. Chem. Res.* 2002, *35*, 695-705; (h) Trost, B. M.; Frederiksen, M. U.; Rudd, M. T. *Angew. Chem. Int. Ed.* 2005, *44*, 6630-6666; (i) Bruneau, C.; Dérien, S.; Dixneuf, P. H. *Topics Organomet. Chem.* 2006, *19*, 295-326; (j) Bruneau, C.; Dixneuf, P. H. *Angew. Chem. Int. Ed.* 2006, *45*, 2176-2203; (k) *Curr. Org. Chem.* 2006, *10(2)*, 113-225 (a thematic issue devoted to this topic).

[3] (a) Anastas, P. T.; Warner, J. C. *Green Chemistry Theory and Practice*; Oxford University Press: Oxford, 1998; (b) Matlack, A. S. *Introduction to Green Chemistry*; Marcel Dekker: New York, 2001; (c) *Handbook of Green Chemistry and Technology*; Clark, J. H., Macquarrie, D. J., Eds.; Blackwell Publishing: Abingdon, 2002; (d) Lancaster, M. *Green Chemistry: An Introductory Text*; RSC Editions: Cambridge, 2002.

[4] (a) Li, C.-J. *Chem. Rev.* 1993, *93*, 2023-2035; (b) Lubineau, A.; Augé, J.; Queneau, Y. *Synthesis* 1994, 741-760; (c) Li, C.-J.; Chan, T. H. *Organic Reactions in Aqueous Media*; John Wiley and Sons: New York, 1997; (d) da Silva, F. M.; Jones, J. *Quim. Nova* 2001, *24*, 646-657; (e) Engberts, J. B. F. N.; Blandamer, M. J. *Chem. Commun.* 2001, 1701-1708; (f) Lindström, U. M. *Chem. Rev.* 2002, *102*, 2751-2772; (g) Li, C.-J. *Chem. Rev.* 2005, *105*, 3095-3165; (h) Andrade, C. K. Z.; Alves, L. M. *Curr. Org. Chem.* 2005, *9*, 195-218; (i) Li, C.-J.; Chen, L. *Chem. Soc. Rev.* 2006, *35*, 68-82.

[5] (a) Kalck, P.; Monteil, F. *Adv. Organomet. Chem.* 1992, *34*, 219-284; (b) Herrmann, W. A.; Kohlpaintner, C. W. *Angew. Chem. Int. Ed. Engl.* 1993, *32*, 1524-1544; (c) *Aqueous Organometallic Chemistry and Catalysis*; Horváth, I. T., Joó, F., Eds.; Kluwer: Dodrecht, 1995; (d) *Aqueous-Phase Organometallic Catalysis: Concepts and Applications*; Cornils, B., Herrmann, W. A., Eds.; Wiley-VCH: Weinheim, 1998; (e) Hanson, B. E. *Coord. Chem. Rev.* 1999, *185-186*, 795-807; (f) *Aqueous Organometallic Catalysis*; Horváth, I. T., Joó, F., Eds.; Kluwer: Dodrecht, 2001; (g) Pinault, N.; Bruce, D. W. *Coord. Chem. Rev.* 2003, *241*, 1-25.

[6] Thematic issues of journals have been devoted to the catalytic applications of water-soluble organometallic compounds: (a) *J. Mol. Catal. A: Chem.* 1997, *116(1-2)*, 1-316;

(b) *Catalysis Today* 1998, *42(4)*, 371-478; (c) *Adv. Synth. Catal.* 2002, *344(3-4)*, 219-451.

[7] (a) Gimeno, J.; Cadierno, V.; Crochet, P. In *Comprehensive Organometallic Chemistry III*; Crabtree, R. H., Mingos, D. M. P., Eds.; Elsevier: Oxford, 2007, Vol. 6, 465-550; (b) Gimeno, J.; Cadierno, V. In *Comprehensive Organometallic Chemistry III*; Crabtree, R. H., Mingos, D. M. P., Eds.; Elsevier: Oxford, 2007, Vol. 6, 551-628.

[8] (a) Rigby, J. H.; Kondratenko, M. A. *Topics Organomet. Chem.* 2004, *7*, 181-204; (b) Dérien, S.; Dixneuf, P. H. *J. Organomet. Chem.* 2004, *689*, 1382-1392.

[9] (a) Joó, F.; Kovács, J.; Kathó, A.; Bényei, A. C.; Decuir, T.; Darensbourg, D. J. *Inorg. Synth.* 1998, *32*, 1-8; (b) Herrmann, W. A.; Kohlpaintner, C. W. *Inorg. Synth.* 1998, *32*, 8-25; (c) Darensbourg, D. J.; Ortiz, C. G.; Kamplain, J. W. *Organometallics* 2004, *23*, 1747-1754.

[10] Suárez, T.; Fontal, B.; Reyes, M.; Bellandi, F.; Contreras, R.; León, G.; Cancines, P. *React. Kinet. Catal. Lett.* 2002, *76*, 161-169.

[11] Anthony, C. R.; McElwee-White, L. *J. Mol. Catal. A: Chem.* 2005, *227*, 113-117.

[12] Díez, J.; Gamasa, M. P.; Lastra, E.; García-Fernández, A.; Tarazona, M. P. *Eur. J. Inorg. Chem.* 2006, 2855-2864.

[13] Le Bozec, H.; Touchard, D.; Dixneuf, P. H. *Adv. Organomet. Chem.* 1989, *29*, 163-247.

[14] Geldbach, T. J.; Brown, M. R. H.; Scopelliti, R., Dyson, P. J. *J. Organomet. Chem.* 2005, *690*, 5055-5065.

[15] Dyson. P. J.; Ellis, D. J.; Laurenczy, G. *Adv. Synth. Catal.* 2003, *345*, 211-215.

[16] (a) Tokunaga, M.; Wakatsuki, Y. *Angew. Chem. Int. Ed.* 1998, *37*, 2867-2869; (b) Tokunaga, M.; Suzuki, T.; Koga, N.; Fukushima, T.; Horiuchi, A. Wakatsuki, Y. *J. Am. Chem. Soc.* 2001, *123*, 11917-11924.

[17] Wong, T. Y. H.; Pratt, R.; Leong, C. G.; James, B. R.; Hu, T. Q. *Chem. Ind.* 2001, *82*, 255-266.

[18] (a) Hernandez, M.; Kalck, P. *J. Mol. Catal. A: Chem.* 1997, *116*, 131-146; (b) Hernandez, M.; Kalck, P. *J. Mol. Catal. A: Chem.* 1997, *116*, 117-130.

[19] Horváth, H.; Laurenczy, G.; Kathó, A. *J. Organomet. Chem.* 2004, *689*, 1036-1045.

[20] Tai, C-C.; Pitts, J.; Linehan, J. C.; Main, A. D.; Munshi, P.; Jessop, P. G. *Inorg. Chem.* 2002, *41*, 1606-1614.

[21] Daguenet, C.; Scopelliti, R.; Dyson, P. J. *Organometallics* 2004, *23*, 4849-4857.

[22] Suzuki, T.; Tokunaga, M.; Wakatsuki, Y. *Org. Lett.* 2001, *3*, 735-737.

[23] Gelpke, A. E. S.; Kooijman, H.; Spek, A. L.; Hiemstra, H. *Chem. Eur. J.* 1999, *5*, 2472-2482.

[24] Ishizaki, T.; Kumobayahi, H. *Eur. Patent Appl.* EP 0544455A1, 1993.

[25] Davis, M. E.; Wan, K. T. *PCT Int. Appl.* WO 95/22405A1, 1995.

[26] Amano, A.; Igarashi, D.; Sayo, N. *PCT Int. Appl.* WO 2005/097811A1, 2005.

[27] Daigle, D. J.; Pepperman Jr., A. B.; Vail, S. L. *J. Heterocycl. Chem.* 1974, *11*, 407-408.

[28] Phillips, A. D.; Gonsalvi, L.; Romerosa, A.; Vizza, F.; Peruzzini, M. *Coord. Chem. Rev.* 2004, *248*, 955-993.

[29] Daigle, D. J. *Inorg. Synth.* 1998, *32*, 40-45.

[30] Allardyce, C. S.; Dyson, P. J.; Ellis, D. J.; Heath, S. L. *Chem. Commun.* 2001, 1396-1397.

[31] Scolaro, C.; Bergamo, A.; Brescacin, L.; Delfino, R.; Cocchietto, M.; Laurenczy, G.; Geldbach, T. J.; Sava, G.; Dyson, P. J. *J. Med. Chem.* 2005, *48*, 4161-4171.

[32] Scolaro, C.; Geldbach, T. J.; Rochat, S.; Dorcier, A.; Gossens, C.; Bergamo, A.; Cocchietto, M.; Tavernelli, I.; Sava, G.; Rothlisberger, U.; Dyson, P. J. *Organometallics* 2006, *25*, 756-765.

[33] Allardyce, C. S.; Dyson, P. J.; Ellis, D. J.; Salter, P. A.; Scopelliti, R. *J. Organomet. Chem.* 2003, *668*, 35-42.

[34] Ang, W. H.; Daldini, E.; Scolaro, C.; Scopelliti, R.; Juillerat-Jeannerat, L.; Dyson, P. J. *Inorg. Chem.* 2006, *45*, 9006-9013.

[35] Dyson, P. J.; Ellis, D. J.; Henderson, W.; Laurenczy, G. *Adv. Synth. Catal.* 2003, *345*, 216-221.

[36] Csabai, P.; Joó, F. *Organometallics* 2004, *23*, 5640-5643.

[37] Dorcier, A.; Dyson, P. J.; Gossens, C.; Rothlisberger, U.; Scopelliti, R.; Tavernelli, I. *Organometallics* 2005, *24*, 2114-2123.

[38] Ang, W. H.; Dyson, P. J. *Eur. J. Inorg. Chem.* 2006, 4003-4018.

[39] Romerosa, A.; Campos-Malpartida, T.; Lidrissi, C.; Saoud, M.; Serrano-Ruiz, M.; Peruzzini, M.; Garrido-Cárdenas, J. A.; García-Maroto, F. *Inorg. Chem.* 2006, *45*, 1289-1298.

[40] Mebi, C. A.; Nair, R. P.; Frost, B. J. *Organometallics* 2007, *26*, 429-438.

[41] Akbayeva, D. N.; Gonsalvi, L.; Oberhauser, W.; Peruzzini, M.; Vizza, F.; Brüggeller, P.; Romerosa, A.; Sava, G.; Bergamo, A. *Chem. Commun.* 2003, 264-265.

[42] Frost, B. J.; Mebi, C. A. *Organometallics* 2004, *23*, 5317-5323.

[43] Bolaño, S.; Gonsalvi, L.; Zanobini, F.; Vizza, F.; Bertolasi, V.; Romerosa, A.; Peruzzini, M. *J. Mol. Catal. A: Chem.* 2004, *224*, 61-70.

[44] Lidrissi, C.; Romerosa, A.; Saoud, M.; Serrano-Ruiz, M.; Gonsalvi, L.; Peruzzini, M. *Angew. Chem. Int. Ed.* 2005, *44*, 2568-2572.

[45] Frost, B. J.; Mebi, C. A.; Gingrich, P. W. *Eur. J. Inorg. Chem.* 2006, 1182-1189.

[46] Mebi, C. A.; Frost, B. J. *Organometallics* 2005, *24*, 2339-2346.

[47] Siele, V. I. *J. Heterocycl. Chem.* 1977, *14*, 337-339.

[48] (a) Majoral, J. P.; Kraemer, R.; Navech, J.; Mathis, F. *Tetrahedron Lett.* 1975, 1481-1484; (b) Jaud, J.; Benhammou, M.; Majoral, J. P.; Navech, J. *Z. Kristallogr.* 1982, *160*, 69-79; (c) Zablocka, M.; Duhayon, C. *Tetrahedron Lett.* 2006, *47*, 2687-2690.

[49] Benhammou, M.; Kraemer, R.; Germa, H.; Majoral, J.-P.; Navech, J. *Phosphorus, Sulfur Relat. Elem.* 1982, *14*, 105-119.

[50] Díaz-Álvarez, A. E.; Crochet, P.; Zablocka, M.; Duhayon, C.; Cadierno, V.; Gimeno, J.; Majoral, J. P. *Adv. Synth. Catal.* 2006, *348*, 1671-1679.

[51] Pringle, P. G.; Smith, M. B. *Platinum Met. Rev.* 1990, *34*, 74-76.

[52] Drieβen-Hölscher, B.; Heinen, J. *J. Organomet. Chem.* 1998, *570*, 141-146.

[53] Cadierno, V.; Crochet, P.; García-Garrido, S. E.; Gimeno, J. *Dalton Trans.* 2004, 3635-3641.

[54] Čubrilo, J.; Hartenbach, I.; Schleid, T.; Winter, R. F. *Z. Anorg. Allg. Chem.* 2006, *632*, 400-408.

[55] Dormond, A.; Baudry, D.; Visseaux, M.; Hepiegne, P. *J. Alloys Compd.* 1994, *213-214*, 1-7.

[56] Goodwin, N. J.; Henderson, W. *Polyhedron* 1998, *17*, 4071-4077.

[57] Downard, A. J.; Goodwin, N. J.; Henderson, W. *J. Organomet. Chem.* 2003, *676*, 62-72.

[58] Sanhueza, J.; Contreras, R.; Valderrama, M. *Bol. Soc. Chil. Quím.* 2002, *47*, 227-233.

[59] Crochet, P.; Díez, J.; Fernández-Zúmel, M. A.; Gimeno, J. *Adv. Synth. Catal.* 2006, *348*, 93-100.
[60] Köckritz, A.; Bischoff, S.; Kant, M.; Siefken, R. *J. Mol. Catal. A: Chem.* 2001, *174*, 119-126.
[61] Sarangapani, M.; Ravinder, V. *Indian J. Chem.* 2002, *41A*, 2060-2064.
[62] Sarangapani, M.; Chandrasekhar, V.; Ravinder, V. *J. Indian Chem. Soc.* 2005, *82*, 497-502.

In: Advances in Organometallic Chemistry Research
Editor: Kenji Yamamoto, pp. 67-79
ISBN: 978-1-60021-779-1

Chapter 3

RUTHENIUM-SILYLENE VERSUS RUTHENIUM-η3-SILANE, IMPLICATIONS FOR HYDROSILYLATION AND OTHER CATALYTIC TRANSFORMATIONS. A THEORETICAL STUDY

Uwe Böhme *
Institut für Anorganische Chemie, TU Bergakademie Freiberg
Leipziger Str. 29, 09596 Freiberg, Deutschland

ABSTRACT

An exciting new mechanism of hydrosilylation had been found in experiments with the ruthenium-silylene complex [Cp*(i-Pr_3P)Ru$(H)_2$Si(H)Ph·OEt_2] [B$(C_6F_5)_4$] by Glaser and Tilley. The proposed catalytic active complex shows surprising bond features. The central unit "Ru$(H)_2$Si" should rather be considered as a ruthenium-η^3-silane and not as a ruthenium-silylene. The ruthenium atom in the catalytic active complex has enough coordinative flexibility to accept the hydrogen atoms in order to activate the silicon atom and to stabilize adducts and transition states. The applicability of ruthenium-η^3-element hydride complexes [Cp(R_3P)Ru-η^3-EH_4]$^{0/+1}$ (E = B, Al, C, Si, Ge) for the insertion of alkenes into E-H bonds is discussed. These insertions should be possible with η^3-complexes of Si, Ge, B, and Al. With B and Al complexes it will probably not work as catalytic reaction, but stoichiometric transformations should be possible.

INTRODUCTION

The hydrosilylation is a reaction which is commonly known and often used in the chemical laboratory and some industrial processes [1,2]. The mechanism for this metal-catalyzed transformation of an organic substrate is widely accepted. Scheme 1 shows the mechanism according to Chalk and Harrod [3,4], which includes the coordination of the

* E-mail: uwe.boehme@chemie.tu-freiberg.de

alkene at the catalyst, oxidative addition of a silane, insertion of the alkene into the transition metal hydrogen bond and finally reductive elimination of the alkylsilane. An exiting new mechanism has been proposed in 2003 by Glaser and Tilley on the basis of experimental work [5,6]. The catalytic reaction is started by adding an alkene to a mixture of $[Cp^*(i\text{-}Pr_3P)Ru(H)_2Si(H)Ph\bullet Et_2O][B(C_6F_5)_4]$ ($Cp^* = \eta^5\text{-}C_5Me_5$), $PhSiH_3$, and benzene or methylene chloride. The ether molecule is easily replaced in solution by other donor molecules. Therefore it can be assumed that the cation $[Cp^*(i\text{-}Pr_3P)Ru(H)_2Si(H)Ph]^+$ represents the catalytic active complex. The authors were able to obtain alkylsilanes from $PhSiH_3$ and different alkenes in 57 to 94 % yield. The intermediates proposed in the catalytic cycle of scheme 2 have not been proved experimentally. A quantum chemical study which supports this mechanism has been published by Beddie and Hall [7].

A closer look at the structure and electronic properties of the catalytic active complex reveals bonding features which might be of importance for the hydrosilylation and other possible transformations. These features will be discussed in the next chapter. Quantum chemical calculations have been performed in order to obtain insight into the bonding features of the catalytic active complex $[Cp^*(i\text{-}Pr_3P)Ru(H)_2Si(H)Ph]^+$ and a number of related complexes $[Cp(H_3P)Ru(H)_2E(H)_2]^{0/+1}$ (E= B, Al, C, Si, Ge).

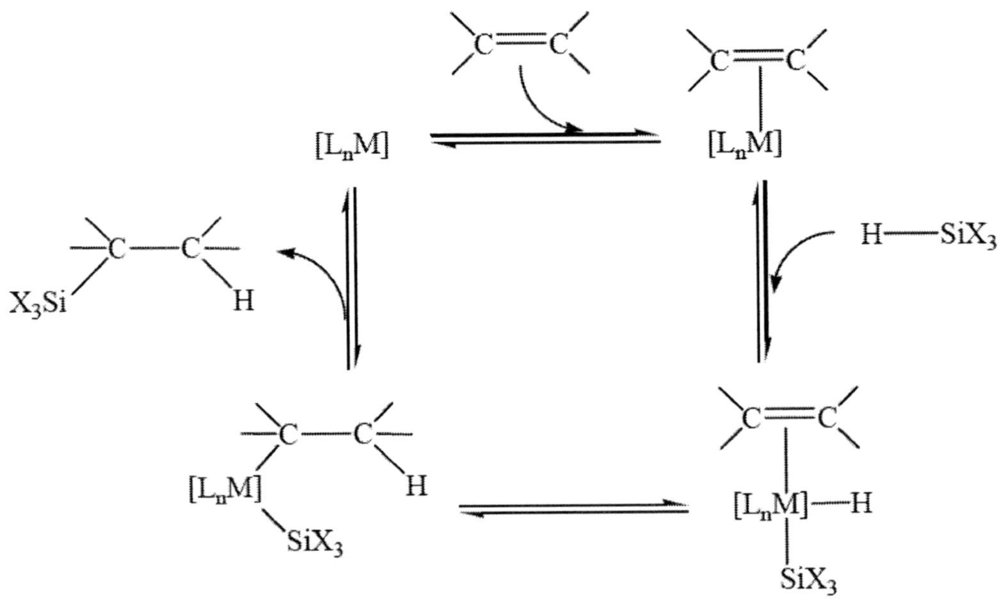

Scheme 1. Mechanism of hydrosilylation according to Chalk and Harrod [3].

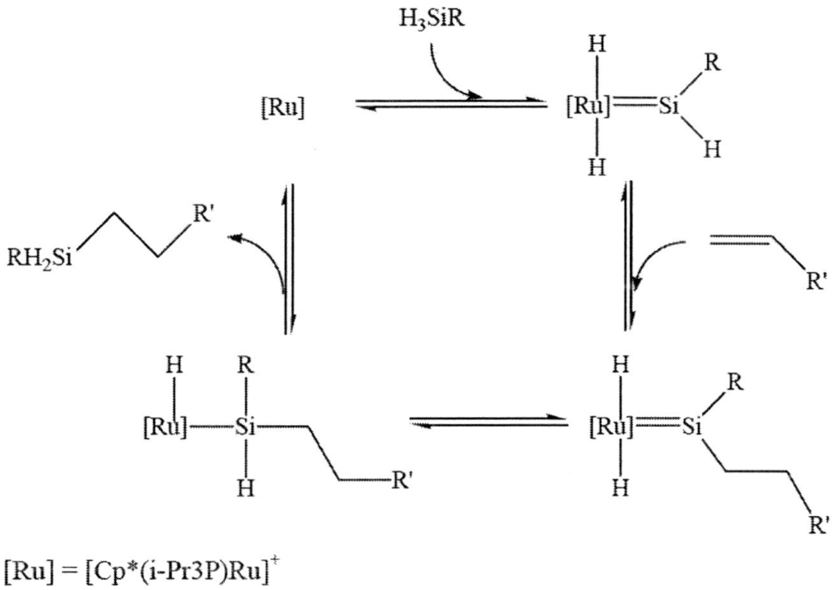

Scheme 2. New Mechanism of hydrosilylation proposed by Glaser and Tilley [5, 7].

STRUCTURE AND PROPERTIES OF $[Cp(H_3P)Ru(H)_2Si(H)Ph]^+$ (1)

The molecular system used for the experiments by Glaser and Tilley contains a pentamethylcyclopentadienyl ligand (Cp*) and a $(i\text{-}Pr)_3P$ ligand. Instead of these large ligands some simplifications have been introduced into the model compounds. In order to obtain the results in a reasonable time cyclopentadienyl (C_5H_5) and phosphane (PH_3) were used for the calculations with the model complex $[Cp(H_3P)Ru(H)_2Si(H)Ph]^+$ (1). All calculations have been performed with the density functional method B3LYP [8,9]. The core electrons of ruthenium were replaced by a pseudo potential, which was supplemented with a suitable basis set [10] and an additional f-polarization function [11]. All other elements obtained the 6-31G(d,p) basis set [12,13]. There were performed full geometry optimizations for all molecules and frequency calculations to verify the minima and transition states with Gaussian 03 [14].

Glaser and Tilley used a ruthenium silylene complex which is stabilized by one molecule of diethylether. The ether molecule is loosely bound to the silicon atom, since it can be easily replaced in solution by THF or benzophenone. The quantum chemical calculation shows that the removal of the ether molecule requires about 8 kJ/mol of Gibbs free energy. At this point arises the question: what is the bonding situation between Ru and silicon in this compound? Glaser and Tilley proposed a ruthenium silylene according to the valence structure A in figure 1. Of course this is a reactive intermediate and it would be rather difficult to obtain an X-ray structure of this compound. The optimized geometry of complex 1 is shown in figure 1. Surprisingly one finds the hydrogen atoms nearly in the middle between Ru and Si. This geometry forms a square arrangement of Ru, Si and two hydrogen atoms. The valence structure with a formal Ru=Si double bound (A) seems to be not appropriate to describe the bonding situation. A very similar arrangement of atoms for the central unit $Ru(H)_2Si$ was obtained by optimization of $[Cp^{*}(Me_3P)Ru(H)_2Si(H)Ph]^+$. This was checked in order to prove that the applied simplification in the model complex does not produce an erroneous geometry.

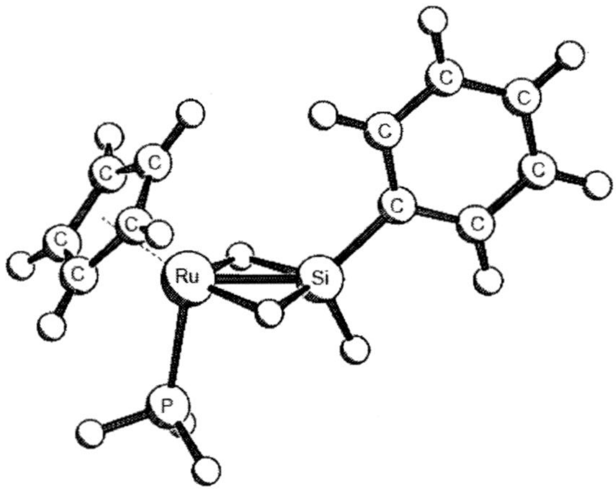

Ru-Si	2.306 Å
Ru-H	1.743 Å
Si-H(bridge)	1.708 Å
Si-H(terminal)	1.481 Å

A B

Figure 1. Optimized geometry of the catalytically active complex 1 (top) and representation of possible valence structures (bottom).

The topological analysis of the electron density distribution is a powerful tool for the analysis of bonding modes and bonding properties in a given molecule. This analysis can be carried out using the atoms in molecules theory (AIM) as developed by Bader and others [15,16,17]. This method partitions the electron density of a molecule ρ(r) into individual non-overlapping atomic fragments by rigorously defined interatomic surfaces [16]. Various properties can be derived from the electron density distribution. The electron density in a molecule is a three dimensional function, which has four types of extrema: maxima, minima and two types of saddle points [17]. The maxima coincide with the nuclear positions. Minima are found in the center of cage molecules and are called cage critical points. Saddle points which are a maximum in one dimension and a minimum in two dimensions are called ring critical points. These are found in the centre of atomic rings. Saddle points which are a maximum in two dimensions and a minimum in one dimension are called bond critical points. When a bond critical point (BCP) is found between two atoms one can draw an atomic interaction line. When the molecule is in a equilibrium geometry, the atomic interaction line is called a bond path. A bond path is not identical with a bond in the sense used by Lewis. A bond path is observed for predominately ionic bonds, covalent bonds, weak hydrogen bonds and even interactions between anions. The Laplacian of the electron density ($\nabla^2\rho$) at the BCP between two atoms gives essential information about the nature of the bond between the involved atoms. The Laplacian is the algebraic sum of the three principal curvatures λ_1, λ_2, and λ_3 of ρ(r). The nature of the atomic interactions depends on the dominant curvatures [18].

Covalent bonds (shared interactions) are dominated by negative curvatures (λ_1, λ_2) with charge concentrated in the internuclear region. The charge concentration is reflected by a large value of ρ(r) and by a negative value of the Laplacian $\nabla^2\rho$. For ionic interactions (closed shell interactions) the positive curvature λ_3 is dominant and charge is depleted at the BCP. The electron density is concentrated at the negatively charged atom and the spatial distribution of the Laplacian is predominately atom-like, that means there is a spherical distribution. The charge depletion at the BCP in this case is characterized by a low value of ρ(r) and by a positive value of the Laplacian $\nabla^2\rho$.

The topological analysis of 1 (figure 2) shows the presence of BCP's between Ru and H, Si and H, and Ru and Si. These BCP's give evidence for an η^3-bound silane unit [19]. The presence of two ring critical points in the square arrangement of atoms (Ru-H-Si-H-) corroborates this hypothesis. Numeric values of the topological analysis are given in table 1. The high value of ellipticity at the BCP between Ru and Si is striking. This hints to a highly delocalized ring system. The electron density at the BCP's between Ru-H, Ru-Si, and Si-H have similar values between 0.583 to 0.603 e/Å^3. Further information about the nature of these bonds can be drawn from the Laplacian $\nabla^2\rho$ (table 1 and figure 4). The Ru-H interaction is mainly of closed shell type, i. e. predominant ionic, as can be seen from the large positive value of the Laplacian (5.151 e/Å^5) and the spherical form of the electron density distribution of the hydrogen atoms in the direction to the ruthenium atom. This is in contrast to the charge concentration at the Si-H bond axes and the negative value of the Laplacian (-2.133 e/Å^5) which hints to a rather covalent interaction between Si and H in the four membered ring.

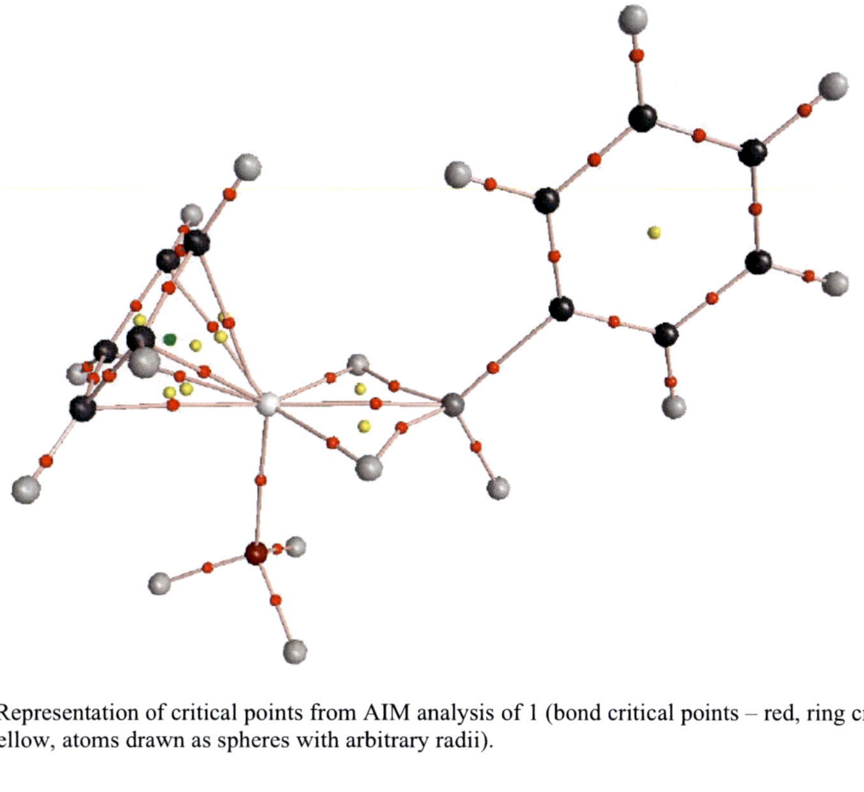

Figure 2. Representation of critical points from AIM analysis of 1 (bond critical points – red, ring critical points – yellow, atoms drawn as spheres with arbitrary radii).

Table 1. Electron density ρ, Laplacian $\nabla^2\rho$, and bond ellipticity ε at selected bond critical points in 1 and 2

	ρ (e/Å^3)	$\nabla^2\rho$ (e/Å^5)	ε
$[Cp(H_3P)Ru(H)_2Si(H)Ph]^+$ (**1**)			
Ru-H	0.603	5.151	0.528
Ru-Si	0.583	-0.631	1.511
Si-H(ring)	0.596	-2.133	0.750
Si-H(terminal)	0.822	3.249	0.010
ring point	0.546	2.337	
diborane (**2**)			
B-H(ring)	0.846	0.586	0.594
B-H(terminal)	1.260	-9.599	0.082
ring point	0.762	-0.650	

One could argue that the silane unit might be bound in an η^2-mode, with bonds between Ru-H, and H-Si, but not between Ru-Si. Let us consider diborane (B_2H_6, 2) as a typical molecule with hydrogen bridges between two boron atoms, but without a bond between the latter ones (figure 3) [20]. Only one ring critical point is in the centre of the four membered cycle in this molecule. Further information can be drawn from the comparison of the laplacian of rho ($\nabla^2\rho$), which is shown as graphic representation in figure 4. The valence charge density of diborane (2) is localized in the ring surface. There is a notable depletion of charge in the centre of the ring in the area around the ring critical point, which looks like a "hole" of charge depletion. The graphical representation of $\nabla^2\rho$ of 1 shows a complete different picture. The valence charge is concentrated in the ring surface and between Ru and Si. Because of the atom size and the unsaturated coordination sphere the ruthenium atom is enabled to form a η^3-coordination with the SiH_2 unit.

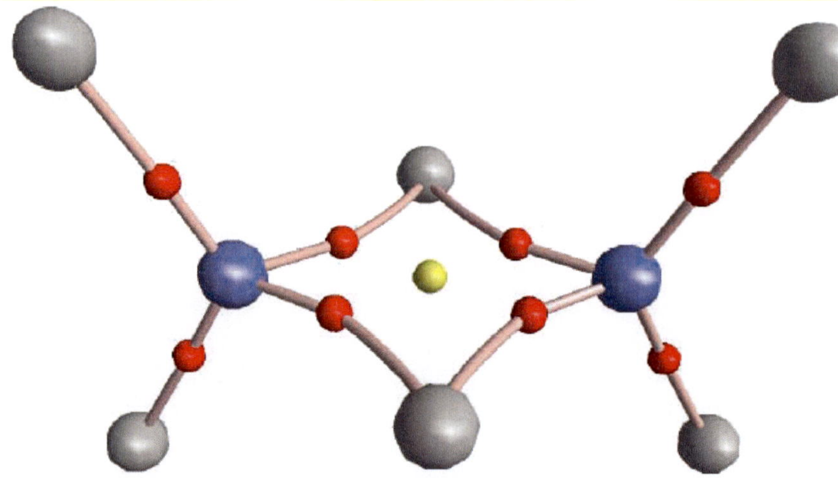

Figure 3. Representation of critical points from AIM analysis of diborane 2 (bond critical points – red, ring critical points – yellow, atoms drawn as spheres with arbitrary radii).

Figure 4. Laplacian of electron density in the ring plane of 1 (top) and diborane 2 (bottom). Positive values of $\nabla^2\rho$ are drawn with dashed lines and represent regions of charge depletion; negative values of $\nabla^2\rho$ are drawn with solid lines and represent regions of charge concentration. Contour values for both representations in atomic units are: 0.001, 0.002, 0.004, 0.008, 0.02, 0.04, 0.08, 0.2, 0.4, 0.8, 2, 4, 8, 20, 40, 80, 200, 400, 800.

According to the topological analysis we can exclude the valence structure with the Ru=Si double bond (A) and recognize the structure with the η^3-coordinated silane (B) as a better description. The valence structure B seems to be rather unusual. Therefore one should look whether there are other complexes with the same structural motif. Indeed there are also some other complexes which have bridging silane ligands with η^3-coordination (figure 5). The ruthenium complexes 3 and 4 with single and double bridged silane ligands have been investigated by Sabo-Ettienne et al. [21,22,23]. Furthermore a number of ruthenium complexes with η^2-bound silane ligands have been characterized with X-ray structure analysis during the last years [24,25,26,27,28,29]. Only one of these complexes, 5 [29], is shown in figure 5 as example. This type of interaction has been found entry into scientific literature as "non-classical" interligand interactions with silicon based ligands [30]. Even examples of triple nonclassical Ru-H-Si interactions, which corresponds to a formal η^4-coordination, has been found recently (6 in figure 5) [31,32]. Highlights, Reviews and quantum chemical calculations about this topic should be mentioned here for the interested reader, [30,33,34].

Now back into the mechanism of the hydrosilylation. After the identification of the catalytic active complex as a η^3-bound ruthenium silane complex there were examined the further steps of the hydrosilylation. The mechanism of this catalytic process was investigated by quantum chemical methods and described in earlier publications [7,35,36], which can be summarized as following: First the alkene coordinates at the silicon atom of the ruthenium-η^3-silane complex (1). The ruthenium atom is able to accept two hydrogen atoms as ligands and keeps the silane in its coordination sphere (7). The release of the hydrogen atoms from the silane unit activates the silicon atom and lowers the activation energy for the insertion of the alkene into the remaining Si-H bond (8). After the insertion of the alkene the hydrosilylated silane is generated and kept in the coordination sphere of the ruthenium complex in the η^3-bonding mode (9). In presence of excess silane R-SiH$_3$ the complex 1 is regenerated and the hydrosilylated silane (10) is expulsed.

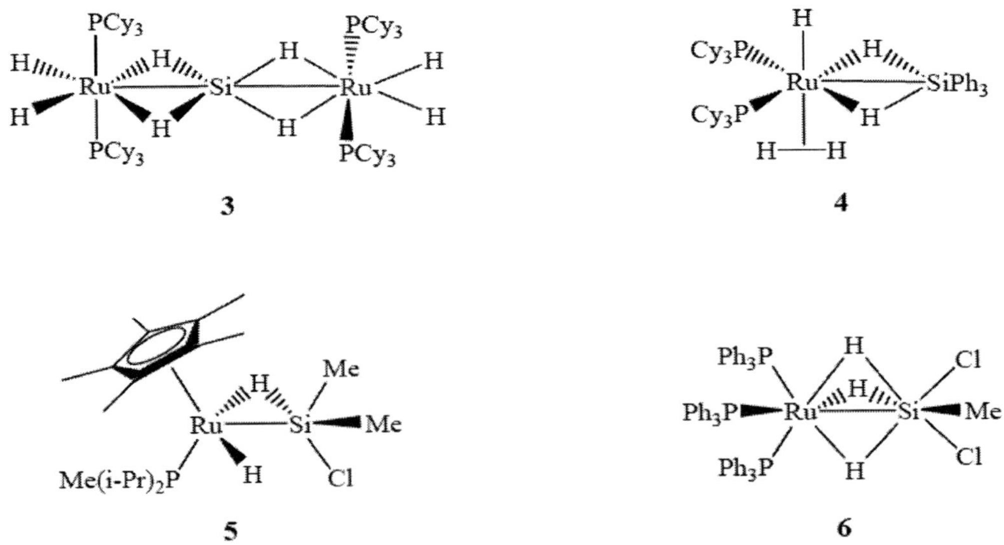

Figure 5. Other Complexes with H-ligands bridging; examples for non-classical interligand interactions (citations see text).

INSERTION REACTIONS INTO OTHER E-H BONDS OF RUTHENIUM H^3-COMPLEXES

With respect to possible applications it would be highly interesting to know whether it is possible to obtain insertion into other E-H bonds of group 13 and 14 elements. This question will be addressed in the forthcoming section. Hydroboration and hydroalumination are widely used processes where the selectivity of such transformations might be of interest. Hydroboration proceeds regioselective (anti-Markownikow) and stereoselective (cis-addition) but it is difficult to control the stoichiometry of this process [37].The hydroalumination is known for its low chemoselectivity with respect to different functional groups [38]. The proposed insertion into E-H bonds of η^3-complexes of ruthenium could open up new perspectives for these reactions.

The rate determining step of the catalytic cycle as outlined in scheme 3 is the insertion of the alkene into the E-H bond. Starting from the η^3-bridged element-ruthenium complex it should be possible to obtain an insertion into the E-H bond via adduct and transition state (see scheme 4). The free energy which is necessary to reach the adduct is denoted as $G_{A\text{-}E}$, the energy to achieve the transition state as $G_{TS\text{-}A}$, and the energy difference between transition state and product as $G_{P\text{-}TS}$. The differences of free energy for the relevant steps are summarized in table 2. The insertion with the Si and Ge compound is possible due to low activation energy. The methane complex exhibits a to large activation energy $G_{TS\text{-}A}$ and a surprisingly low activation energy for the formation of the adduct ($G_{A\text{-}E}$). The adduct of $[Cp(H_3P)RuCH_4]^+$ and ethene (11) is shown in figure 6 in comparison with the adduct of $[Cp(H_3P)RuGeH_4]^+$ with ethene (12). The ethene molecule in 11 is coordinated at the ruthenium atom and not at the carbon atom, which explains the lowering of the total energy. But 11 is not the adduct which is necessary for an insertion into C-H bonds. Therefore one has to conclude that the insertion reaction will not take place with $[Cp(H_3P)RuCH_4]^+$ and similar compounds with coordinated hydrocarbons. On the right side of figure 6 is shown the "ordinary" adduct of the Ge complex (12) for comparison, where the alkene is coordinated to the heteroatom (E = Ge).

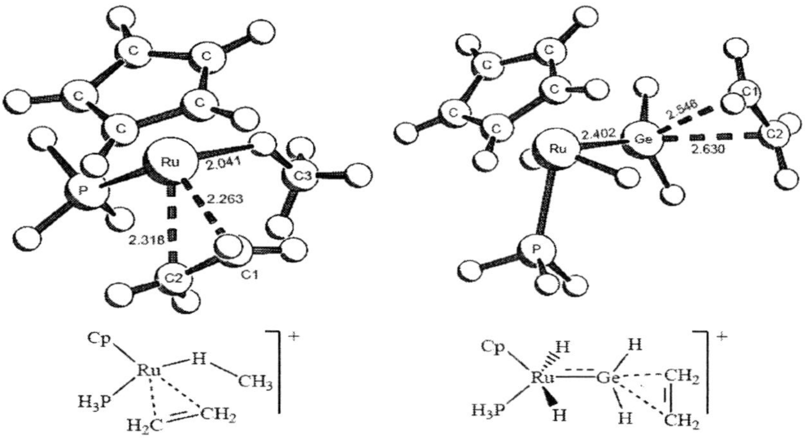

11 12

Figure 6. Optimized geometries of ethene adducts of $[Cp(H_3P)RuCH_4]^+$ (left) and $[Cp(H_3P)RuGeH_4]^+$ (right).

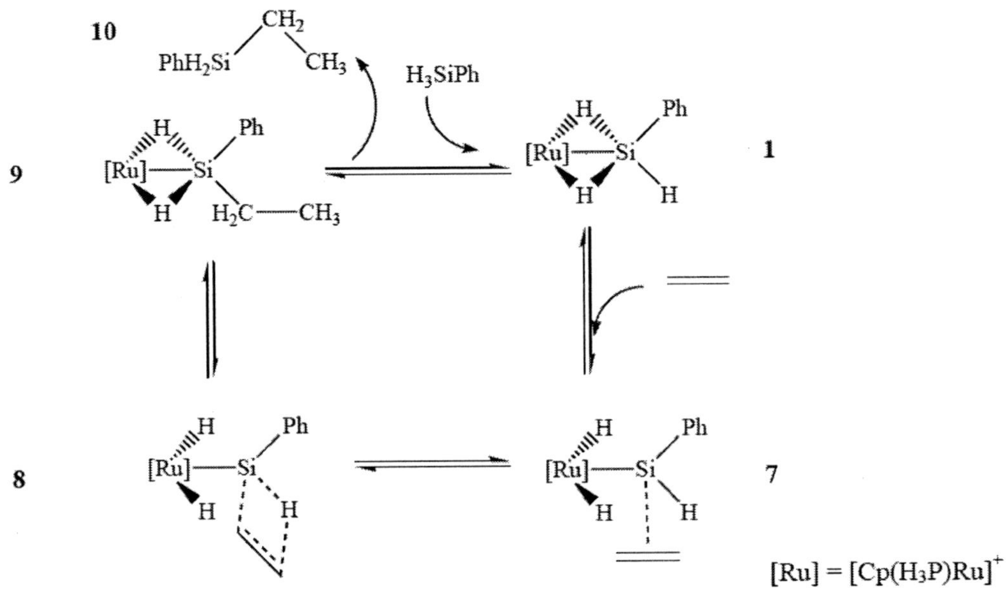

Scheme 3. Calculated Mechanism of Hydrosilylation [34].

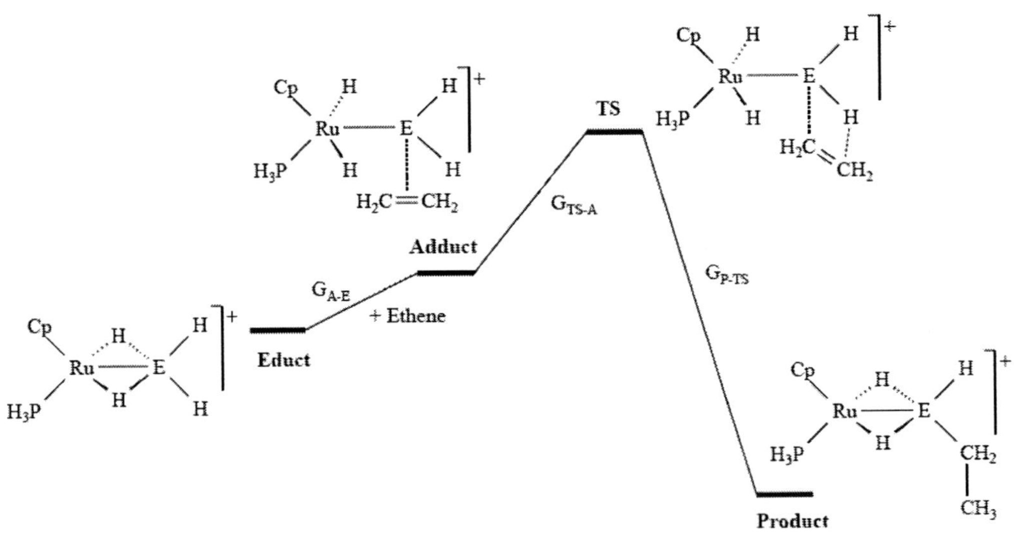

Scheme 4. Mechanism of Insertion into other E-H Bonds (for elements E of group 14 the complexes carry one positive charge).

Table 2. Differences of Gibbs free energy for the insertion into E-H bonds of $[Cp(H_3P)RuEH_4]^{0/+1}$. The designation of energy differences is according to Scheme 4 (ΔG in kJ/mol)

E	$G_{A\text{-}E}$	$G_{TS\text{-}A}$	$G_{P\text{-}TS}$
C	-43.0	221.7	-218.8
Si	18.9	42.4	-146.5
Ge	1.0	55.0	-138.8
B	23.1	125.4	-198.9
Al	30.3	80.0	-168.1

Table 2 contains also energy values for the insertion of alkenes into Ru-η^3-complexes of the group 13 elements B and Al. The same procedure as for the group 14 elements was used here and it is possible to conclude because of the energy differences that we could expect hydroalumination and hydroboration. The activation energies for these insertions are higher than for the non-catalyzed reactions [39]. Therefore the catalytic reaction will not proceed in presence of free borane (BH_3) or aluminium hydride, but a stoichiometric reaction should be possible.

CONCLUSION

Glaser and Tilley have found a new type of catalyst for the hydrosilylation of alkenes. The investigation of the electronic properties of the catalytic active complex reveals that it is a ruthenium complex with a η^3-coordinated silane unit. It is possible to develop a reasonable mechanism for the hydrosilylation with the ruthenium η^3-silane complex as key intermediate. The ruthenium atom in the complex 1 has enough coordinative flexibility to accept the hydrogen atoms in order to stabilize transition states and adducts. The silicon atom undergoes substantial activation, when the ruthenium atom "takes over" the hydrogen atoms.

Insertion into E-H bonds should be possible also with η^3-complexes of germanium, boron, and aluminium. With boron and aluminium it will probably not work as catalytic reaction, since the insertion proceeds already without catalyst. But it should be possible to obtain selectively mono-alkylboranes, or dialkylboranes with two different substituents in stoichiometric reactions.

ACKNOWLEDGEMENTS

The author thanks the Computing Centre of the TU Bergakademie Freiberg for disk space and computation time and the Bundesland Sachsen for financial support.

REFERENCES

[1] Marciniec, B. (Hrsg.): *Comprehensive Handbook on Hydrosilylation*, Pergamon Press, Oxford 1992.

[2] Ojima, I.: The hydrosilylation reaction; in *The Chemistry of Organic Silicon Compounds*, Editors: Patai, S.; Rappoport, Z.; Wiley, Chichester, 1989, 1479.

[3] Chalk, A. J.; Harrod, J. F. *J. Am. Chem. Soc.* 1965, *87*, 16-21.

[4] Duckett, S. B.; Perutz, R. H. *Organometallics* 1992, *11*, 90-98.

[5] Glaser, P. B.; Tilley, T. D. *J. Am. Chem. Soc.* 2003, *125*, 13640-13641.

[6] Brunner, H. Angew. Chem. 2004, 116, 2805; Angew. Chem. Int. Ed. Engl. 2004, 43, 2749.

[7] Beddie, C.; Hall, M. B. *J. Am. Chem. Soc.* 2004, *126*, 13564-13565.

[8] Becke, A. D. *J. Chem. Phys.* 1993, *98*, 5648-5652.

[9] Lee, C.; Yang, W.; Parr, R. G. *Physical Rev.* 1988, *B37*, 785-789.

[10] Hay, P. J.; Wadt, W. R. *J. Chem. Phys.* 1985, *82*, 299-310.

[11] Ehlers, A. W.; Böhme, M.; Dapprich, S.; Gobbi, A.; Höllwarth, A.; Jonas, V.; Köhler, K. F.; Stegmann, R.; Veldkamp, A.; Frenking, G. *Chem. Phys. Lett.* 1993, *208*, 111-114.

[12] Hariharan, P. C.; Pople, J. A. *Theoret. Chim. Acta* 1973, *28*, 213-222.

[13] Francl, M. M.; Pietro, W. J.; Hehre, W. J.; Binkley, J. S.; Gordon, M. S.; DeFrees, D. J.; Pople, J. A. *J. Chem. Phys.* 1982, *77*, 3654-3665.

[14] Gaussian 03, Revision C.02, M. J. Frisch, G. W. Trucks, H. B. Schlegel, G. E. Scuseria, M. A. Robb, J. R. Cheeseman, J. A. Montgomery, Jr., T. Vreven, K. N. Kudin, J. C. Burant, J. M. Millam, S. S. Iyengar, J. Tomasi, V. Barone, B. Mennucci, M. Cossi, G. Scalmani, N. Rega, G. A. Petersson, H. Nakatsuji, M. Hada, M. Ehara, K. Toyota, R. Fukuda, J. Hasegawa, M. Ishida, T. Nakajima, Y. Honda, O. Kitao, H. Nakai, M. Klene, X. Li, J. E. Knox, H. P. Hratchian, J. B. Cross, C. Adamo, J. Jaramillo, R. Gomperts, R. E. Stratmann, O. Yazyev, A. J. Austin, R. Cammi, C. Pomelli, J. W. Ochterski, P. Y. Ayala, K. Morokuma, G. A. Voth, P. Salvador, J. J. Dannenberg, V. G. Zakrzewski, S. Dapprich, A. D. Daniels, M. C. Strain, O. Farkas, D. K. Malick, A. D. Rabuck, K. Raghavachari, J. B. Foresman, J. V. Ortiz, Q. Cui, A. G. Baboul, S. Clifford, J. Cioslowski, B. B. Stefanov, G. Liu, A. Liashenko, P. Piskorz, I. Komaromi, R. L. Martin, D. J. Fox, T. Keith, M. A. Al-Laham, C. Y. Peng, A. Nanayakkara, M. Challacombe, P. M. W. Gill, B. Johnson, W. Chen, M. W. Wong, C. Gonzalez, and J. A. Pople, Gaussian, Inc., Wallingford CT, 2004.

[15] Bader, R. F. W., *Atoms in Molecules*, Clarendon Press, Oxford 1994.

[16] Gillespie, R. J., *J. Chem. Educ.* 2001, 78, 1688.

[17] Explanation taken from: Gillespie, R. A. , Popelier, P. L. A., *Chemical Bonding and Molecular Geometry*, Oxford University Press, New York 2001.

[18] Explanation taken from: Kocher, N., Selinka, C., Leusser, D., Kost, D., Kalikhman, I., Stalke, D., *Z. Anorg. Allg. Chem.* 2004, *630*, 1777-1793.

[19] AIM analyses have been performed with XAim and AIM2000: XAim was developed by Jose Carlos Ortiz and Carles Bo, Universitat Rovira i Virgili, Tarragona, Spain. Biegler-König, F; Schönbohm, J.; Bayles, D; AIM2000 - A Program to Analyze and Visualize Atoms in Molecules, submitted to *J. Comp. Chem.* 2001, 22, 545-559.

[20] The topology of diborane has been investigated earlier: Bader, R. F. W.; Essén, H. *J. Chem. Phys.* 1984, *80*, 1943-1960. The data for diborane have been recalculated for this work in order to have them comparable to 1.

[21] Atheaux, I.; Donnadieu, B.; Rodriguez, V.; Sabo-Etienne, S.; Chaudret, B.; Hussein, K.; Barthelat, J.-C. *J. Am. Chem. Soc.* 2000, *122*, 5664-5665.

[22] Ben Said, R.; Hussein, K.; Barthelat, J.-C.; Atheaux, I.; Sabo-Etienne, S.; Grellier, M.; Donnadieu, B.; Chaudret, B. *Dalton Trans.* 2003, 4139-4146.

[23] Hussein, K.; Marsden, C. J.; Barthelat, J.-C.; Rodriguez, V.; Conejero, S.; Sabo-Etienne, S.; Donnadieu, B.; Chaudret, B. *Chem. Commun.* 1999, 1315-1316.

[24] Campion, B. K.; Heyn, R. H.; Tilley, T. D. *Chem. Commun.* 1992, 1201-1203.

[25] Freeman, S. T. N.; Lemke, F. R.; Brammer, L. *Organometallics* 2002, *21*, 2030-2032.

[26] Lachaize, S.; Sabo-Etienne, S.; Donnadieu, B.; Chaudret, B. *Chem. Commun.* 2003, 214-215.

[27] Ayed, T.; Barthelat, J.-C.; Tangour, B.; Pradère, C.; Donnadieu, B.; Grellier, M.; Sabo-Etienne, S. *Organomeallics.* 2005, *24*, 3824-3826.

[28] Osipov, A. L.; Vyboishchikov, S. F.; Dorogov, K. Y.; Kuzmina, L. G.; Howard, J. A. K.; Lemenovskii, D. A;. Nikonov, G. I. *Chem. Commun.* 2005, 3349-3351.

[29] Osipov, A. L.; Gerdov, S. M.; Kuzmina, L. G.; Howard, J. A. K.; Nikonov, G. I. *Organometallics.* 2005, *24*, 587-602.

[30] Nikonov, G. *J. Organomet. Chem.* 2001, *635*, 24-36.

[31] Yardy, N. M.; Lemke, F. R.; Brammer, L. *Organometallics* 2001, *20*, 5670-5674.

[32] Hübler, K.; Hübler, U.; Roper, W. R.; Schwerdtfeger, P.; Wright, L. J. *Chem. Eur. J.* 1997, *3*, 1608-1616.

[33] Nikonov, G. *Angew. Chem.* 2001, *113*, 3457-3459; *Angew. Chem. Int. Ed. Engl.* 2001, *40*, 3353-3355.

[34] Choi, S.-H., Lin, Z. *J. Organomet. Chem.* 2000, *608*, 42-48.

[35] Beddie, C.; Hall, M. B. *J. Chem. Phys.* 2006, *110*, 1416-1425.

[36] Böhme, U. *J. Organomet. Chem.* 2006, *691*, 4400-4410.

[37] Zaidlewicz, M. in *Comprehensive Organomet. Chem.*; Wilkinson, G.; Stone, F. G. A.; Abel, E. W.; Ed.; Pergamon Press, Oxford, U.K.; 1982;Vol. 7; pp 143-160.

[38] Elschenbroich, C.; *Organometallchemie*, B. G. Teubner Verlag, 5th Editon, Wiesbaden, 2005, 116.

[39] Activation energies for the non-catalyzed reactions $EH_3 + H_2C{=}CH_2 \rightarrow [TS] \rightarrow H_2E\text{-}CH_2\text{-}CH_2$ are 17.0 kJ/mol for E = B and 48.4 kJ/mol for E = Al (B3LYP/6-31G* level of theory).

In: Advances in Organometallic Chemistry Research
Editor: Kenji Yamamoto, pp. 81-96
ISBN: 978-1-60021-779-1

Chapter 4

ACTIVATION OF ALKYNES BY CAMPHOR COMPLEXES

M. Fernanda and N.N. Carvalho
Centro de Química Estrutural, Complexo I, Instituto Superior Técnico
Universidade Técnica de Lisboa; Av. Rovisco Pais, 1049-001 Lisboa, Portugal

ABSTRACT

The search for catalysts that promote or accelerate chemical processes is of utmost relevance for the syntheses of organic species otherwise difficult or even impossible to obtain. Higher activities, lower costs and cleaner processes are the actual driving forces for the development of new catalysts that wider the scope of compounds or improve methods to obtain the existing ones.

Within this purpose in the last decades several transition metals were studied as catalysts for promotion of ring closure due to the relevance of those processes for the syntheses of natural products or biological active species.

The formation of aromatic or heterocyclic systems from linear precursors such as alkynes[1,2] continues to be a challenge to fulfil more economical and environmental friendly synthetic processes.

1. CAMPHOR LIGANDS

Traditional medicine has used camphor as anesthetic for local pain relief since it is ready absorbed through the skin. More recent uses include plasticizer for cellulose nitrate, insect repellent or beauty creams but scarce examples exist of the use of camphor to address actual challenges of medicine. An exception is the recent study of camphorate platinum compounds that revealed as promising antitumor agents[3]. The main difficulty for the use of camphor in conjunction with metals for biological or industrial purposes is the low ability of camphor to bind metals. In order to overcome that difficulty, without substantial structural modifications in the camphor skeleton, R groups have been introduced in positions 1, 2 or 3 (figure 1).

In contrast with medicine, many examples exist of the use of camphor or camphor derivatives[4] as chiral auxiliaries for asymmetric synthesis[5-8] or catalysis[9,10] in metal mediated processes, although, the characteristics of the metal-ligand interaction quite often remained almost unknown. Till now, just one camphor uranyl adduct had its structure confirmed by X-ray analysis[11].

A tentative to overcome the low ability of camphor to bind metals was the introduction of phosphorus[12,13] nitrogen or sulfur[14] substituents in position 2 and/or 3 of the camphor skeleton (see figure 1 for numbering) that promote coordination to the metal. Camphorsulfonic acid has also been used under this purpose, since the sulfur atom enables coordination to metals affording complexes stable enough to be characterized. Some camphor derived compounds have been used as catalysts in asymmetric hydrogenation[13], Pauson-Khand[14], Diels-Alder[15] or C-C[16] coupling and recently essayed as luminescent materials for electronic devices[17]. Complexes that can be fully characterized are frequently less active in organic synthesis or catalysis than those that can not. Nevertheless, they commonly allow some insight into the properties of the ligands that contribute to a better knowledge of the processes.

In order to study camphor type species as ligands and get some data into the characteristics of camphor derived complexes and their activity as catalysts we addressed the study of the coordinative ability of camphorimine (L1), camphorsulfonimine (L2) and dialkyne camphor (L3) species (figure 1) to transition metals of group 10.

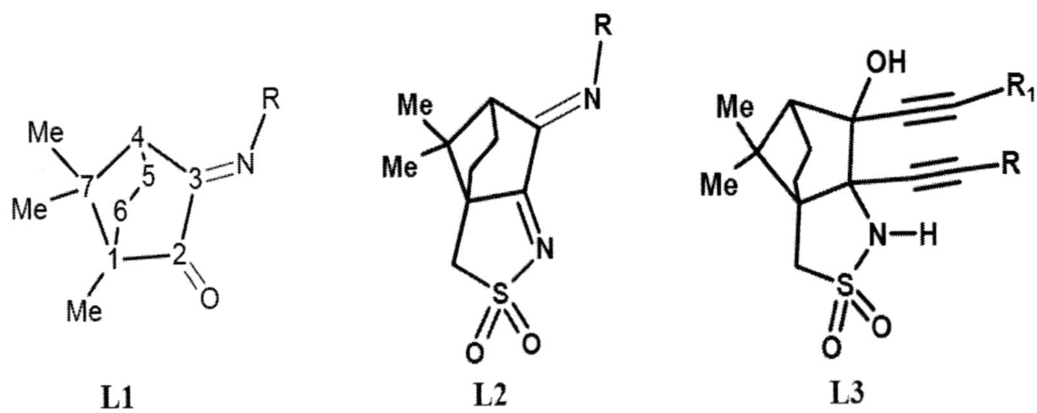

Figure 1. Camphorimine (L1), showing numbering scheme; camphorsulfonimine (L2); camphor diynes (L3).

Camphor derivatives of type L1 were obtained by condensation of the corresponding amine or hydrazine with camphorquinone using a strategy based on ancient procedures[18] after some experimental improvement in order to get suitable yields for the synthesis of species 3-$RNC_{10}H_{14}O$[19,20] (Scheme 1) .

Camphorquinone (2) is a very useful precursor for the synthesis of several camphor derived nitrogen coordinating species[10,21], since it is ready obtained from camphor (1) by oxidation with SeO_2.

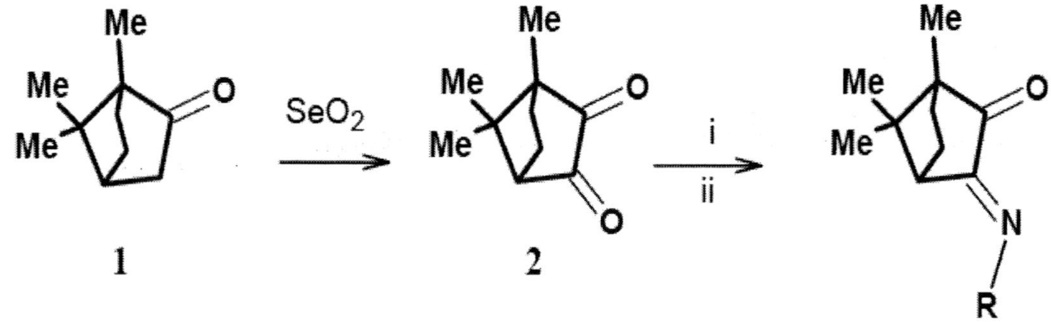

1 2

Scheme 1. Schematic synthesis of camphorimine derivatives: i –RNH_2 (1a), ii – H_2NNRR_1 (1b).

Camphorsulfonimines L2 are obtained from 3-oxo-camphorsulphonimine (4) by reaction with the appropriate amine or hydrazine[19] as the final step of an overall process that starts from the commercial available camphorsulfonic acid (3). In the process the acid is successively converted into a chloride (i) followed by an amine species (ii) that cyclizes to form a sulfonimine species (iii) which is then oxidized with SeO_2 to afford 3-oxo-camphorsulphonimine (iv)[22] (Scheme 2).

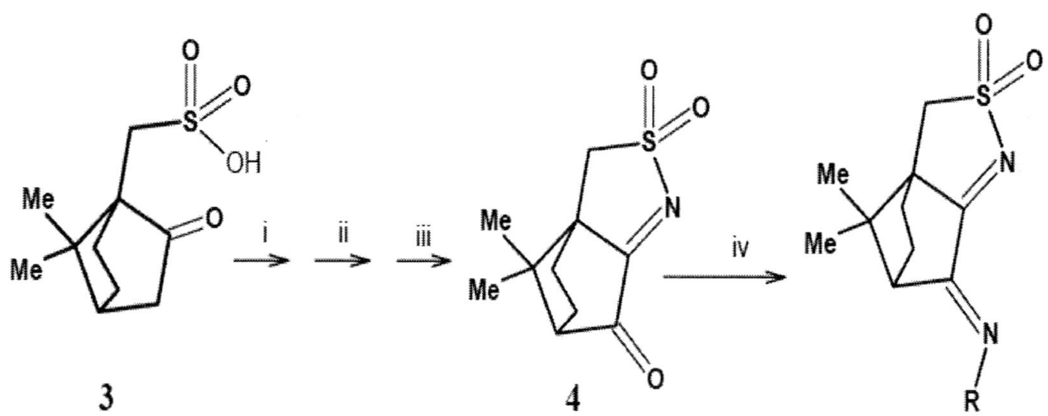

3 4

Scheme 2. Schematic synthesis of camphorsulfonimines; i- SO_2Cl; ii- ammonia solution; iii- SeO_2; iv- RNH_2.

Camphorsulphonimide diynes L3 are obtained from 3-oxo-camphorsulphonimine (4) by reaction with the appropriate deprotonated terminal alkyne.[23,24] Strict control of the experimental conditions allows the isolation of mono-substituted species of type L4 or L5, Scheme 3. The characteristics of the R group play an important role for the type of mono-alkyne species formed, L4 or L5, as a general trend L5 forms preferably to L4.

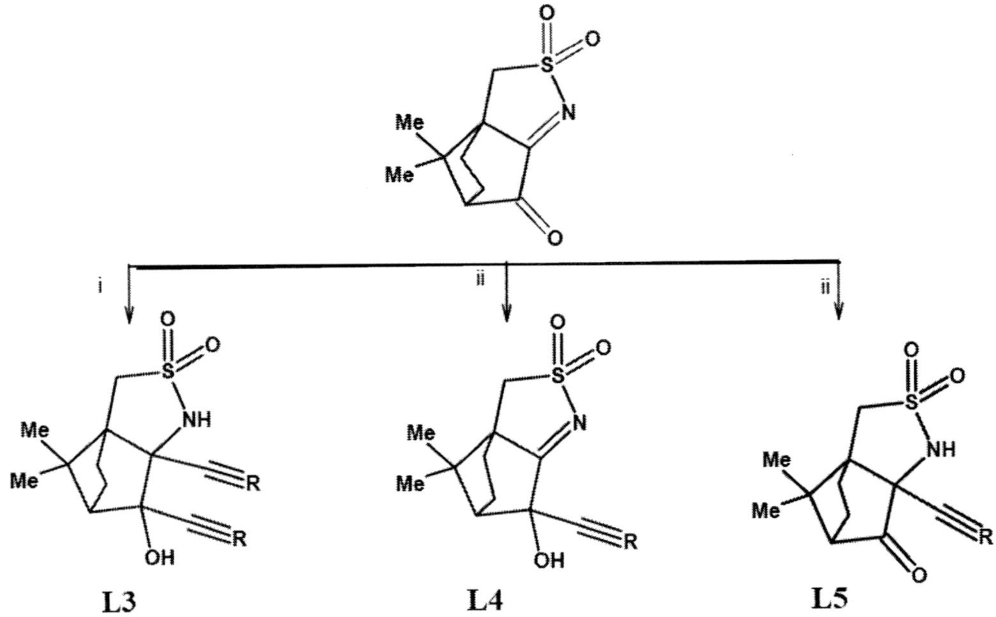

Scheme 3. Schematic syntheses of camphor sulfonimide mono- or di-ynes. i: > 2 LiC≡CR ; ii: slightly under one equivalent of LiC≡CR.

2. COMPLEXES

The camphorimine complexes derived from L1 ligands display square planar geometries with the L1 ligands occupying mutually *trans* positions. Camphorimines L1 coordinate transition metals (e.g. Pd, Pt) through the nitrogen atom of the imine group (figure 2 (b)) as confirmed by structural analysis performed by X-rays in *trans*-$[MCl_2L_2]$ (M=Pd or Pt, L1: R=NMe_2 and M=Pd, L1:R=*iso*-Pr). The ORTEP[20] drawing of the complex *trans*-$[PtCl_2(Me_2NNC_{10}H_{14}O)_2]$ (L1 type) is displayed in figure 2 (a) as well as the schematic structure of the complex (figure 2 (b)).

All complexes whose structures were analyzed by X-rays show the non-coordinating oxygen atom of the keton group (figure 2) at a distance of the metal that is in the order of the sum of the van der Waals radii of the oxygen and metal atoms. Consequently, the complexes *trans*-$[MCl_2L_2]$ (L= L1) have some *pseudo* octahedral character which is believed to be a relevant structural feature to their catalytic properties.

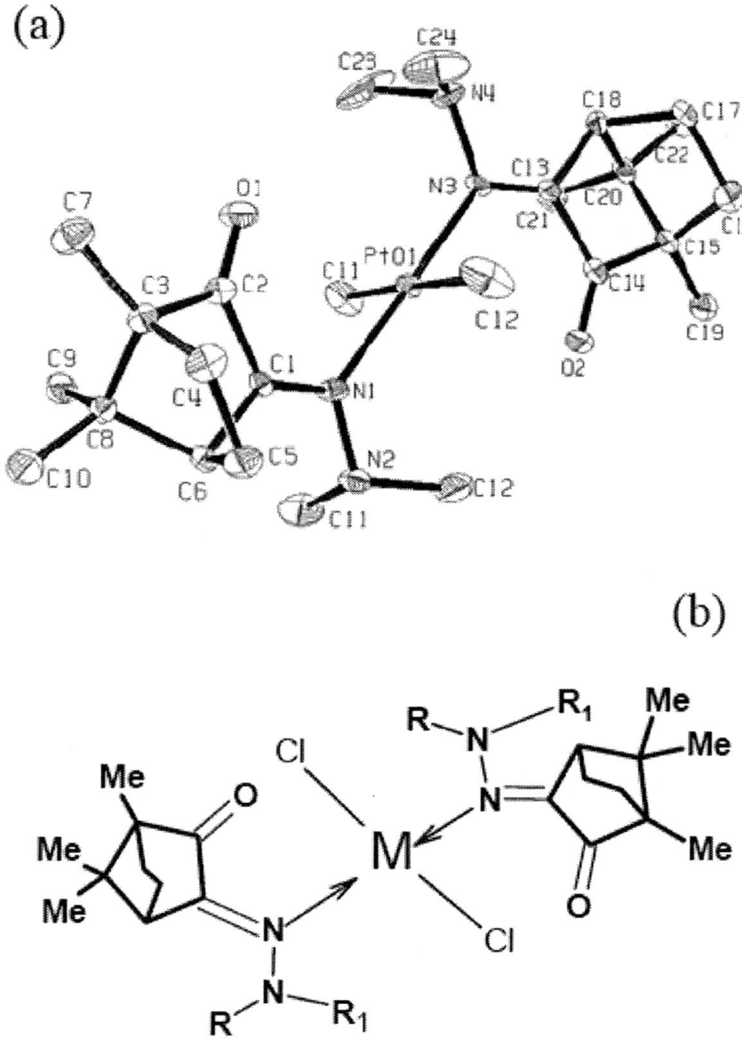

Figure 2. (a) – ORTEP drawing of *trans*-[$PtCl_2(Me_2NNC_{10}H_{14}O)_2$]; (b) - Schematic structure of complexes *trans*-[$MCl_2(L1)_2$] .

In contrast with L1 the camphor species L2 coordinate the Pd or Pt metals through the nitrogen atom of the sulfonimine group at position C-2 even if an imine group is available at position C-3 (figure 3). This tendency was confirmed by structural analysis performed at *cis*-[$PdCl_2${3-dimethylhydrazino)camphorsulfinimine}$_2$]. In the complex the two L2 ligands occupy mutually *cis* positions within an arrangement that reduces competitive electron withdrawing by the two camphor derived ligands. In Pd or Pt complexes there is a tendency of the camphor species L1 to arrange in *trans* position while species L2 arrange in *cis* position as verified by infra-red spectroscopy even if no X-ray data was available, *i.e.* the IR spectra of the complexes derived from L1 display just one M-Cl stretching band while those of complexes derived from L2 species display two M-Cl stretching bands (often very close to each other) in the range of 330-350 cm^{-1} in agreement, respectively with *trans* and *cis* arrangements.

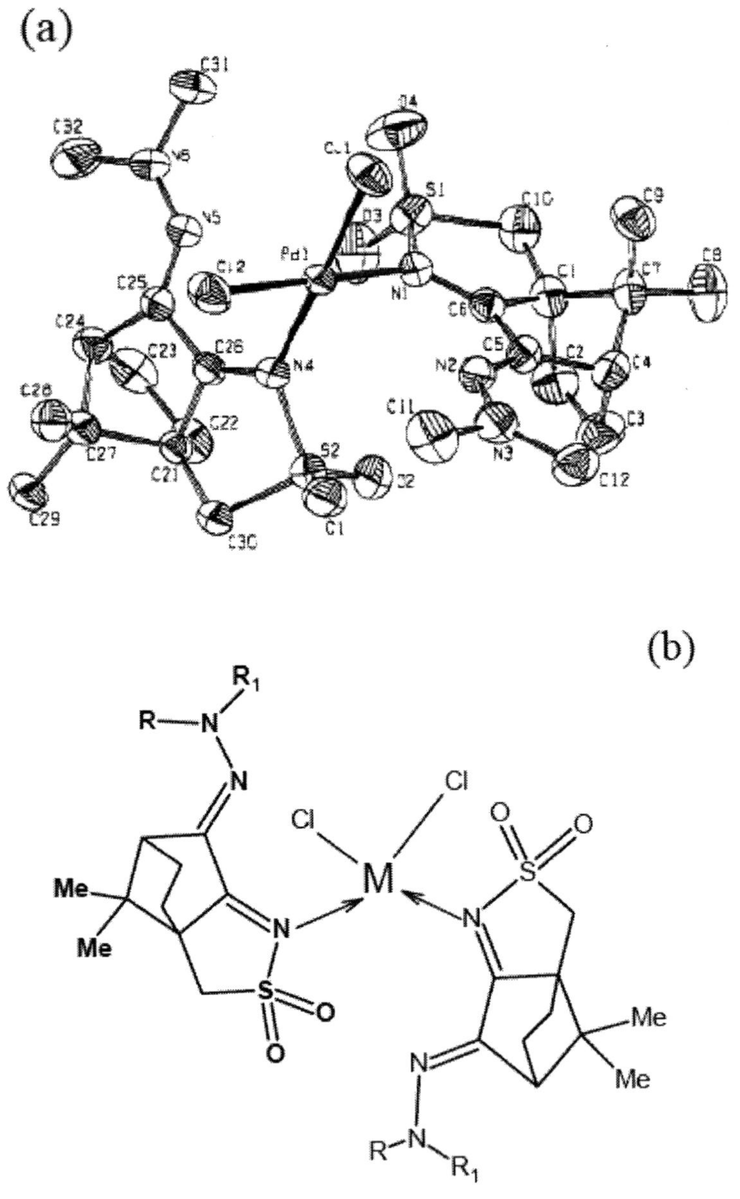

Figure 3. (a) ORTEP drawing of *trans*-[$PdCl_2$(3-dimethylhydrazino) camphorsulfinimide}$_2$]; (b) - Schematic structure of complexes *trans*-[MCl_2L_2] (L=type 2 ligand).

The geometry of complexes derived from L2 ligands are square planar such as that of complexes derived from ligands L1 but no evidence exists for an intramolecular interaction between the imine nitrogen atom (at C-3) and the metal that could be related with that one established between the oxygen of the keton group and the metal in L1 derived complexes. Although, L1 and L2 resemble each other in many aspects they have considerably different coordinative properties that induce structural and electron differences that affect the properties of the complexes.

The coordinative characteristics of camphor di-alkyne species L3 are still poorly known, since no crystals suitable for structural analysis were obtained. Studies of L4 and L5 as ligands for transition metals are under progress.

3. ACTIVATION OF ALKYNES

Transition metals are often used as auxiliaries to promote organic reactions otherwise difficult. In some cases the processes become *truly* catalytic and useful under synthetic purposes. Cascade reactions are of particular interest since several steps can be accomplished without experimental interruptions, thus improving the yields and accounting for time and atom economy.

Catalytic activation of unsaturated molecules by transition metals is a rather convenient process for the synthesis of some organic species difficult to obtain through classical organic strategies.[25] Among the unsaturated molecules, alkynes are attractive versatile precursors for the synthesis of organic species through metal mediated processes, since they display high coordinative ability to most of the transition metals forming alkynyl, vinylidene or other type of complexes[26-29] that in same cases trigger catalytic reactions such as coupling[30], cyclization[26,31,32] or oligomerization.[33-36]

Assembling of aromatic rings from linear alkynes is a process of utmost relevance for synthetic chemistry[37] that has been known since Repp[38] in the fifties discovered that palladium dichloride promotes the cyclo oligomerization of alkynes. Latter, Maitlis[39] studied the process in more detail and sort out a mechanism proposal for the cyclization process. Recently Yamamoto[1] extended the scope of possibilities to use alkynes as building molecules for organic syntheses. The strategy to promote the cyclo trimerization of alkynes in order to produce high substituted aromatic compounds[2,40] has been used with efficiencies tuned by the characteristics of the catalyst but also of the alkyne.

Within our work we verified that in very mild experimental conditions palladium camphorimine complexes *trans*-$[PdCl_2L_2]$ (L1 or L2) react with alkynes leading to C-C bond formation. As a general trend complexes derived from L1 are efficient towards alkynes cyclo trimerization, whether complexes derived from L2 promote their oligomerization.

The catalytic activity of complexes derived from L1 strongly depends on the characteristics of the R group bound to the imine (figure 1). Among all the complexes essayed as catalysts till now, *trans*-$[PdCl_2L_2]$ (L, R=NMe_2) revealed as the best catalyst for cyclo trimerization of alkynes.

The formation of 1,3,5-trimethyl-2,4,6-triphenylbenzene from 1-phenyl-1-propyne, reaction 1, catalyzed by *trans*-$[PdCl_2L_2]$ (L1: R=NMe_2) occurs with high *regio* selectivity (higher than 99%) and activity (A = 1.1×10^4 gT/(mol [Pd] x [(C≡C)]x/h). The process is also efficiently promoted in other internal alkynes such as diphenyl acetylene or dimethyl acetylenedicarboxylate, however, the activity and selectivity of camphorimine (L1) palladium complexes towards promotion of cyclotrimerization of terminal alkynes strongly depend of the characteristics of the alkyne.[41]

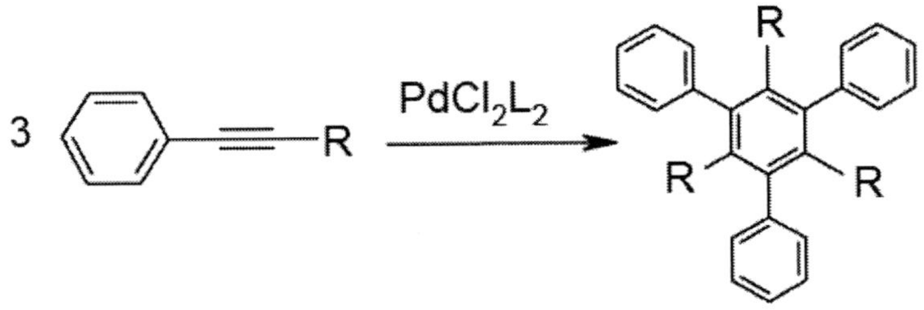

(1)

As a general trend the catalytic process works better for internal than for terminal alkynes, except for terminal bulky (R=Bu^t) or electron withdrawing (R=CO_2Me) alkynes that also undergo cyclo trimerization with good efficiency and selectivity.

The mechanism of cyclization is still unclear. However, some experimental details obtained by ^{1}H and ^{13}C NMR indicate that the process involves the coordination of the alkyne to the metal in addition to the rearrangement of the ligand. The experimental data obtained by NMR suggests that there is no release of the camphorimine ligand during the process.

cis-[$PdCl_2L_2$] (L2) and *trans*-[$PdCl_2L_2$] (L1) are conceptually analogous complexes, since they have ligands derived from (+)-camphor that coordinate to the palladium through a nitrogen atom of an imine group. These aspects contributed to expect that they would behave similarly with respect to induce reactivity in alkynes. However, it was demonstrated that their catalytic properties are rather different.

cis-[$PdCl_2L_2$] (L2), in contrast with *trans*-[$PdCl_2L_2$] (L: L1) do not catalyse the cyclotrimerization of internal or terminal alkynes. Instead, *cis*-[$PdCl_2L_2$] (L: L2) complexes promote alkyne oligomerization. [31]

The considerably different catalytic properties of palladium complexes derived from L2 or L1 are supported on the coordinative and electronic characteristics of the two types of ligands. Ligand L2 coordinates the metal by the nitrogen atom of sulfonimine a fact that drives the *cis* geometry of the complexes to avoid electron competition by the two ligands. L1 ligands which are less electron withdrawing allow the complexes to acquire a *trans* geometry that enables the interaction of the oxygen atom of the keton group with the metal site thus giving to the complex some octahedral character that may be relevant for the catalytic process.

At complexes derived from L2 (figure 1) the molecules are more crowded in consequence of the *cis* coordination than those derived from L1, at least at one of the sides. Furthermore, due to one additional ring, L2 ligands display lower versatility than L1. These are structure aspects that may be relevant for the different catalytic activities of palladium L1 or L2 derived complexes.

In what concerns the electronic characteristics, the lower electron density of *trans*-[PdL_2L_2] (L1: R=NMe_2) compared to *cis*-[PdL_2L_2] (L2: R=NMe_2) is revealed by the redox properties of the complexes[42], *e.g.* the reduction potential Pd(II) → Pd(I) is *ca.* 300 mV less negative for complexes derived from L1 than those derived from L2 (analogous R group). Lower electron density in addition to high coordinative versatility may give to the L1 palladium complexes a better ability to interact with alkynes conceivably promoting the coordination of more than one substrate molecule thus allowing cyclization. Such a process seem less favourable using *cis*-[$PdCl_2L_2$] (L: L2) as catalyst due to higher electron density and higher steric constraints. Thus, promotion of oligomerization instead of cyclization of alkynes by L2 palladium complexes seems a more feasible process, although there are several aspects to be confirmed before the mechanism is fully understood.

4. ACTIVATION OF ALKYNOLS

The search for catalysts that enhance the synthesis of heterocycle species is of high relevance for organic synthesis under the topics of natural products[43] or biologically active species[44,45].

Alkynes with additional alcohol groups can in principle be used as precursors for oxygen-containing heterocycles such as furans[46], pyrans[47], ketals[48] or spiroketals.[49]

Within our studies we verified that camphor palladium systems that activate alkynes towards cyclotrimerization can also promote the cyclization of 3-pentyn-1-ol, 4-pentyn-1-ol,[50] or 5-hexyn-1-ol.[51]

In the process cyclization *plus* dimerization of the alkynols occur affording reasonably stable ketal-type species (5 or 7) that can be isolated and characterized by spectroscopic techniques (IR and NMR). In mild catalytic conditions, *i.e.* room temperature, *trans*-$[PdCl_2L_2]$ (L1: R=NMe_2 or Pr^i) (2-5%) promote the cyclization of 4-pentyn-1-ol or 3-pentyn-1-ol affording respectively, 2-methyl-2-pent-4-ynyloxy-tetrahydrofuran (5) or 2-methyl-2-pent-3-ynyloxy-tetrahydrofuran (7) (Scheme 4).

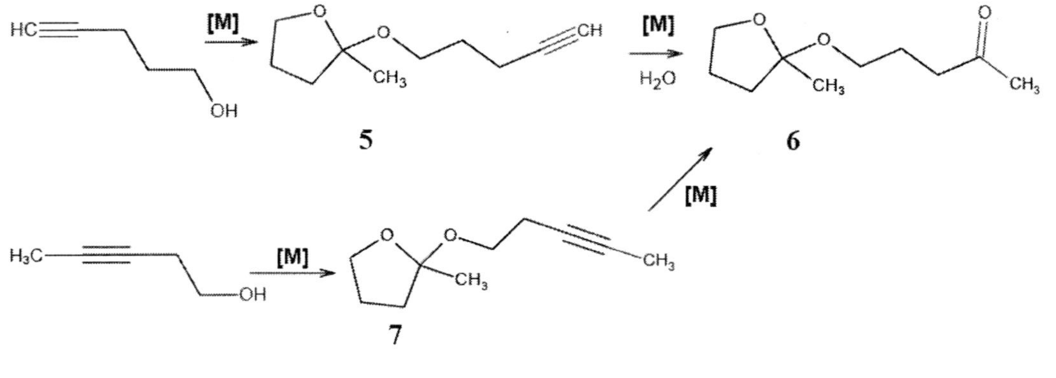

Scheme 4. Cyclization of terminal or internal pentynols promoted by *trans*-$[PdCl_2L_2]$ (**L1**: R=NMe_2 or Pr^i).

Species 5 or 7 further react with water affording 5-(2-methyl-tetrahydrofuran-2-yloxy)-pentan-2-one (6) in the presence of catalytic amounts of *trans*-$[PdCl_2L_2]$ or $PdCl_2$.

The hydration process is considerably faster than the cyclization using $PdCl_2$ as catalyst thus rending difficult the isolation of species 5 or 7, even if just traces of water exist. Complexes *trans*-$[PdCl_2L_2]$ considerably accelerate the heterocycle formation and delay the hydration process compared to $PdCl_2$ thus allowing the isolation of the ketal species (5 or 7) in good yields.

The kinetics and yield for the cyclization of 4-pentyn-1-ol by *trans*-$[PdCl_2L_2]$ (R=Ph) is similar to that of $PdCl_2$, while that of *trans*-$[PdCl_2L_2]$ (R=*iso*-Pr) is two times faster. Moreover, *trans*-$[PdCl_2L_2]$ (R=*iso*-Pr) as catalyst is much more selective to direct the formation of the ketal, since just traces of the cyclic keton 5 form in the absence of added water.

From complexes *trans*-[$PdCl_2L_2$] the best catalyst to drive ketal stabilization is the derived from camphorimine L1 (R=NMe_2) that compared with $PdCl_2$ or *trans*-[$PdCl_2L_2$] (R=Ph) accelerates the ketal formation by *ca.* twenty five times, thus making the control of hydration much more efficient. *trans*-[$PdCl_2L_2$] (R=NMe_2) is also the best catalyst for the cyclization of 3-pentyn-1-ol without interference of traces of water.

These results show that internal and terminal pentynols can be catalytically activated by palladium camphorimine complexes towards furan ring formation with concomitant dimerization. Hydration of the new alkyne species (5 or 7) is achieved with selectivity that depends on the imine R group of the ligand (L1).

Two products could in principle form by hydration of the triple bonds at 5 or 7, i.e. by addition of OH^- group to the terminal carbon atom (anti-Markovnikov) or to the internal carbon atom (Markovnikov). Mercury compounds[52] selectively promote hydration of terminal alkynes by a Markovnikov type mechanism in contrast with internal non-symmetric alkynes in which a random non selective process is activated that leads to several *regio*-isomers with respect to the position of the carbonyl group in the chain. In contrast to mercury compounds *trans*-[$PdCl_2L_2$] catalyses the hydration of the internal non-symmetric 3-pentyn-1-ol with high selectivity affording species 6 as the only product of hydration.

By hydrolysis, 5-(2-methyl-tetrahydrofuran-2-yloxy)-pentan-2-one (6) is converted into 5-hydroxy-2-pentanone (8), conceivably through formation of an instable linear intermediate that could not be observed by NMR (Scheme 5).

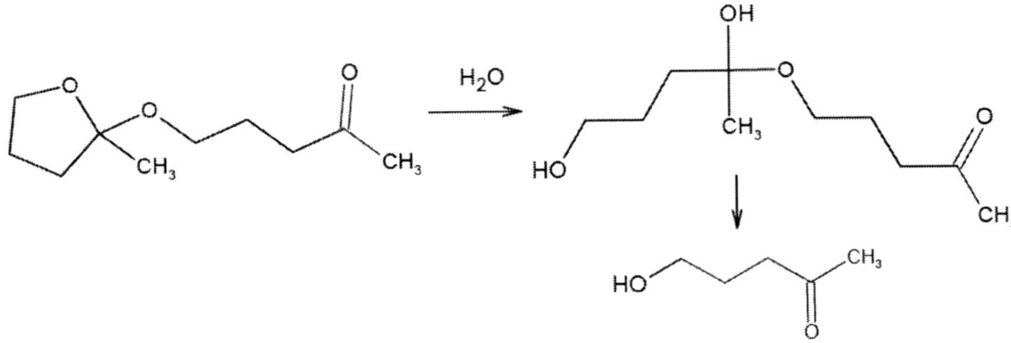

8

Scheme 5. Hydrolysis of 5-(2-methyl-tetrahydrofuran-2-yloxy)-pentan-2-one through a linear ketal (a proposal).

The selective conversion of internal or terminal pentynols into linear ketones using *trans*-[$PdCl_2L_2$] catalysts (Scheme 4 and 6) can be accomplished within a cascade process or step by step depending on the reaction conditions.

Preliminary results in the activation of 5-hexyn-1-ol by *trans*-[$PdCl_2L_2$] (R= NMe_2) indicate that cyclization of the alkynol is catalytically promoted forming 2-methyl-tetrahydro-pyran-2-ol (9, Scheme 6). No evidence exists for dimerization products similar to those obtained by reaction with pentynols, instead oligomerization occurs. Further studies are in progress.

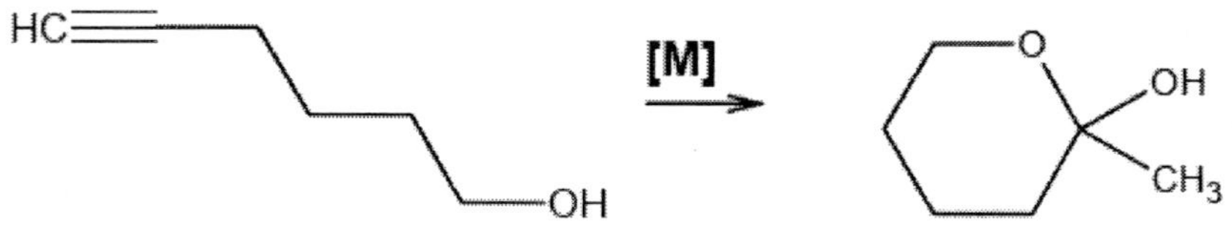

9

Scheme 6. Cyclization of 5-hexyn-1-ol promoted by *trans*-[$PdCl_2L_2$] (L1: R= NMe_2).

Several proposals were made concerning the mechanism of cyclization of alkynols[48, 53,54]. A theoretical study of the mechanism of 4-pentyn-1-ol cyclization by W(0) complexes showed that the substrate coordinates the metal through the alkyne group followed by intramolecular attack of the alcohol to the *exo* carbon atom affording an heterocyclic intermediate that rearranges to the furan type product[55] which is then released. No such type of heterocycle was obtained by us using *trans*-[$PdCl_2L_2$] complexes as catalysts. Instead, a second molecule reacts affording species 5 or 7 thus suggesting that at least at some stage a different mechanism operates.

From NMR data it is believed that coordination of the triple bond to the metal site starts the catalytic process. Whether a second alkyne molecule coordinates the metal or cyclization occurs first, could not be verified by experimental means neither did the role of the camphorimine ligand. In fact, naked $PdCl_2$ also promotes the formation of the furan ring, although the process is much slower than that using *trans*-[$PdCl_2L_2$] (L1: R=NMe_2) as catalyst.

The solvent conceivably has an effect in what concerns the release of the heterocycle or further reaction of it with an incoming alkynol molecule, since in water heterocycle formation from 4-pentyn-1-ol is catalysed by palladium phosphine catalysts without dimerization while ketal type species (Scheme 4) were obtained in DMSO.[53]

DFT calculations are in progress to try to rationalize the mechanism involved in coordination of alkynol to palladium camphorimine complexes and the intermediates that lead to heterocycle formation. Meanwhile, the versatility of the camphorimine ligands that can adopt a *E* or *Z* configuration may be a relevant feature to the catalytic behaviour of the L1 camphorimine derived complexes.

Recent studies[20] demonstrated that (*E*)-camphorimine (L1: R=NHMe) undergoes rearrangement to the (*Z*) isomer upon coordination in *trans*-[MCl_2L_2] (M= Pt or Pd) thus allowing the establishment of intramolecular hydrogen bonds in each one of the two coordinated camphorimines which corresponds to a stabilization of the complex by *ca.* 16 Kcal[20], figure 4.

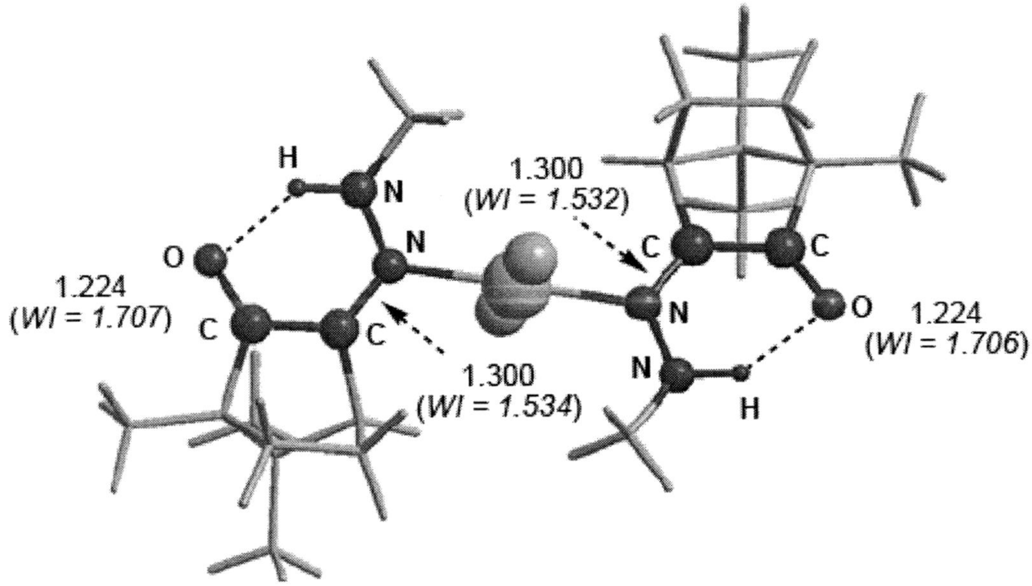

Figure 4. Optimized geometry of *trans*-[$PtCl_2(3\text{-}MeHNNC_{10}H_{14}O)_2$] displaying intra-ligand hydrogen bonding, C=O and C=N bonds bond distances (Å) and Wiberg indices (WI, italics).

The stabilization of *trans*-[MCl_2L_2] (R=NHMe) due to ligand isomerization does not inhibit its ability to catalyse the cyclization 4-pentyn-1-ol but renders the formation of **(5)** slower than using as catalyst the related complex *trans*-[MCl_2L_2] (R=NMe_2). This behaviour is consistent with the need to overcome a higher activation barrier to start the process.

Within the study by NMR of the catalytic process that leads to heterocycle formation from alkynols the rearrangement of the ligand is evident no matter the R group of the camphorimine, thus further studies are necessary to establish whether E/Z isomeration is involved in the catalytic process.

5. Ring Rupture / Expansion

Camphor diynes of type L3 (figure 1) display low ability to stabilize transition metals and consequently just a few complexes (Fe, Cu and Pd) could be prepared[24] and none had its structure solved by X-ray analysis, thus no coordinative details could be obtained. Formally, camphor diynes L3 can bind to metals through several different atoms since they have at least four potentially coordinating groups. In fact, coordination to Pt(II) by the alcohol group at position C-3 (see figure 1 for numbering) of the camphor skeleton or the triple bond of one of the alkynes is proposed in the mechanisms that lead to ring expansion from five to eight (11) or five to six members (12), respectively, of the camphor skeleton (Scheme 7).

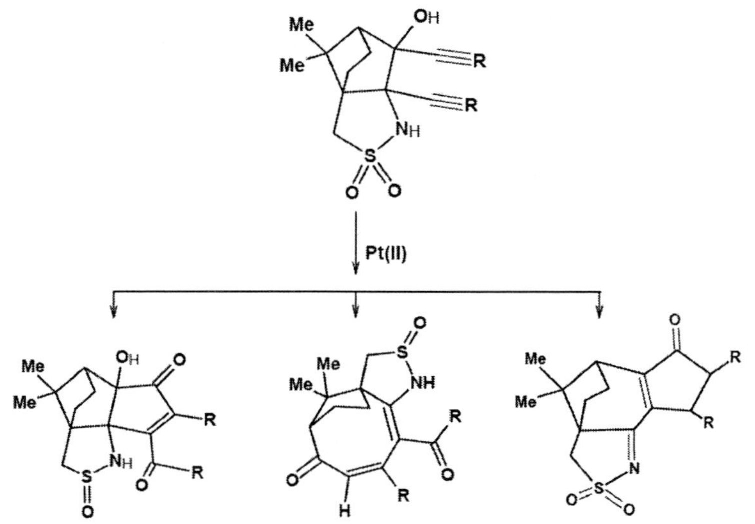

10 11 12

Scheme 7. Species obtained by catalytic activation with $PtCl_2$ or $[PtCl_2(NCPh)_2]$.

Coordination to Ti(IV) by the deprotonated nitrogen atom of the sulfonamide and the deprotonated oxygen atom of the alcohol groups forming a five members chelate is considered to initiate the process of annulation of a five members ring to the positions C-2 and C-3 of the camphor skeleton (10) catalyzed by $TiCl_4$ [56] (Scheme 7). Although, no evidence or proposal exists for coordination of sulfur or oxygen atoms of the sulfonamide group of L3 to transition metals that possibility also exists, despite the fact that in protic medium the species detected are cationic with the charge localized on sulfur atom. Perrhenate (ReO_4^-) revealed to be able to stabilize species (13) formed by proton attack at L3 (R=Ph)[24] (Scheme 8) in contrast with anions derived from Brønsted acids such as HCl or CF_3COOH. In the process, proton addition to the carbon atom of the alkyne group closer to the alcohol group (at C-3) is followed by the annulation of a five membered ring to the camphor skeleton in addition to binding of the oxygen atom of the sulfonamide to the alkyne bound to C-3. The process is stabilized by the formation of a new heterocycle that transfers the positive charge from the carbon atom of the alkyne to the sulfur atom (13, Scheme 9). Cationic species of this type also involved in the reaction of L3 with halogens.[57]

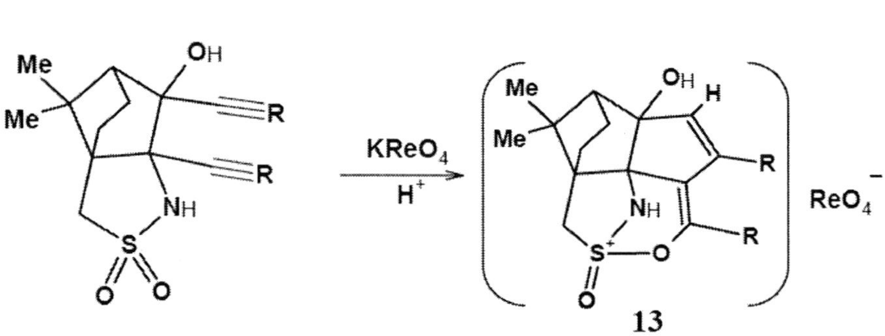

13

Scheme 8. Stabilization by ReO_4^- of the sulfur cation obtained by annulation of a five members ring to camphor.

In the absence of strong electrophiles Pt(II) or Pd(II) promote C2-C3 bond breaking in camphor diynes (L3, R=Ph, n-butyl) within a process that allows ring expansion. Platinum catalysts are considerably more efficient than palladium catalysts.

$PtCl_2$ or $[PtCl_2(NCPh)_2]$ catalyze intra molecular alkyne coupling leading to formation of a new C-C bond and rupture of the camphor skeleton C2-C3 bond (see figure 1 for numbering) within a cascade reaction that ends up with the expansion of the camphor skeleton from five to eight members ring affording a species of the taxoloato type[16,24] (11, Scheme 7) in addition to a minor oxidation product (10). The mechanism of ring expansion remains uncertain, although coordination to platinum through the hydroxyl group was assumed. Hydrogen transfer to the α carbon of the alkyne group at position 3[16], reduction of sulphur atom and C-C coupling of the two vicinal alkynes completes the process.

Formation of species 12 (R=H) involves the coordination of the triple bond of one of the alkyne groups of L3 to platinum affording a η^1-alkyne complex. Oxygen transfer between vicinal carbon atoms through an epoxy intermediate in addition to formation of an alkylidene hydride complex triggers C-C bond formation between the two alkynes in position 2 and 3 of the camphor skeleton (Scheme 9).

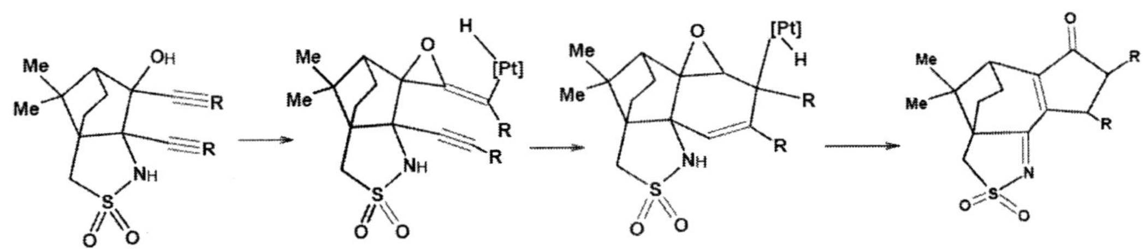

Scheme 9. Mechanism for the annulation of a five members ring to camphor catalyzed by Pt(II).

The results obtained till now show that the characteristics of the R groups at the alkynes have a predominant effect in the processes. Pt (II) compounds are convenient catalysts to promote ring expansion or annulation of rings otherwise no achieved.

A number of other possible intramolecular rearrangement of vicinal alkynes at position C-2 and C-3 of camphor dyines remain to achieve.

Acknowledgements

We acknowledge financial support by Fundação para a Ciência e Tecnologia (FCT) and FEDER under POCI 2010 (Project POCI/QUI/58119/2004).

References

[1] Y. Yamamoto, *Curr. Org. Chem.* 2005, *9*, 503.

[2] V. Gevorgyan, U. Radhakrishnan, A. Takeda, M. Rubina, M. Ribin, Y. Yamamoto, *J. Org. Chem.* 2001 *66*, 2835-41.

[3] L. Wang, S. Gou, Y Chen, Y. Liu, *Bioorg. Med. Chem. Lett.* 2005, *15*, 3417

[4] C-I Cheng, C-T Shei , K. Sung, *Chirality*, 2007, *19*, 235 and references therein.

[5] a) S.R. Davis, M.C. Mitchell, C.P. Cain, P.G. Devitt, R.J. Taylor, T.P. Kee, *Syn Commun*, 2004, *34*, 1191. b) S.E. denmark, I. Rivera, *J. Org. Chem.* 1994, 59, 6887. c) K. Tanaka, H. Ushio, Y. Kawabata, H. Suzuki, *J. Chem. Soc., Perkin Trans. 1*, 1991, 1445. d) M.P. Bonner, E.R. Thornton, *J. Am. Chem. Soc.*, 1991, *113*, 1299. e) W. Oppolzer, P*ure and Appl. Chem.*, 1990, *62*, 1241. f) W. Oppolzer, P*ure and Appl. Chem.*, 1988, *60*, 39; W. Oppolzer, *Tetrahedron*, 1987, *43*, 1969.

[6] X. Verdaguer, J. Vásquez, G. Fuster, V. Bernardes-Génisson, A.E. Greene, A. Moyano, M.A. Pericàs, A. Riera, *J. Org. Chem.* 1998, *63*, 7037

[7] C. Garcia, L.K. LaRochelle, P.J. Walsh, *J. Am. Chem. Soc.*, 2002, *124*, 10970

[8] M. Kitamura, S. Suga, R. Noyori, *J. Am. Chem. Soc.*, 1986, *108*, 6071

[9] H-J. Knölker, *Chem. Rev*, 2000, *100*, 2941;

[10] Z. Xu, C. Chen, J. Xu, M. Miao, W. Yan, R. Wang, *Org. Lett.*, 2004, *6*, 1193

[11] S. Kannan, M.R.A. Pillai, V. Venugopal, P.A. Droege, C.L. Barnes, *Inorg. Chimica Act*a, 1997, *254*, 113.

[12] B.L. Shaw, N. Iranpoor, S.D. Perera, M. Thornton-Pett, J.D. Vessey, *J. Chem. Soc., Dalton Trans.* 1998, 1885. b) R.J. Cross, L.J. Farrugia, P.D. Newman, R.D. Peacock, D. Stirling; *J. Chem. Soc., Dalton Trans.* 1996, 41. c) R. Clarke, M. Gahagan, R.K. Mackie, D.F. Foster; D.J. Cole-Hamilton, M. Nicol, A.W. Montford, *J. Chem. Soc., Dalton Trans.* 1995, 1221. d) D.A. Knight, D.J. Cole-Hamilton, D.C. Cupertino, M. Harman, M.B. Hursthouse, *Polyhedro*n, 1992, *11*(16), 1987. e) S.D. Perera, B.L. Shaw, M. Thornton-Pett, *J. Chem. Soc., Dalton Trans.* 1992, 999. f) D.A. Knight, D.J. Cole-Hamilton, D.C. Cupertino, *J. Chem. Soc., Dalton Trans.* 1990, 3051.

[13] R. Kadyrov, I.Z. Ilaldinov, J. Almena, A. Monsees, T.H. Riermeier, *Tett. Lett.*, 2005, *46*, 7397

[14] a) X. Verdaguer, M.A. Pericàs, A. Riera, M. A. Maestro, J. Mahía, *Organometallics*, 2003, *22*, 1868. b) I. Marchueta, E. Montenegro, D. Panov, M. Poch, M.A. Pericàs, X. Verdaguer, A. Moyano, A. Riera, *J. Org. Chem.* 2001, *66*, 6400. c) J. Balsells, P.J. Walsh, *J. Am. Chem. Soc.*, 2000, *122*, 3250

[15] W. Oppolzer, C. Chapuis, D. Dupuis, M. Guo, *Helv. Chim. Acta*, 1985, *68*, 2100

[16] M. Lampropoulou, R. Herrmann, G. Wagner, *Tetrahedron*, 2004, *60*, 4635

[17] J. Kavitha, S-Y Chang, Y. Chi, J-K Yu, Y-H. Hu, P-T. Chou, S-M. Peng, G-H. Lee, Y-T. Tao, C-H. Chien, A.J. Carty, *Adv. Funct. Mater.* 2005, *15*, 223

[18] M.O. Foster, T. Thornley, *J. Chem. Soc.* 1909, *95*, 942.

[19] M.F.N.N. Carvalho, L.M.G. Costa, A.J.L. Pombeiro, A. Schier, W. Scherer, S.K. Harbi, U. Verfürth, R. Herrmann, *Inorg. Chem.* 1994, *33*, 6270.

[20] M.F.N.N. Carvalho, A.S.D. Ferreira, J.L. Ferreira da Silva, L.F. Veiros, *Collect. Czech Chem. Commun.* 2007, 72 (5-6), 649.

[21] G-I, Cheng, C-T Shei, K. Sung, *Chirality.* 2007, *19*, 235

[22] M.A. ReyChler, *Bull. Soc. Chim. Paris*, 1898, 120

[23] G. Wagner, C. Hei□, U. Verfürth, R. Herrmann, *Z. Naturforsch,* 1996, *51b*, 1655.

[24] M.F.N.N. Carvalho, A.J.L. Pombeiro, G. Wagner, B. Pedersen, R. Herrmann, *Z. Naturforsch.* 1999, *54b*, 725.

[25] B. M. Trost, *Angew. Chem., Int. Ed.* 1995, *34*, 259.

[26] N.J. Long, C.K. Williams, *Angew. Chem., Int. Ed.* 2003, *43* 2586.

[27] Y. Li, M. Yang, *J. Mol. Catal. A: Chemical.* 2002, 18 4161.

[28] V.W.-W. Yam, L. Zhang , C.-H. Tao , K.M.-C. Wong, K.-K. Cheung, *J. Chem. Soc. Dalton Trans.* 2001, 7, 1111.
[29] K. Ohe, K. Miki, S. Uemura, *J. Synth.Org.Chem.*, 2004 62, 978.
[30] P. Siemsen, R.C. Livingston, F. Diederich, *Angew. Chem. Int. Ed.* 2000, *39*, 2632.
[31] M.F.N.N Carvalho, F.M.T. Almeida, A.M. Galvão, A.J.L. Pombeiro, *J.Organometal. Chem.* 2003, *679*, 143.
[32] R.A. Begum, N. Chanda, T.V.V. Ramakrishna, P.R. Sharp, *J. Am. Chem. Soc.*, 2005, *127*, 13494
[33] Y. Li, M. Yang, *J. Mol. Catal. A: Chemical*, 2002, 184, 161
[34] P. Xue, H.R.Y. Sung, I.D. Williams, G. Jia, *J. Organometal. Chem.* 2006, *691*, 1945
[35] N. Panziera, P. Pertici, I. Fratoddi, A. La Groia, M.V. Russo, *J. Organometal. Chem.*, 2006, *691*, 2648
[36] N. Ohshiro, F. Takei, K. Onitsuka, S. Takahashi, *J. Organometal. Chem.*, 1998, *569*, 195
[37] a) N. Morohashi, K. Yokomakura, T. Hattori, S. Miyano, *Tetrahedron Lett.* 2006 *47,* 1157. b) Y. Yamamoto, *J. Synthetic Org. Chem.* 2005, *63,* 112-121. c) J.J. Li, G.W. Gribble, Palladium in Heterocyclic Chemistry, Pergamon Oxford, 2000. d) S. Saito, Y. Yamamoto, *Chem. Rev.* 2000, *100*, 2901-15.
[38] a) W. Reppe, O. Schilichting, K. Klager, T. Toepel, *Liebigs Ann. Chem.* 1948, *560*, 3. b) W. Reppe, W.J. Schweckendiek, *Liebigs Ann. Chem.* 1948, *560*, 104
[39] P.M. Maitlis, *J.Organometal. Chem.*, 1980, *200*, 161
[40] F.C. Pigge, Z. Zheng, *Tett. Lett.* 2001, *42*, 8259
[41] M.F.N.N Carvalho, M.T. Duarte, R. Herrmann, *Collect. Czech. Chem. Comm.* 2006, *71(3),* 302.
[42] M.F.N.N. Carvalho, Portugaliæ Electrochimica Acta, 2004, 22, 3
[43] a) J.T. Gupton, *Top Heterocycl Chem.* 2006, *2*, 53. b) B.M. Trost, M.R. Machacek, B. D. Faulk, *J. Am. Chem. Soc.* 2006, *128*, 6745
[44] T. Brown, H. Holt Jr, M. Lee, *Top Heterocycl Chem.* 2006, *2*, 1
[45] M. Daneshtalab, *Top Heterocycl Chem.* 2006, *2*, 153
[46] G. Zeni, R. Larock, *Chem. Rev.* 2004, *104*, 2285
[47] B. Liu, J.K. De Brabander, *Org. Lett.* 2006, *8*, 4907
[48] K. Utimoto, *Pure Appl. Chem.* 1983, *55*, 1845
[49] C.V. Ramana, R. Mallik, R.G. Gonnade, M.K. Gurjar, *Tetrahedron Lett.* 2006, *47*, 3649
[50] M.F.N.N Carvalho, A.S.D. Ferreira, R. Herrmann, *J.Organometal. Chem.*, 2006, *691,* 4124.
[51] M.F.N.N Carvalho, unpublished results
[52] R. Paul, S. Tchelitcheff, *Bull. Soc. Chim. France.* 1953, 20, 417
[53] D.W. Lucey, J.D. Atwood, *Organometallics*, 2002, *21*, 2481 and references therein.
[54] J. Muzart, Tetrahedron, 2005, *61*, 4179
[55] T. Sordo, P. Campomanes, A. Diéguez, F. Rodriguez, F.J. Fananás, *J. Am. Chem. Soc.* 2005, *127*, 944
[56] G. Wagner, C. Hei□, U. Verfürth, R. Herrmann, *Z. Naturforsch.* 1996, *51b*, 1655
[57] G. Wagner, R. Herrmann, A. Schier, *J. Chem. Soc., Perkin Trans.*, 1997, 701

In: Advances in Organometallic Chemistry Research ISBN: 978-1-60021-779-1
Editor: Kenji Yamamoto, pp. 97-138

Chapter 5

THE CHEMISTRY OF ORGANOIMIDO DERIVATIVES OF POLYOXOMETALATES - ORGANIC CHEMISTRY AS A USEFUL TOOL FOR THE FUNCTIONALIZATION OF POMS AND THE PREPARATION OF NOVEL POM-BASED ORGANIC-INORGANIC HYBRID MOLECULAR MATERIALS

***Yongge Wei*[1, 2], *Pingfan Wu*[3], *Yuan Wang*[2] *and Meicheng Shao*[2]**
[1] Department of Chemistry, Tsinghua University, Beijing 100080, China
[2] State Key Laboratory for Structural Chemistry of Unstable and Stable Species, College of Chemistry and Molecular Engineering, Peking University, Beijing 100871, China
[3] Department of Bioengineering, Hubei University of Technology, Wuhan 430068, China

ABSTRACT

The chemistry of organoimido derivatives of polyoxometalates (POMs), an innovative emerging organic chemistry of polyoxometalates, has received increasing interest in the supramolecular chemistry and chemistry of materials recently. Organoimido derivatives of POMs are valuable building blocks for constructing novel nanostructured organic-inorganic hybrid molecular materials with the so-called ''value-adding properties'' and possible synergistic effects, which could result in unique catalytic, photo and electronic properties dramatically different from the corresponding parent materials, since the π electrons in the organic component may extend their conjugation to the inorganic framework and dramatically modify the electronic structure and redox properties of the corresponding parent POMs. The surface modification of POMs, that is, the substitution of one or more terminal or bridged oxo groups with organoimido ligand, is the fundamental method to synthesize organoimido derivatives of POMs via common organic chemistry reactions. In this chapter of the book, the synthetic chemistry of organoimido derivatives of POMs using different organic reagents will be reviewed, of which the novel DCC-dehydrating protocol to prepare organoimido derivatives of POMs recently developed by us and co-authors will be concentrated. As it can be expected, the reaction chemistry of organoimido derivatives of POMs stands for the fascinating future of the chemistry of organoimido derivatives of POMs since it opens not only a new road to the chemical modification of POMs, but also an exciting research arena where a variety of hybrid materials containing covalently bonded POM clusters and organic conjugated segments can be prepared in a more controllable and rational manner. Hence, in the rest of this chapter, Pd-catalyzed coupling reactions of the functionalized organoimido derivatives of POMs will be introduced in detail for this purpose. Especially, different rationally assembled hybrid molecules and polymers where the organic molecules and inorganic clusters are bonded together by an imido nitrogen atom will then be highlighted.

1. INTRODUCTION

Polyoxometalates (POMs), formed by the early transition metal ions such as vanadium, molybdenum and tungsten in their high oxidization states with bridged oxide anions, are a unique class of well-defined inorganic nanocluster compounds with much diversity in size, composition, structure and function, which can be, in fact, considered as the "oligomeric" or molecular states and model compounds of the corresponding "polymeric" metal oxides with extended solid state structure [1]. Since the first salt of POM, ammonium 12-molybdophosphate, $(NH_4)_3[PMo_{12}O_{40}]$ was reported by Berzelius in 1826, as a yellow precipitate from the aqueous solution of phosphoric acid and ammonium molybdate [2], the chemistry of POMs has gained dramatic development and is remarkable for its significance in quite diverse disciplines including catalysis, medicine and materials science [3- 5]. Especially, during the last decade, a lot of nanoscale huge clusters of polyoxometalates and their assemblies have been prepared and structurally characterized by our research team at Peking University [6- 9], A. Müller's group at University of Bielefeld [10- 24] and others [25- 28]. As examples, in figures 1-3, there are shown three typical nanoscale polyoxomolybdates, *i. e.*, tyre-like $\{Mo_{176}\}$ (1) [7, 12], ball-like $\{Mo_{132}\}$ (2) [9, 15], and chain-like $\{Mo_{38}\}$ (3) self-assembled from $\{Mo_{36}\}$ [6], synthesized by A. Müller *et al.* and us, respectively. These high nuclearity huge clusters are not only ideal theoretical models for mesocopic physics but also provide us an exciting research arena for nanochemistry and materials science— The nanotyres 1 are potential building blocks for fabricating novel nanotubes if they are arranged to stack along their symmetrical axes. The nanoballs 2 have a spherical cavity of *ca* 8 nm^3 with accessible openings, indicating that they are a good candidate for nanocontainers and may be used in controllable drug-releasing. The nanochains 3 show significantly red-shif in its UV-vis spectra, which implies that they are potential applications in making unique nanoscale molecular quantum wells or one-dimensional conducting materials. Furthermore, T. B. Liu *et al.* have recently observed and confirmed, according to laser light scattering studies, that in their solution, such nanoclusters of polyoxometalates spontaneously assemble into almost monodisperse, well-defined vesicle-like nanoaggregates with spherical structures [29, 30], via an unusually slow self-association process [31]; and more interestingly, in the NaCl-containing aqueous solution, dissolution and precipitation, two opposite processes, of these hydrophilic nanoanions automatically and subsequently occur without changing external conditions [32].

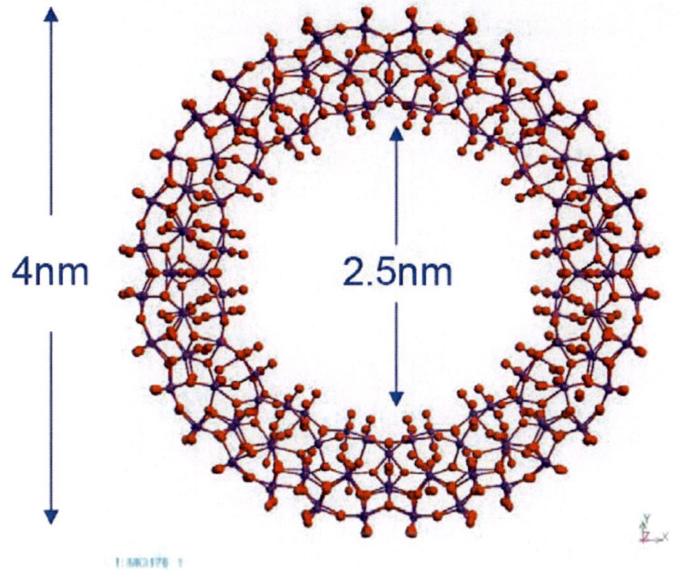

Figure 1. Structure of the nanotyre 1 with 176 Mo atoms in a cluster anion (view parallel to the C_8 axis of the cluster). Also see Ref. 7 and 12.

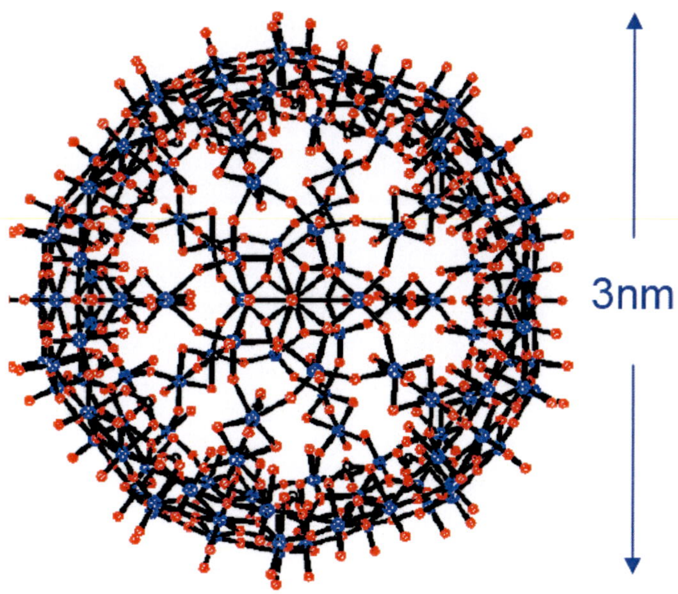

Figure 2. A view of the $\{Mo_{132}\}$ nanoball **2** along its C_5 axis. Also see Ref. 9 and 15.

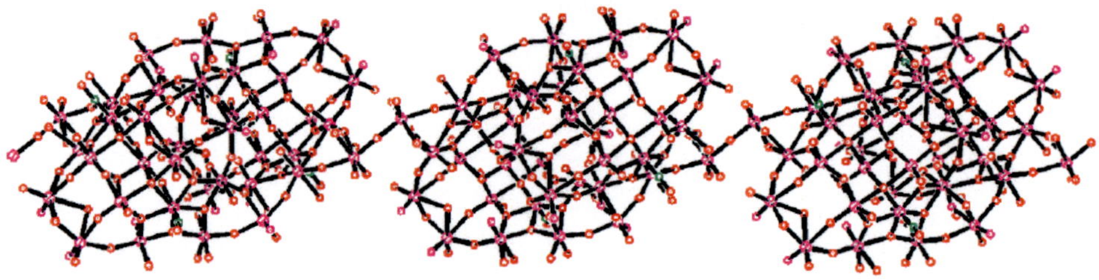

Figure 3. Drawing of the nanochain 3 extended along the 21 screw axis. Also see Ref. 6.

Parallel to the rapid and great progress made on these nanosized inorganic clusters of metal-oxide anions, studies on conjugated organic and polymeric materials have also exploded and flourished [33- 36], since such molecular materials have exciting technological applications in various photic and electronic devices, for instance, organic light emitting diodes (OLEDs) [37- 38], organic field-effect transistors (OFETs) or organic thin film transistors (OTFTs) [39- 42], and solid-state lasers [43, 44]. The fact that the 2000's Nobel Prize in chemistry was bestowed upon three preeminent scientists working on conjugated polymers is the ultimate testimony to the important developments in organic conjugated systems [45- 48].

Although they are vastly different in molecular structures, POMs and conjugated organic molecules/ polymers are both electrically active materials with similar electrical and optical properties such as photochromism, electrochromism, and conductivity [34, 49, 50]. The underlying mechanisms of these properties are, however, different for the two types of materials, with dπ electrons responsible for the inorganic POM clusters, and delocalized pπ electrons responsible for the organic counterpart. While both areas have been enjoying considerable success, there has been little success in bringing these two types of materials together through covalent bonds to make novel POM-based organic-inorganic hybrids [51] owing to the lack of reliable synthetical chemistry tools to functionalize POMs and prepare their organic derivatives in high throughput. As could be expected, however, such organic-inorganic hybrid materials will not only combine the advantages of organic materials, such as good processability and fine-regulable structure and electronic properties [34], with those of inorganic POM clusters, such as good chemical stability and strong electron acceptability [1], to produce so-called ''value-adding properties'' [52], but also may bring exciting synergistic effects due to the close interaction of delocalized organic pπ orbits with the inorganic POM cluster's dπ orbits. With their unique structures and potential novel properties, originating from the functional nanobuilding blocks of POMs [51], these sorts of hybrid materials are extremely interesting not only to synthetic chemists and materials scientists, but also to theoretical and experimental physicists.

Organically functionalized POMs, *i.e.*, species where one or some oxo or $\{Mo_x\}^{n+}$ groups have been replaced with organic functional groups, have been extensively studied and now form the largest subclass of POM derivatives [53], since the pioneering work at du pont, of which alkylated derivatives or esters of Keggin-structure phosphomolybdic and phosphotungstic acid, such as $(n\text{-}Bu_4N)_3[PMo_{12}O_{39}(OMe)]$ (4) and$(n\text{-}Bu_4N)_3[PW_{12}O_{39}(OMe)]$ (5), were prepared by Knoth et al in the early 1980s [54, 55]. There are a lot of reasons for the current interest in these derivatized POMs [53]: 1) Due to the perceived structural analogies of POMs to the surfaces of metal oxides, they can be viewed as soluble analogues of metal oxides and therefore are of special interest as model compounds for investigating the reactions and properties of metal oxides. Given the versatility of metal oxides in catalyzing organic conversions, and the difficulty in determining the intimate mechanism of these reactions, the study of the stoichiometric reactivity of well-defined surface POMs might provide an understanding of the elementary steps of heterogeneous reactions, particularly with respect to surface-bound intermediates. Therefore, the characterization of organic derivatives of POMs is relevant to the molecule-level modeling of catalytic reactions occurring at metal oxide surfaces. Furthermore, POM-supported transition-metal catalysts represent a new class of oxide-supported catalyst materials that can be fully studied at the atomic level, both structurally and mechanistically. 2) Derivatization may result in the activation of surface oxo groups of POMs, thus giving more chances to be modified. 3) Functionalization can conduce to the stabilization of otherwise labile POM frameworks in the derivatives, providing novel building blocks for the assembly of larger systems. 4) The construction of interconnected POM networks could be built up through the incorporation of organic groups with functionality and polysubstituted derivatives could be utilized in the context of the development of novel dendrimers. 5) Such derivatives can probably provide multifunctional oxidation or acidification catalysts that display selective recognition of substrates, thus higher selectivities, and might develop targeting of POMs in antiviral and anticancer chemotherapy.

Among many organic derivatives of POMs, organoimido derivatives have attracted increasing interest and recently, a lot of arylimido derivatives including polymers of the Lindqvist hexamolybdate cluster, $[Mo_6O_{19}]^{2-}$, have been synthesized continuously [56], since the π electrons in the organic component of such derivatives may extend their conjugation to the inorganic POM framework via the N atoms of the imido groups, thus result in strong d–pπ interactions and dramatically modify the electronic structure and redox properties of the corresponding parent POMs. In addition, organoimido derivatives of POMs with a remote active functional group can be exploited as building blocks to conveniently and controllably fabricate the complicated covalently-linked POM-based organic-inorganic hybrids, including the nano-dumbbells, polymeric chains and even networks of POMs [56]. This modular building block approach brings rational design and structure regulation into the synthesis of organic-inorganic hybrid molecular materials and has just been reviewed by one of our co-workers at UMKC [56]. In this chapter, concentrating on the novel DCC-dehydrating protocol to prepare organoimido derivatives of POMs recently developed in this laboratory and others, various synthetic approaches to organoimido derivatives of POMs using different organic reagents will be reviewed. As it can be expected, the reaction chemistry of organoimido derivatives of POMs stands for the fascinating future of the chemistry of organoimido derivatives of POMs since it opens not only a new road to the chemical modification of POMs, but also an exciting research arena where a variety of hybrid materials containing covalently bonded POM clusters and organic conjugated segments can be prepared in more controllable and rational manner. Hence, in the rest of this chapter, Pd-catalyzed coupling reactions of the functionalized organoimido derivatives of POMs will be introduce in details. Furthermore, various rationally assembled hybrid molecules and polymers where the organic molecules and inorganic clusters are bonded together by an imido nitrogen atom will then be highlighted.

2. Synthetic Chemistry of Organoimido Derivatives of POMs

As an important subclass of oxo-imido complexes of early transition metals, organoimido derivatives of POMs have only been synthesized and characterized very recently. They were mentioned first by J. Zubieta *et al* in an outstanding communication in 1988 [57], however, there was lack of a reliable and detailed structural characterization in this paper, probably due to the purification problem of their products. Three years later, using a different synthetic methodology, E. A. Maatta's group finished the synthesis and single crystal X-ray structural determination of the first organoimido derivative of POMs [58]. Since then, the organoimido derivatives of POMs have been extensively investigated and a number of organoimido derivatives of the Lindqvist hexamolybdate ion, $[Mo_6O_{19}]^{2-}$, have been reported [53]. However, little work has been conducted on other POMs including the other Lindqvist POMs, the Keggin POMs, and especially, the recently discovered nanoscale huge clusters of POMs such as compounds 1-3 [6-28], the synthesis of their organoimido derivatives is still a formidable challenge. Furthermore, although the bridging oxo groups of POMs can be replaced, in principle, with isoelectronic organoimido ligands, the substitution of terminal oxo groups features largely in the reported organoimido derivatives of POMs, in which the imido groups attached to only one metal center in a linear fashion and the N-atom in the imido groups can be considered sp hybridized as σ-π donor ligand to form one σ and two π bonds with the bound metal atom.

In general, organoimido derivatives of POMs can be obtained in the following ways [53]:

- Imido metathesis via Wittig-like (net) [2 + 2] exchange reactions of M=O bonds with imido-releasing agents, including isocyanates, phosphinimines, or sufinylamines.
- The α-hydrogen transfer reactions with amines.
- The displacement of labile ligands from an imido-containing metal complex by active oxometalates, followed by aggregation, or by coordination-unsaturated lacunary POMs.
- The addition or displacement reactions of active functional groups at the imido ligands from existing organoimido precursors of POMs.

Apparently, the first two strategies create the new metal-nitrogen multiple bonds and the latter ones deal with other chemical bonds from known metal imido complexes. In this section of the chapter, the synthetic methodologies in respect to the former three strategies will be surveyed. The fourth strategy will be introduced in Section 3 because it is involved in the reaction chemistry of organoimido derivatives of POMs.

2.1. Organoimido Derivatives of the Lindqvist Hexamolybadte

The hexamolybdate ion, $[Mo_6O_{19}]^{2-}$, is amongst the most well-known POM clusters for its thermal and chemical robust and ease to be prepared [59], which has the so-called Lindqvist structure [60]. As it is depicted in figure 4, the Lindqvist structure consists of a central oxyanion surrounded in an octahedral cage formed by six metal atoms. All the six metal atoms also have an octahedral environment. Besides the central oxygen atom, they each coordinate triply to one terminal oxygen atom, forming a terminal metal-oxo group (M≡O), and share an additional four doubly-bridging oxygen atoms (μ_2-O atoms) with neighboring metal atoms. Generally speaking, a Lindqvist ion has a superoctahedral structure approach to O_h point group and features its six terminal metal-oxo groups aligned along the Cartesian axes. For the hexamolybdate, these molybdyl groups (Mo≡O) are reactive enough for the terminal oxygen atoms to be directly replaced by various nitrogenous species, for example, diazenido [61], diazoalkyl [62], and imido groups [58]. In the case of imido derivatives of the hexmolybdate, so far, three types of reactions have been developed. These include reactions with phosphinimines [Eq. (1)] [58, 63], isocyanates [Eq. (2)] [64, 65], and primary amines [Eq. (3)] [66- 68], which are given in the following equations, respectively.

$$(n\text{-}Bu_4N)_2[Mo_6O_{19}] + x\ Ph_3P{=}NAr \longrightarrow (n\text{-}Bu_4N)_2[Mo_6O_{19\text{-}x}(NAr)_x] + x\ Ph_3P{=}O \quad (1)$$

$$(n\text{-}Bu_4N)_2[Mo_6O_{19}] + x\ RNCO \longrightarrow (n\text{-}Bu_4N)_2[Mo_6O_{19\text{-}x}(NAr)_x] + x\ CO_2 \quad (2)$$

$$(n\text{-}Bu_4N)_2[Mo_6O_{19}] + x\ ArNH_2 \longrightarrow (n\text{-}Bu_4N)_2[Mo_6O_{19\text{-}x}(NAr)_x] + x\ H_2O \quad (3)$$

With these approaches, the six terminal oxo (in certain case, some bridging-oxo groups) in the hexamolybdate cluster can be partially or completely substituted with organoimido ligands. Nowadays, a large number of monosubstituted [58, 63- 78], disubstituted [63, 65, 66, 71, 76, 79- 83] and polysubstituted [65, 71, 84, 85] organoimido derivatives of the hexamolybdate have been synthesized and structurally characterized.

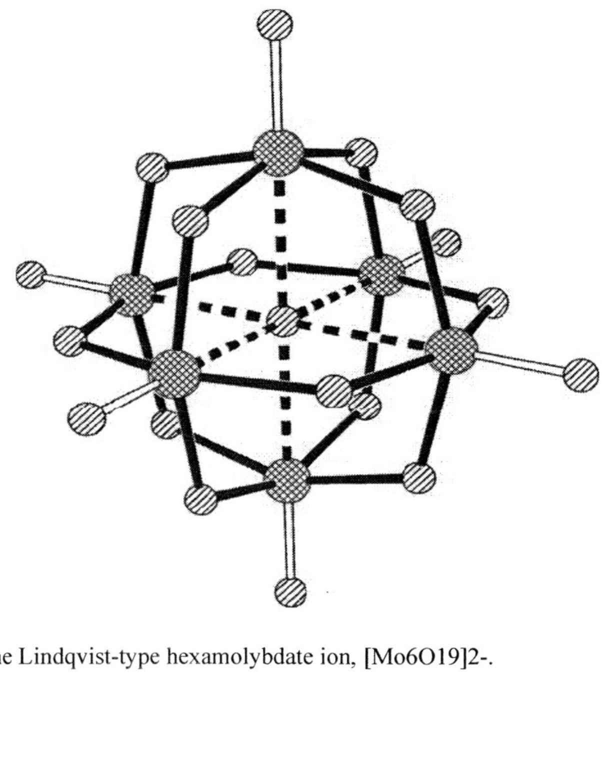

Figure 4. Structure of the Lindqvist-type hexamolybdate ion, [Mo6O19]2-.

2.1.1. Monosubstituted Derivatives

Monofunctionalized organoimido derivatives of the hexamolybdate were firstly obtained by E. A. Maatta *et al* with phosphinimines as imido-releasing reagents [Eq. (1)] in 1991 [58]. They discovered that the reaction of equimolar amounts of $(n\text{-}Bu_4N)_2[Mo_6O_{19}]$ (6) and $Ph_3P{=}Ntol$ (prepared from the reaction of p-tolyl azide and Ph_3P in ether) in strictly anhydrous pyridine at 85 °C yields $(n\text{-}Bu_4N)_2[Mo_5O_{18}(MoNC_6H_4CH_3)]$ (7) within 2 days. Two years later, A. Proust and co-workers founded that this type of reaction also works well in refluxed anhydrous acetonitrile under dried nitrogen gas for 60 hours and synthesized the monophenylimido derivative, $(n\text{-}Bu_4N)_2[Mo_6O_{18}(NPh)]$ (8) [63]. Although these reactions involve the not-easily accessible phosphinimine reagents, proceed slow, and the triphenylphosphine oxide byproduct is very difficult to separate from the desired reaction products, they nonetheless demonstrate that direct functionalization can be carried out on the hexamolybdate salt itself in a straightforward manner. Moreover, phosphinimines are powerful reagents to introduce C=N bond into organic compounds from carbonyl substrates via the famous aza-Wittig reactions in classic organic syntheses [86], the reactions thus imply that the molybdyl in the hexamolybdate behaves like a carbonyl of aldehydes, ketones or esters in some extent and may have the chemistry similar to that of a carbonyl.

The phosphinimine approach is very useful for the synthesis of arylimido monosubstituents of the hexamolybdate cluster. Recently, an interesting styrylimido-hexamolybdate complex, $(n\text{-}Bu_4N)_2[Mo_6O_{18}(NC_6H_4CH{=}CH_2)]$ (9), shown in figure 5, which is the important building block in fabricating polystyrene compositions with the POM as a pendant group, has also been prepared in *ca* 90% high yield by this approach [72]. However, attempts to make the alkylimido analogues using this approach are unsuccessful. For instance, the reaction of $(n\text{-}Bu_4N)_2[Mo_6O_{19}]$ (6) with $Bu^n{}_3P{=}NBu^t$ failed to give the tertbutylimdo derivative [64] because phosphinalkylimines are more reactive than the corresponding phosphinarylimines and result in the reduction and rearrangement of the hexamolybdate cluster.

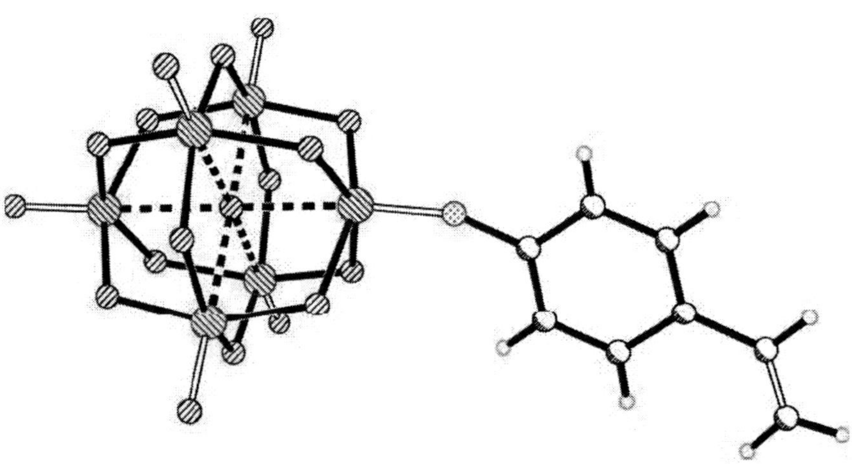

Figure 5. Ball-and-stick model of the styrylimido-hexamolybdate complex 9.

The second approach [Eq. (2)] to produce organoimido derivatives of the hexamolybdate was first set up by R. J. Errington and co-workers a little over a decade ago [64], who have shown that both of the corresponding monoimido-substituted derivatives could be afforded by reactions of alkyl and aryl isocyanates with $[Mo_6O_{19}]^{2-}$ at elevated temperatures in acetonitrile or benzonitrile, and reported the first structurally characterized tertbutylimdo derivative of the hexamolybdate, $(n\text{-}Bu_4N)_2[Mo_6O_{18}(NBu^t)]$ (10) [64]. Maatta and co-workers have subsequently developed this approach successfully using dry pyridine as the solvent [65]. They discovered that alkyl isocyanates are more reactive with the hexamolybdate clusters in dry pyridine and the reaction between them can occur at room temperature and finish in 2 to 3 days [69, 71]. Nevertheless, the reaction of aryl isocyanates with $[Mo_6O_{19}]^{2-}$ in dry pyridine still needs to be carried out at high temperature (110 °C) and must be refluxed under strictly anhydrous conditions for a prolonged time [69, 71]. It is worthwhile to point out here that this reaction is also analogous to that of isocyanates with aldehydes or ketones to form imines [87], which again suggests the comparability between the molybdyl and the carbonyl.

Using this improved route, E. A. Maatta *et al* have further synthesized a lot of monosubstituted organoimido derivatives [65, 69- 71], including the charge-transfer ferrocenylimido complex, $(n\text{-}Bu_4N)_2[(FcN)Mo_6O_{18}]$ (11), FcN = $(\eta^5\text{-}C_5H_5)Fe(\mu\text{-}\eta^5\text{:}\sigma\text{-}C_5H_4N)$, (figure 6) [70] and the organodi-imido-bridged bis(hexamolybdate) derivative, $(n\text{-}Bu_4N)_2[(Mo_5O_{18})Mo{\equiv}N\text{-}Z\text{-}N{\equiv}Mo\ (O_{18}Mo_5)]$ (12), Z = 1,4-cyclo-C_6H_{10}, (figure 7) [69]. Compound 11 features the covalent attachment of an oxidizable ferrocenyl donor to a reducible POM acceptor system. Compound 12 stands for the fist example of the hexamolybdate complexes of diimido ligands, which are coordinated to two hexamolybdates and links them as a bridge.

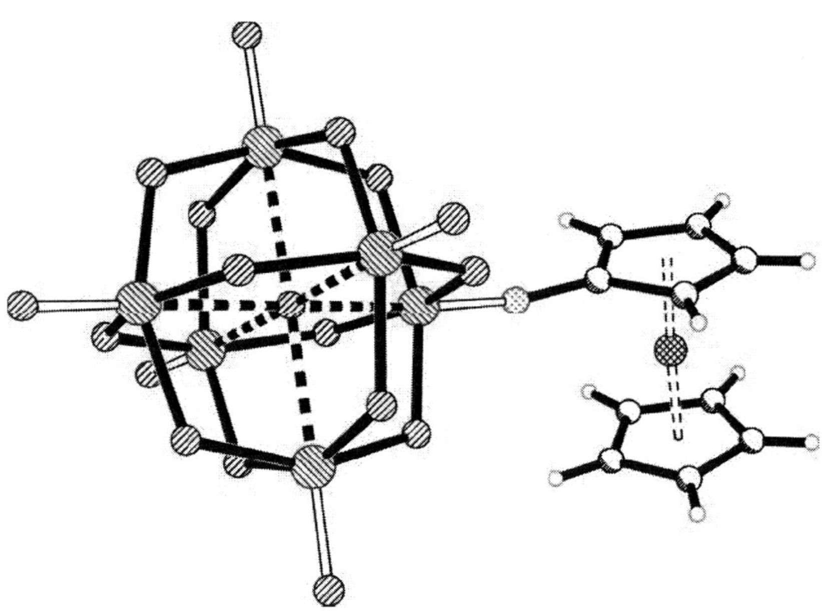

Figure 6. Ball-and-stick model of the charge-transfer ferrocenylimido-hexamolybdate complex 11.

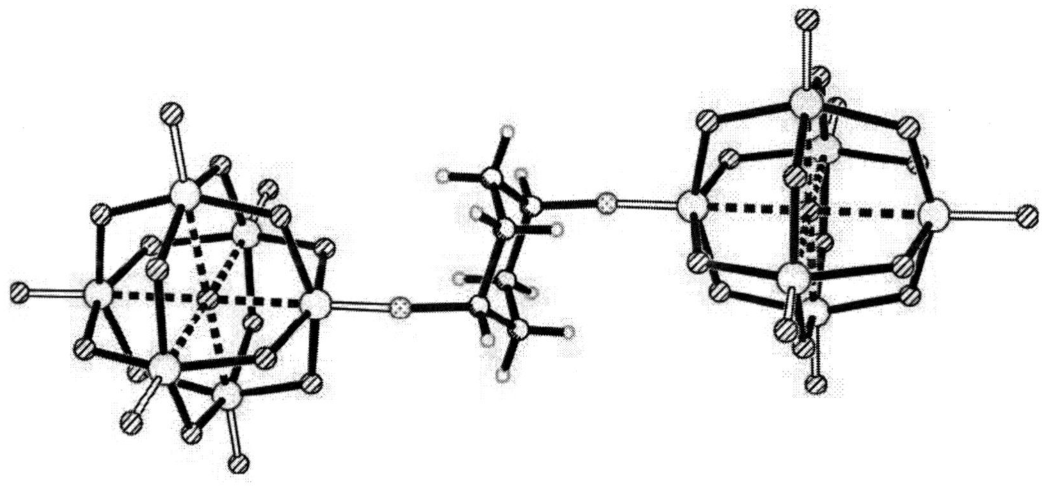

Figure 7. Ball-and-stick model of the organodi-imido-bridged bis(hexamolybdate) derivative 12.

The last approach given in Eq(3) was also put forward by R. J. Errington and co-workers firstly. They observed that in the presence of triethylamine, an aromatic primary amine, for example 1,4-diaminobenzene, reacted directly with the hexamolybdate ions in benzonitrile solvent at *ca* 150 °C [66]. Albeit this reaction occurs under rather harsh conditions and only a mixture that includes mono- and di-functionalized derivatives, and various oligomers is obtained, from which pure imido derivatives are difficult to isolate from each other, it indeed provides us a chance to prepare organoimido derivatives with cheaper and more accessible arylamines as the imido-releasing reagents. Additionally, this reaction confirms further the similarity between the molybdyl and carbonyl groups since it extremely resembles the Schiff-base reaction taken place between primary amines with aldehydes or ketones, which involves of nucleophilic addition of the amine to the carbonyl followed by elimination of water [88].

Several years later, R. A. Roesner et al improved this approach with dry pyridine as the solvent instead of benzonitrile. However, only aromatic amines with the strongly electron-donating substituents, such as the alkoxy groups, promise this reaction, under heating and dry N_2-gas protecting conditions, to create the corresponding mono-imido derivatives of the hexamolybdate in moderate yield [67].

We have now a long time interest in the chemistry of derivatives of POMs [6, 8, 9, 68, 73- 84], including new synthetic methodologies for the organoimido derivatives [68, 73- 83]. In the light of the pioneering investigations of E. A. Maatta and R. J. Errington's groups on this field, we strongly feel that the molybdyl is, under certain cases, one of the inorganic analogues, in some extent, of the carbonyl in organic chemistry [88], and some organic synthetic techniques can be extended into the chemistry of derivatives of POMs. For the primary amine approach, for example, if the resulting water is able to remove from the reactive mixture successively, just like common dehydrating reactions in organic chemistry, the reaction of Eq(3) will proceed rapidly and irreversibly, and complete thoroughly at the end.

Five years ago, together with our co-workers at UMKC, we discovered that using carbodiimides, which are among the most widely-used reagents in common organic synthesis [89-91], such as DPC (N,N'-diisopropylcarbodiimide) and DCC (N,N'-dicyclohexylcarbodiimide), as the dehydrating reagent, the reaction of $[Mo_6O_{19}]^{2-}$ with aromatic primary amines can be dramatically promoted, and have developed a novel DCC-dehydrating synthetic protocol to afford arylimido derivatives of POMs in the context of the primary amine approach [68]. As shown in Eq(4), in the presence of one equivalent of DCC, one equivalent of aromatic primary amines reacts smoothly with an excess of $(n\text{-}Bu_4N)_2[Mo_6O_{19}]$ (6) under dry N_2 gas and refluxing acetonitrile, pyridine or hot N,N'-dimethylformamide, to give the corresponding mono-substituted imido derivatives usually in excellent yield of more than 90%. Not only a variety of aromatic primary amines occur this reaction, but also it is usually completed in less than 12 hours, and pure *crystalline* products can be readily obtained with convenient bench manipulations. Moreover, this reaction can be carried out in the open air without nitrogen protection if slightly over one equivalent of DCC (1.2 equivalent) is added. Best of all, the reaction also tolerates additional remote functional groups on the aromatic ring, including iodo, ethynyl and terpyridyl groups. Figures 8-10 show the X-ray structures of three imido derivatives containing such three functional groups, $(n\text{-}Bu_4N)_2[Mo_6O_{18}(NAr)]$, Ar = 4-iodo-2,6-dimethyl-C_6H_2 (13) [68], 4-ethynyl-2,6-dimethyl-C_6H_2 (14) [68], 4-(4'-terpyridylethynyl)-2,6-dimethyl-C_6H_2 (15) [73], respectively, of which compounds 13 and 14 both have two stable conformational isomers differing by a rotation of the aromatic ring relative to the hexamolybdate skeleton. All the three complexes are valuable building blocks to construct POM-based organic-inorganic hybrid molecular materials in our controllable molecule-module-assembly strategy [56].

$$[Mo_6O_{19}]^{2-}\text{ (excess)} + ArNH_2 + DCC \longrightarrow [Mo_6O_{18}(NAr)]^{2-} + (cyclo\text{-}C_6H_{11}NH)_2CO \quad (4)$$

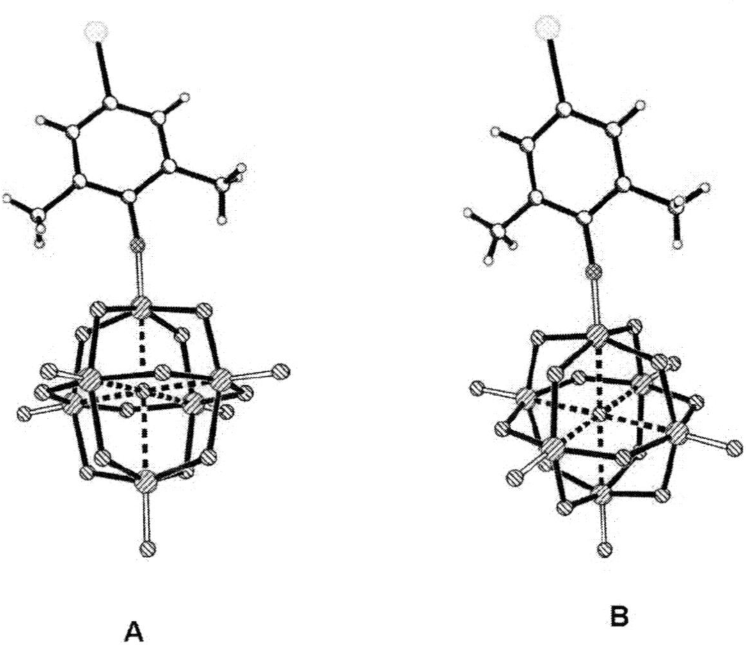

A B

Figure 8. Structure of iodo-substituted derivative 13. A and B stand for its two conformational isomers differing by a rotation of the aromatic ring relative to the hexamolybdate skeleton.

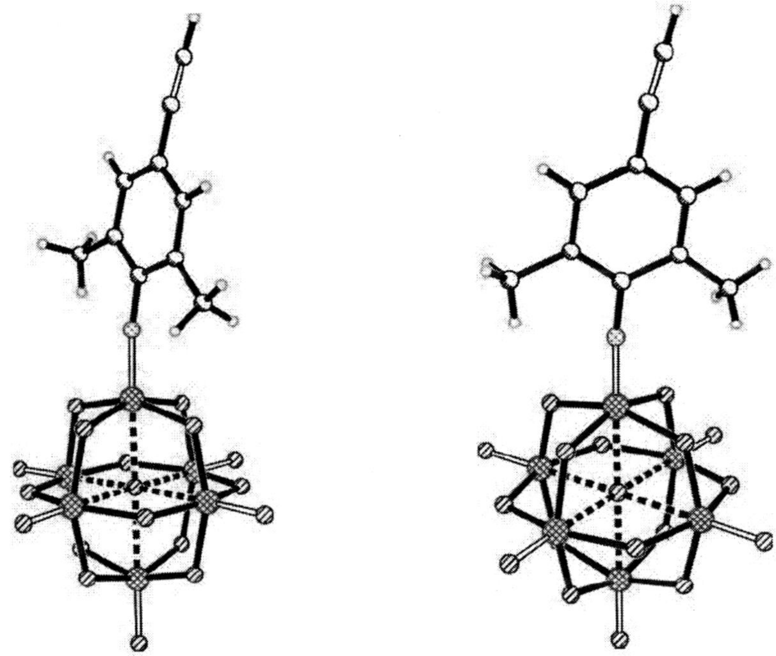

Figure 9. Structure of ethynyl-functionalized derivative 14. A and B stand for its two conformational isomers differing by a rotation of the aromatic ring relative to the hexamolybdate skeleton.

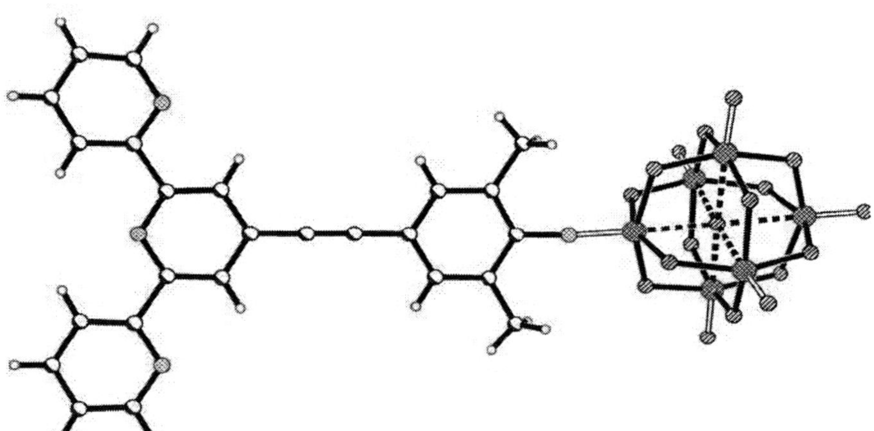

Figure 10. Ball-and-stick drawing of the organoimido hexamolybdate compound 15 containing a terpyridyl group.

In our investigations, other drying or dehydrating agents have also been tested instead of carbodiimides, including magnesium sulfate, calcium chloride, 4Å zeolite, phosphorus pentoxide, *N'N*-carbonyldiimidazole or even calcium hydride. However, they do not show similar activation effect like that of DCC [68]. This suggests that DCC does not simply serve as a dehydration agent to remove the water generated in the reaction. The likely role of DCC is presumedly to activate the terminal Mo-O bond and hence to increase the electrophilicity of the molybdenum atom, similar to its activating effect on the carboxyl group in amide or peptide synthesis [90]. In addition, an extremely excess of DCC is harmful to this reaction because DCC can undergo the side reactions directly with the molybdate ions and the initially formed products; for example, addition of eight equivalent of DCC, reaction of 2,6-dimethylaniline with $(n\text{-Bu}_4\text{N})_2[\text{Mo}_6\text{O}_{19}]$ (6) under 12 hours of refluxing in dry acetonitrile affords, in place of the expected imido products, only an octamolybdtate compound, $(cyclo\text{-C}_6\text{H}_{11}\text{NH}_3)_4$ $\beta\text{-}[\text{Mo}_8\text{O}_{26}]$ (16), which is usually the major byproduct found in this reaction [68].

Recently, we have succeeded in extending this synthetic route from aniline homologs and their derivatives to other aromatic primary amines containing fused rings, including naphthylamine, anthramine, phenanthramine and aminopyrene. For instance, 1-naphthylamine undergoes this reaction effectively, in *ca* 70% yield, to afford the corresponding imido derivative (figure 11), $(n\text{-Bu}_4\text{N})_2[\text{Mo}_6\text{O}_{18}(\text{NC}_{10}\text{H}_7\text{-1})]$ (17) [74].

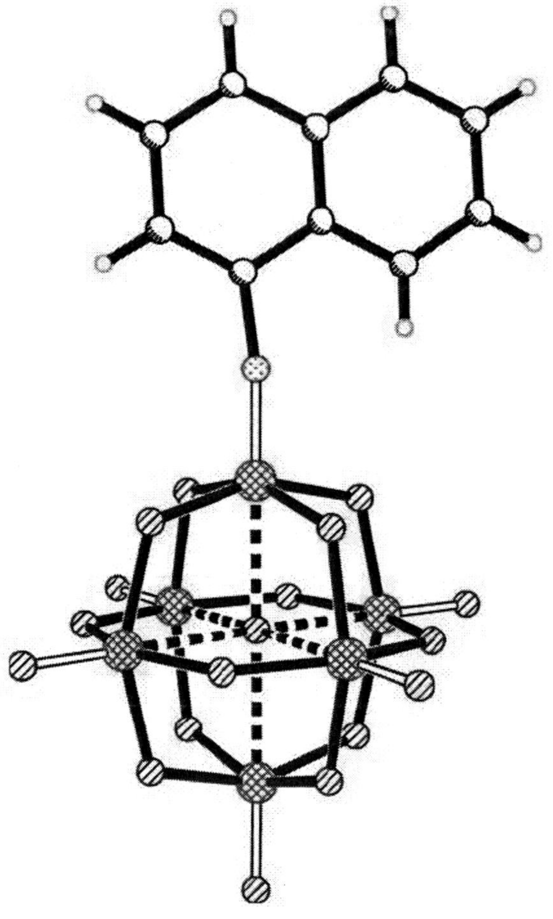

Figure 11. Ball-and-stick showing of the 1-naphthylimido hexamolybdate 17.

It should be pointed out here that, aromatic amines only with electron-withdrawing substituents such as nitro, fluoro and chloro groups do not undertake this reaction, neither do the aminopyridines, due to the weak nucleophilicity or basicity and easy oxidization of these kinds of amines [75]. We have also failed to prepare the alkylimido derivatives using this route with the corresponding aliphatic primary amines, such as benzylamine, since they violently reduce the hexamolybdate ion, usually resulting the single electron reduction produce of the parent cluster, $[Mo_6O_{19}]^{3-}$ [68, 75]. However, much effort to broaden the applicability of the DCC-dehydrating protocol to a broad spectrum of amines is still being taken in our laboratory.

Moreover, in our recent attempts to functionalize the octamolybdtates with organoimido groups, we discovered that in the presence of DCC, a proton can dramatically speed up the reaction of α-$[Mo_8O_{26}]^{4-}$ with aromatic primary amines under much milder conditions and, in the meanwhile, monofunctionalized organoimido derivatives of $[Mo_6O_{19}]^{2-}$ are selectively synthesized in high purity and moderate yield with easy workup [75]. This acid-assisting route has allowed the facile synthesis of a large number of mono-imido hexamolybdates, particularly, including those even bearing electron-withdrawing substituents such as bromo, chloro and fluoro groups [75- 78]. As shown in Eq(5), on addition of DCC (2 mmol), a variety of halogenated aromatic amines and their hydrochloride salts, in equivalent amounts (1.34 mmol), react with $(n\text{-}Bu_4N)_2\cdot\alpha\text{-}[Mo_8O_{26}]$ (1.0 mmol) smoothly, under nitrogen in hot acetonitrile (10 mL), obtaining the corresponding monosubstituted imido derivative in *ca* 50% yield after about 6 hours refluxing. Two structurally characterized halogenated imido derivatives, $(n\text{-}Bu_4N)_2[Mo_6O_{18}(NAr)]$, Ar = *p*-chlorophenyl (18) [75], *p*-bromophenyl (19) [77], respectively, are exemplified in figures 12 and 13. Nitroanilines also undergo this reaction, however, we have not successfully separated the corresponding derivatives from the reaction mixtures yet due to their decomposition in solution.

$$[Mo_8O_{26}]^{4-} + ArNH_2 + ArNH_3Cl \xrightarrow[\text{Refluxing, 6h}]{\text{2DCC, CH}_3\text{CN}} [Mo_6O_{18}(NAr)]^{2-} \quad (5)$$

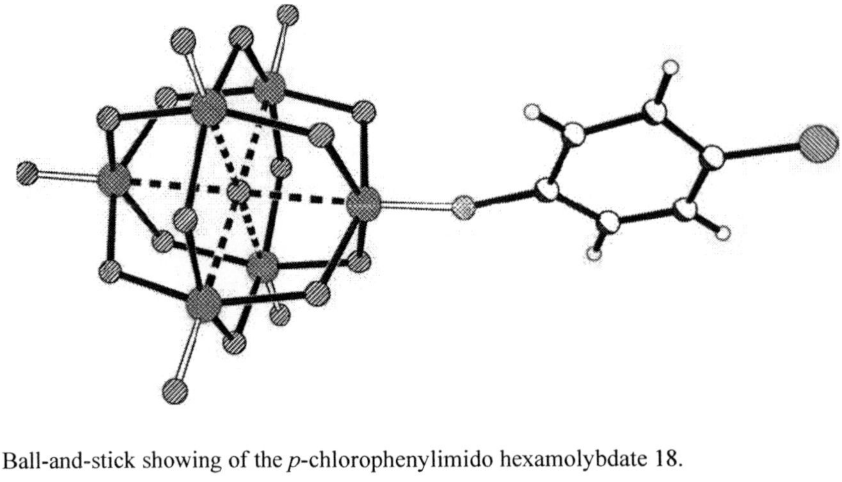

Figure 12. Ball-and-stick showing of the *p*-chlorophenylimido hexamolybdate 18.

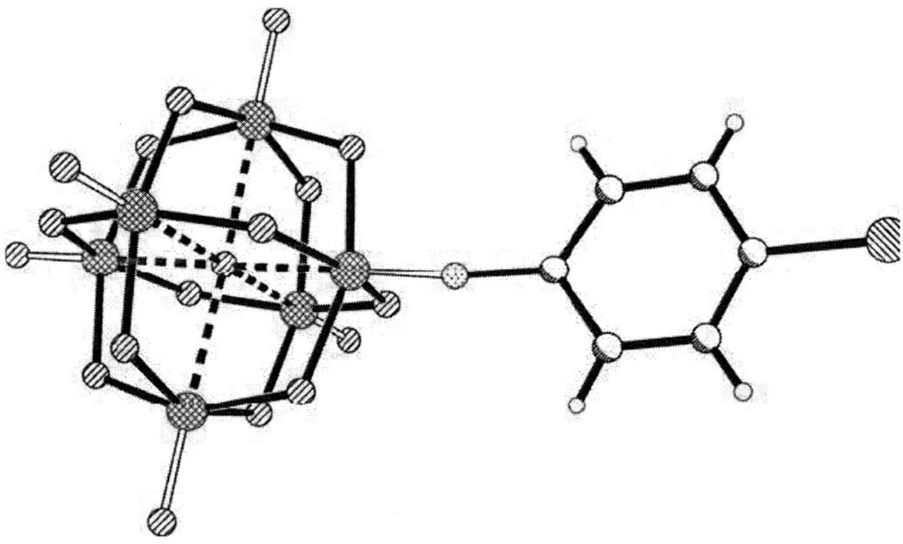

Figure 13. Ball-and-stick showing of the *p*-Bromophenylimido hexamolybdate 19.

It is worth pointing out that, the proton is crucial to the above route for the formation of monofunctionalized derivatives. It was observed that, in the absence of hydrochloride salts–which offer the protons, activated aromatic amines containing electron-donating substituents such as alkyl, alkoxy and alkylamino groups, react with α-$[Mo_8O_{26}]^{4-}$ to yield only the bifunctionalized arylimido derivatives of hexamolybdate. On the other hand, under the same conditions, no dehydrating reaction occurs between α-$[Mo_8O_{26}]^{4-}$ and inert aromatic amines with electron-withdrawing substituents such as chloro, fluoro and nitro groups, and the amines are oxidized instead. Furthermore, increasing the amount of hydrochloride salts can speed up the reaction, and correspondingly, it can be carried out even at room temperature. While more work needs to be done to shed light on the detailed reaction mechanism, the imaginable role of the proton is to complex with DCC and hence to strengthen the electrophilic ability of DCC to attack the oxo group of Mo-O group. Additionally, it also promotes the conversion of the α-octamolybdate into the hexamolybdate through a degradation and re-assembly process since in an acidic organic solvent, the hexamolybdate ions and their derivatives are much more stable than the α-octamolybdate ions.

There are a lot of spectroscopic techniques having been used to characterize the organoimido derivatives. In the low wavenumber region of the IR spectra ($\widetilde{\nu}$ <1000 cm^{-1}), the monofunctionalized organoimido derivatives display similar characteristic patterns of the parent hexamolybdate anion: there are two very strong bands of the Mo-O_t stretching and Mo-O_b-Mo asymmetric stretching vibrations at *ca* 955 (±5) and 795 (±5) cm^{-1}, respectively. Different from the parent cluster, there is also a strong shoulder peak appeared near 975 (±5) cm^{-1}, around the Mo-O stretching vibrations. This band may derive from the Mo-N stretching vibration [92] and is a diagnostic character for mono-organoimido substitution.

Compared to the ^{1}H NMR spectra of corresponding free amines, the protons attached to the α-carbon atoms or benzene rings in the monofunctionalized alkylimido and arylimido derivatives, respectively, all exhibit significantly downfield chemical shifts, indicating that the shielding nature of $[Mo_5O_{18}(Mo{\equiv}N\text{-})]^{2-}$ is much weaker than that of the amino group NH_2- and in agreement with the strong electron acceptability of POM clusters.

While the UV/Vis spectra of monofunctionalized alkylimido derivatives resemble that of the hexamolybdate parent intimately in the low energy region, those of monofunctionalized arylimido derivatives show considerable changes very distinctly due to their different molecular and electronic structures. The lowest energy electronic transition at 325nm in $[Mo_6O_{19}]^{2-}$, which was assigned to a L→M charge-transfer transition from the terminal oxygen non-bonding π-type HOMO to the molybdenum π-type LUMO [92], is bathochromically shifted by more than 20 nm in the monofunctionalized arylimido derivatives and becomes significantly more intense. The observed bathochromic shifts imply that the Mo-N π-bond is formed and delocalized over the organic conjugated π systems in such arylimido derivatives. In other words, there are strong electronic interactions between the hexamolybdate cluster and the conjugated organic ligands. In addition, the extent of bathochromic shifts in the reported mono-arylimido derivatives is also consistent with the length of conjugated segments and the trend in the π-donor ability of arylimido ligands.

Electrochemical data and ^{95}Mo, ^{17}O, and ^{14}N NMR data have also been recorded for a range of mono-imido derivatives of the hexamolybdate [63, 71]. The following trends have been observed [53]: (i) All species display a one-electron reduction process, the potential of which undergoes a cathodic shift in the order O < NAr < NR and, for arylimido derivatives, in the order of increasing electron-donating properties of the substituent. (ii) The $[^{95}Mo{\equiv}NR]$ resonance is shifted upfield in the order O < NAr < NR, and in the series of arylimido complexes, in the order of increasing electron-attracting properties of the substituent. (iii) The ^{17}O NMR spectra each are globally shielded with respect to that of $[Mo_6O_{19}]^{2-}$. These data are consistent with the trend in π-donor ability of the ligands, O < NAr < NR. For arylimido derivatives, molybdenum and nitrogen shieldings increase with increase in the $(\sigma + \pi)$ acceptor ability of the substituent.

The X-ray crystal structures of all the mono-organoimido derivatives have some common features: They resemble that of their hexamolybdate parent, but one terminal oxo group has been replaced by a organoimido substituent. The short Mo-N bond lengths range from 1.68-1.75 Å and nearly linear linkages of Mo-N-C are consistent with a substantial degree of triple bond character, although the angles sometimes deviate from linearity more than expected due to the force of the crystal packing [75]. The central oxygen atoms become close to the imido-bearing molybdenum atoms since the *trans* influence of an imido group is weaker compared to a terminal oxo group [76]. While the bond length involving the bridging-oxygen atoms shows considerable changes, there are no appreciable variations in the terminal bond lengths, raging from 1.66 to 1.72 Å. The pattern of short/long trans bond alternation in the three mutually orthogonal Mo_4O_4 belts observed in the original parent cluster still remains in the Mo_4O_4 belts perpendicular to the Mo-N bonds. In addition, the supramolecular "pairing self-assembly" of the cluster anions through π-π stacking between parallel phenyl rings on two neighboring clusters has also been observed in some mono-arylimido derivatives [75-77], which results the formation of dimers in their crystalline states (figure 14).

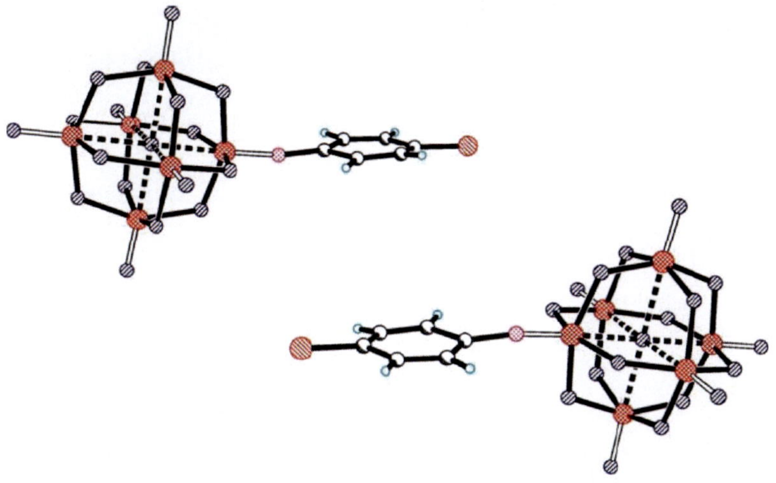

Figure 14. Dimer formation of the cluster anion in the solid of compound 18.

2.1.2. Disubstituted Derivatives

A characteristic feature of the derivatization of the hexamolybdate with organoimido ligands is the capacity for polyfunctionalization. Of which the difunctionalization is the easiest and most common. A. Proust et al noticed that, the *cis*-diimido derivative formed even when only one equivalent of Ph_3P=NPh was exploited to react with $[Mo_6O_{19}]^{2-}$ [53]. However, using two equivalent of Ph_3P=NPh, it still gave a mixture of un-reacted hexamolybdate anions and, mono- and *cis*-diimido derivatives. The isolation of the successive derivatives is also difficult because of they cocrystallize usually in the crystalline state [63].

The first example of well characterized di-organoimido derivative was reported by E. A. Maatta and co-workers [65, 71]. They found that the reaction of $(n\text{-}Bu_4N)_2[Mo_6O_{19}]$ (6) with two equivalent of 2,6-diisopropylphenyl isocyanate in refluxing pyridine for 8 days occurs to form the *cis*-disubstituted imido hexamolybdate, $(n\text{-}Bu_4N)_2[cis\text{-}Mo_6O_{17}(NAr)_2]$, Ar = 2,6-diisopropylphenyl (20), in 67% yield. It is obvious that this is a poor route in efficiency. Moreover, they also mentioned that the crude products contain an admixture of higher- and lower-substituted hexamolybdates, which makes the separation of large amounts of the desired difunctionalized product tedious and impractical, if not impossible.

In the past years, we also checked the possibility of our DCC-dehydrating protocol in the preparation of di-organoimido derivatives of the hexamolybdate. In most case, attempts to use the same route as for the mono-substituted derivatives but with stoichiometric ratio of two equivalent aromatic primary amines failed to obtain the disubstituted derivatives in high yields. Instead, a mixture of various polysubstituted hexamolybdates are generated since there are six equally reactive sites on a hexamolybdate ion, and thus repeated recrystallization is needed for purification of the anticipant product. However, an attempt to functionalize the α-isomer of octamolybdate ion, $\alpha\text{-}[Mo_8O_{26}]^{4-}$, with this reaction routine resulted in the selective synthesis of difunctionalized arylimido derivatives of hexamolybdate with good yield and high purity [79].

This novel route is shown in Eq(6), which allows the efficient preparation of a number of di-imido hexamolybdates more conveniently with easy and simple bench operations [79-81]. For example, the refluxing reaction of $(n\text{-}Bu_4N)_2{\cdot}\alpha\text{-}[Mo_8O_{26}]$ with two equivalent of DCC and two equivalent of 4-iodo-2,6-diisopropylaniline or 4-ethynyl-2,6-dimethylaniline in dry acetonitrile for 6 to 12 hours affords the corresponding di-organoimido hexamolybdates, (*n*-$Bu_4N)_2[cis\text{-}Mo_6O_{17}(NAr)_2]$, Ar = 4-iodo-2,6-diisopropyl-C_6H_2 (21) [79], 4-ethynyl-2,6-diisopropyl-C_6H_2 (22) [81], respectively, in good yields of 60-80%. The X-ray structures of compounds 21 and 22 are illustrated in figures 15 and 16, respectively. They both have the *cis* disubstituted structures with two activate remote functional groups at the end, which make them very important building bocks in fabricating novel organic-inorganic hybrids.

$$[Mo_8O_{26}]^{4-} + 2ArNH_2 \xrightarrow[\text{Refluxing, 6-12h}]{\text{2DCC, CH}_3\text{CN}} [cis\text{-}Mo_6O_{17}(NAr)_2]^{2-} \quad (6)$$

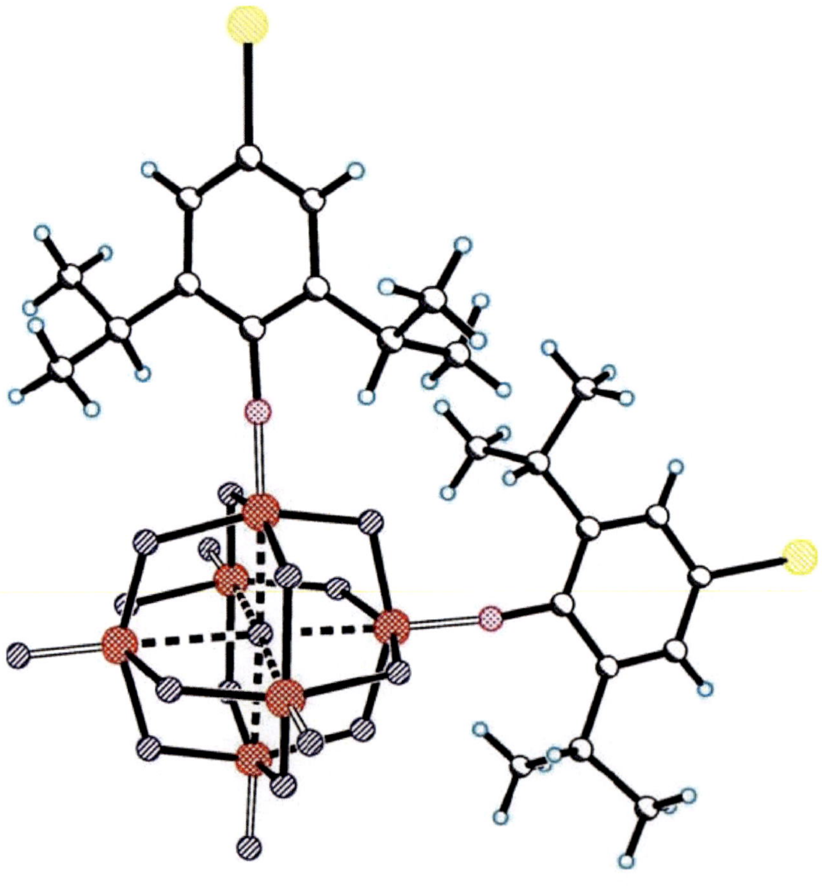

Figure 15. Ball-and-stick structure of the cis-disubstituted hexamolybdate imido derivative 21.

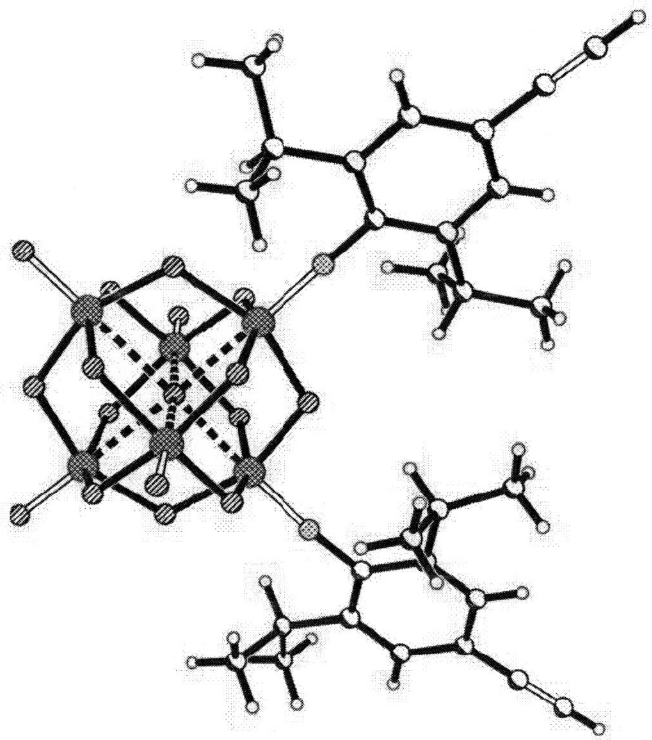

Figure 16. Ball-and-stick structure of the cis-didisubstituted hexamolybdate imido derivative 22.

In principle, there are, *cis* and *trans*, two position isomers in the terminal substituted di-imido hexamolybdates. If the steric and electronic effects of the imido groups are ignored, both of them can be formed in the same reaction, theoretically at a *cis* and *trans* ratio of 4:1, since there are four *cis* sites and only one *trans* terminal oxo reaction sites in a mono-imido derivative. While the steric repulsion is considered, which, in fact, is very common in organic substitution reactions, the *trans*-isomer will be more than 25% and becomes the principal product. But contrary to expectations, except for the *cis*-isomers, the *trans*-disubstituted hexamolybdates have scarcely been detected in the above reaction routes. It is seems that the disubstituted hexamolybdate with aryliimdo prefers to form a *cis*-isomeric derivative. Such a result is amazing, which suggests that the steric effect is less important here and on the other hand, the present of an extant imido substituent exerts an activating effect at proximal molybdyl groups. Indeed, a recent density functional theory study reveals that the *cis*-disubstituted derives are more stable in energy than the corresponding *trans*-isomers [93].

While the above routes give only the *cis*-disubstituted derives, R. J. Errington and co-workers discovered that, using their aromatic primary amines route [Eq(3)], a *trans*-disubstituted hexamolybdate, $(n\text{-}Bu_4N)_2[trans\text{-}Mo_6O_{17}(NC_6H_4NH_2)_2]$ (23), the unique example reported in the literature so far [66], was formed predominantly in the reaction between $(n\text{-}Bu_4N)_2[Mo_6O_{19}]$ (6) and two equivalent of 1,4-phenylenediamine. Moreover, ^1H NMR also confirmed the formation of the corresponding *cis*-isomer in the resulting products. Their work implies that the *trans*- disubstituted hexamolybdate is a kinetic stable product under certain supported conditions and can be isolated indeed. It also intensively suggests that there is a obvious different mechanism of the directly functionalization of the hexmolybdate with aromatic primary amines from those with phosphinimines or isocyanates and the aromatic primary amines approach is potential in the *trans*-disubstituted hexamolybdate derivatives.

Keeping the above idea in mind, we have had a renewed check very recently on our DCC-dehydrating protocol for the direct functionalization of the hexamolybdate with stoichiometric ratios of reactants [82- 83]. Indeed, for some activated aromatic primary amines containing electron-donating substitents, such as 2,6-dimethylaniline and *ortho*-anisidine, in the presence of one equivalent of DCC, the reaction of them and 0.5 equivalent of $(n\text{-}Bu_4N)_2[Mo_6O_{19}]$ (6) in heated acetonitrile can afford both the *cis*- and *trans*-disubstituted imido complexes in reasonable yields by a careful control of the refluxing time [Eq(7)].

$$[Mo_6O_{19}]^{2-} + 2ArNH_2 \xrightarrow[\text{Refluxing, 6-24h}]{\text{2DCC, CH}_3\text{CN}} [Mo_6O_{17}(NAr)_2]^{2-} \quad (7)$$

In general, a prolonged refluxing time promotes the formation of the *cis*-isomer, for example, the compound, $(n\text{-}Bu_4N)_2[cis\text{-}Mo_6O_{17}(NAr)_2]$, Ar = 2-methoxy-$C_6H_4$, (24) [82], was obtained from 24 hours refluxing reaction; On the other hand, the *trans*-isomer can be isolated from a relative short-time reaction system, for example, the compound, $(n\text{-}Bu_4N)_2[trans\text{-}Mo_6O_{17}(NAr)_2]$, Ar = 2,6-dimethyl-$C_6H_3$, (25) [83], was acquired from 6 hours refluxing reaction. In figures 17 and 18, the X-ray structures of compound 24 and 25 are shown, respectively. Compared to the parent hexamolybdate, compound 24 has a smaller optical band gap of 1.75 eV due to the incorporation of two strong electron-donating anisidido groups, which suggests that compound 24 is a potential molecular semiconductor and the optical band gap of POM-based semiconducting materials could be tuned exactly and controllably via chemical modification using suitable organic ligands. The chemical and physical properties of compound 25 are still underway in our laboratory.

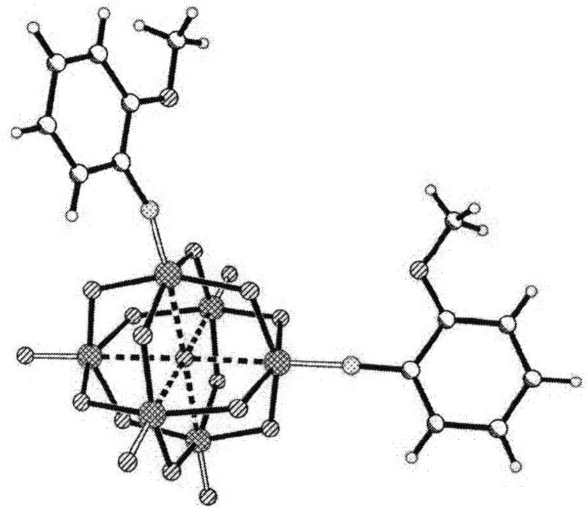

Figure 17. Ball-and-stick structure of the cis-didisubstituted hexamolybdate imido derivative 24.

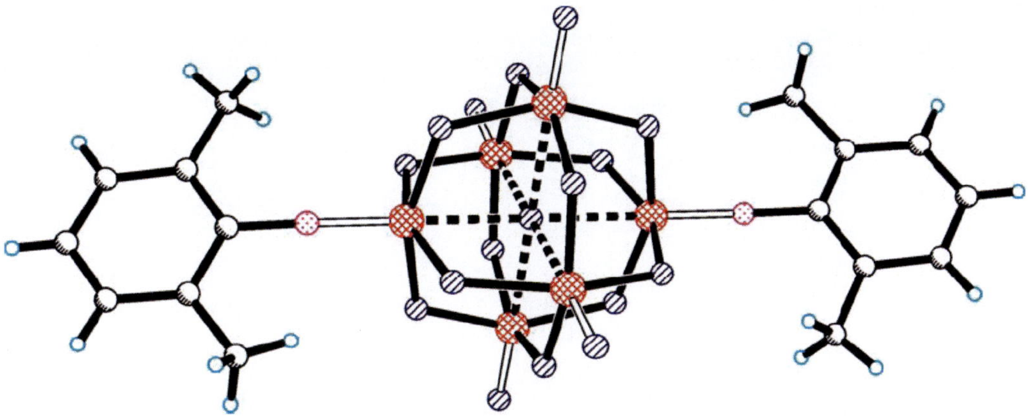

Figure 18. Ball-and-stick structure of the trans-didisubstituted imido complex 25.

Similar to their monosubstituted analogues, all the reported diimido hexamolybdates have of their own common features and some general trends in the structures and spectroscopies: (i) The bond lengths of [Mo≡NAr] triple bond are within the range of 1.71 to 1.75 Å, together with nearly linear linkages of Mo-N-C angles close to 180°. For the *cis*-isomers, one of the angles sometimes deviate from linearity more than expected due to the aggregation of the cluster anions through π-π stacking and the force of the crystal packing [80]. (ii) The central oxygen atoms inside the hexamolybdate cage are drawn closer to the imido-bearing Mo atoms, while the four bridge oxygen atoms are pushed away from the imido-bearing Mo atoms. (iii) The short-long bond alternation in the in the Mo_4O_4 belts of the parent hexamolybdate disappears in the *cis*-isomers. However, in the trans-isomers, this pattern is still kept in the belt perpendicular to the long axis of the cluster along the two Mo-N bonds. (iv) The patterns of supramolecular self-assembly due to π-π stacking or/and C-H...O hydrogen bonding are very popular. In particular, the I-O interaction has also been observed in compound 21 (figures 19). (v) In their IR spectra, the Mo-N stretching vibrations at 968 (±2) cm^{-1}, appear as a strong shoulder peak of the Mo-O stretching vibrations at 945 (±2) cm^{-1}. Compared to the corresponding mono-imido derivatives, there are obviously blue-shifts of *ca* 5 and 10 cm^{-1} for the Mo-N and Mo-O stretching vibrations, respectively. Correspondingly, the $Mo\text{-}O_b\text{-}Mo$ asymmetric stretching vibrations at *ca* 795 (±5) cm^{-1} of the mono-imido and parent hexamolybdates are significantly blue-shifted more than 20 cm^{-1} and appear as a broad and strong band at *ca* 775 cm^{-1}. (vi) Compared to the corresponding mono-imido derivatives, the lowest energy electronic transition of L→M charge-transfer is further red-shifted by *ca* 5 nm due to the incorporation of the other arylimido ligand. In the meanwhile, ^1H NMR of the aromatic protons shows visible upfield shifts, and the reduction potentials are also observed to shift cathodically due to the increase of electron density in the cluster anions accompanying the incorporation of the second arylimido ligand.

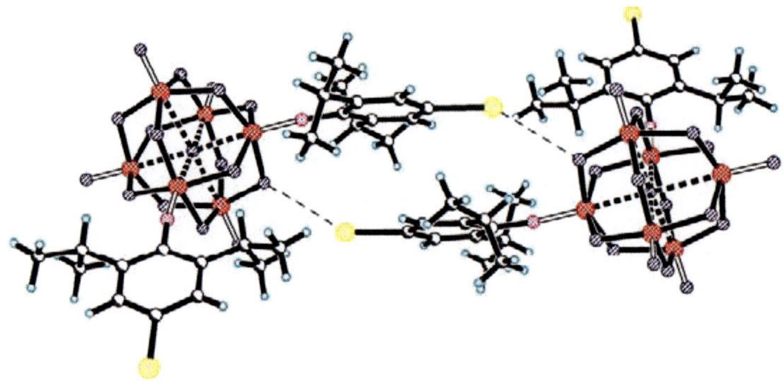

Figure 19. A view of the I-O interaction in a dimer of compound 21 in the crystalline state.

2.1.3. Polysubstituted Derivatives

The polysubstituted hexamolybdates, where three to six terminal oxo groups replaced by 2,6-diisopropylphenylimido ligands, have also been reported by E. A. Maatta and co-workers for nearly a decade [65, 71, 84]. They discovered that, by carefully controlling the molar ratios of 2,6-diisopropylaniline isocyanate over the hexamolybdate cluster , multiple-functionalized hexamolybdates, $[Mo_6O_{(19-x)}(NAr)_x]^{2-}$, Ar = 2,6-diisopropylphenyl, x = 3 (26) [71], 4 (27) [65, 71], 5 (28) [65, 71], 6 (29) [84] could be synthesized in refluxing pyridine after a prolonged reaction time of 13, 12, 11 and 21 days for 26, 27, 28, and 29, respectively. Of which the X-ray molecular structures of compounds 26 and 29 are exemplified in figures 20 and 21, respectively. While all these polysubstituted derivatives have been isolated and characterized, the nonselective nature of the reaction, which always results in an admixture of various multiple functionalized products, makes their separation process rather difficult. Apparently, the rational synthesis of polysubstituted hexamolybdates is a formidable challenge for a chemist.

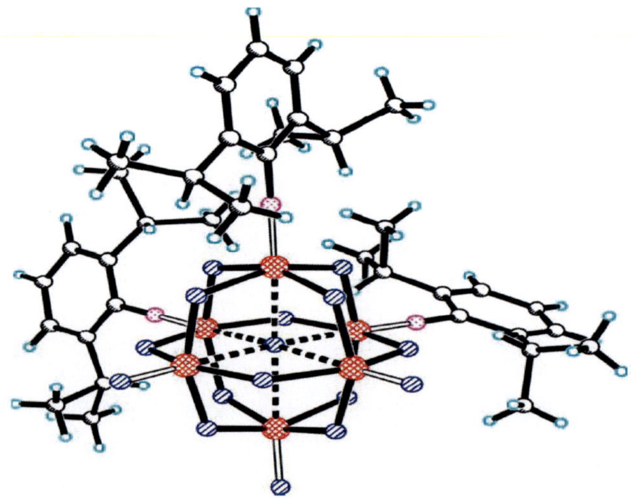

Figure 20. X-ray molecular structure of the trisubstituted hexamolybdate 26.

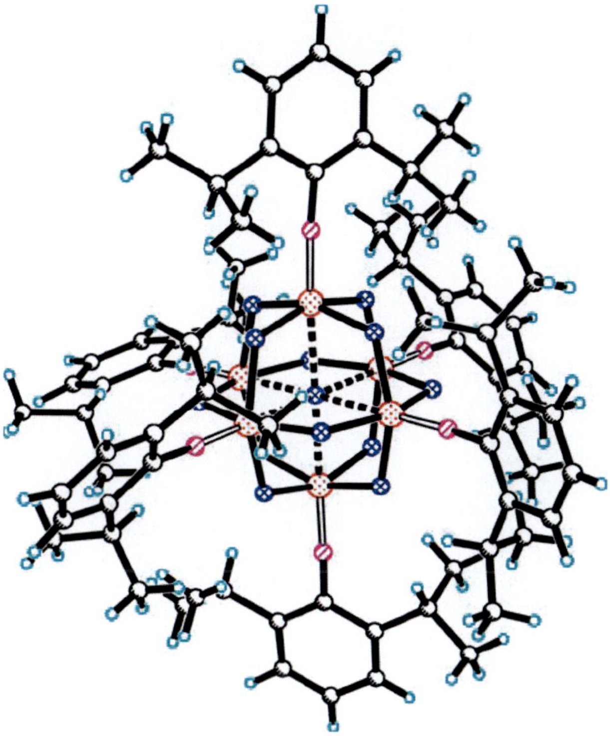

Figure 21. X-ray molecular structure of the hexasubstituted hexamolybdate 29.

We have also interested in the preparation of the polysubstituted hexamolybdates in the context of our DCC-dehydrating protocol using aromatic primary amines as the imido-releasing agents. In a recent attempt to synthesizing the *fac*-tri-2,6-dimethylanilido hexamolybdate [Eq(8)], an analogue of compound 26, contrary to our expectation, the as-resulted tri-imido derivative, $(n\text{-}Bu_4N)_2[cis\text{-}Mo_6O_{16}\ \mu_2\text{-}(NAr)(NAr)_2]$, Ar = 2,6-dimethyl-$C_6H_3$, (30) is not exclusively terminal-oxo replaced complex, and instead, one of bridging-oxo groups is substituted.

$$[Mo_6O_{19}]^{2-} + 3ArNH_2 \xrightarrow[\text{Refluxing, 24h}]{\text{3DCC, CH}_3\text{CN}} [Mo_6O_{16}(NAr)_3]^{2-} \tag{8}$$

As is illustrated in figure 22, the compound 30 has a feature that the μ_2-bridging oxygen atom sharing by the two imido-bearing Mo atoms in a *cis*-diimido hexamolybdate is substituted with a μ_2-bridging organoimido ligand. It suggests that such oxygen atoms in the *cis*-diimido hexamolybdates have been doubly activated by the neighboring imido groups and become more negative (nucleophilic) than other oxygen atoms, resulting in the easy electrophilic attack by DCC. Compound 30 stands for the first example of a bridging oxygen atom in POMs replaced by the bridging imido groups. It breaks the myth that only the terminal oxo groups can be directly replaced with imido ligands before and brings us a wonderful prospect in the chemistry of organoimido derivatives of POMs.

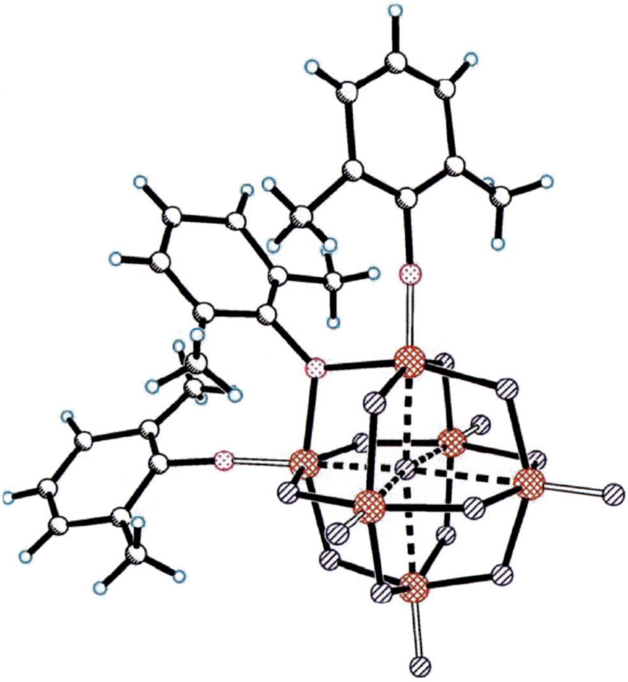

Figure 22. Ball-and-stick model of the trisubstituted hexamolybdate 30.

2.4. Organoimido Derivatives of other Lindqvist Poms

While there is fruitful in the synthetic chemistry of hexamolybdate, related studies on other POMs keep scarcely well developed. So far, there only has been obtained quite limited success in extending the above approaches to the other Lindqvist POMs.

Attempts to directly functionalize the hexatungstate anion, $[W_6O_{19}]^{2-}$, another important Lidquvist homopolyoxometalate besides the hexamolybdate, have failed to utilize all the above-mentioned approaches, since $[W_6O_{19}]^{2-}$ does not react at all with phosphinimines, isocyanates or aromatic primary amines under the present developed conditions. It seems that compared to the molybdyl groups in the hexamolybdate, the tungstyl groups in the hexatungstate is rather unreactive. The only one example of imidosubstituted hexatungstates, $(n\text{-}Bu_4N)_2[W_6O_{18}(NAr)]$, Ar = 2,6-diisopropyl-$C_6H_3$, (31) was obtained about ten years ago by E. A. Maatta et al via the reaction of $(n\text{-}Bu_4N)_2[WO_4]$ with 2,6-diisopropylphenyl isocyante in either 1,2-dichloroethane or pridine, but the yield is very low (*ca* 10% based on $n\text{-}Bu_4N)_2[WO_4]$) [94]. They thought that the plausible mechanism involves the initial formation of the imdo-tungstate complex, $[WO_3(NAr)]$, which then aggregates with additional $[WO_4]^{2-}$, in a self-assembly manner, to yield the target product. In addition, it is probably the proton in acidic contaminant, not water mentioned by Maatta *et al*, that involves in the assembly of the hexanuclear cluster, just like the above-mentioned acid-assisting route developed by us for the mono-imido hexamolybdates [75]. The deliberate addition of a suitable acid will perhaps bring us an improvement of this route.

Although the hexatungstate cluster is less labile, replacement of tungstyl groups with reactive molybdyl groups will exert an activate effect on the cluster. Surely, we discovered that the pentatungstenmolybdate ion, $[MoW_5O_{19}]^{2-}$, a Lidquvist heteropolyoxometalate, can be directly functionalized with aromatic primary amines using the DCC-dehydrating protocol [95].

$$[MoW_5O_{19}]^{2-} + ArNH_2 \xrightarrow[\text{Refluxing, 12h}]{\text{1.5DCC, CH}_3\text{CN}} [W_5O_{18}(MoNAr)]^{2-} \quad (9)$$

As shown in [Eq(9)], on addition of one and a half equivalents of DCC, the reaction of one equivalent of 2,6-dimethylaniline or 4-iodo-2,6-dimethylaniline with one equivlant of $(n\text{-}Bu_4N)_2[MoW_5O_{19}]$ does occur in hot acetonitrile under nitrogen, and in *ca* 60% yields affords the mono-imido derivatives [95], $(n\text{-}Bu_4N)_2[W_5O_{18}(MoNAr)]$, Ar = 2,6-dimethyl-$C_6H_3$, (32), 4-iodo-2,6-dimethyl-C_6H_2, (33), respectively. The X-ray crystal structure of compound 33 is given in figure 23. Indeed as expected, the terminal oxygen atom bonded to the molybdenum atom is selectively replaced by an imido substituent. This result opens a general road for the synthesis of organo imido derivatives of less reactive POM clusters, namely, by replacement one of the inert terminal metal-oxo groups with a functionalizable Mo-O group.

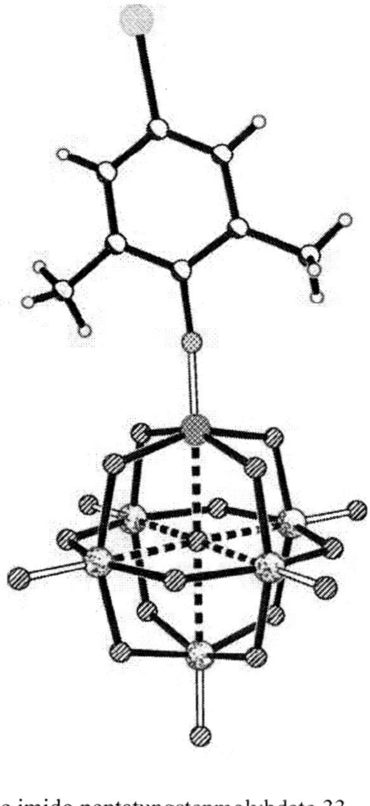

Figure 23. Ball-and-stick model of the imido pentatungstenmolybdate 33.

2.5. Organoimido Derivatives of Keggin POMs

Keggin-type POMs have a general formula of $[XMO_{40}]^{n-}$ and feature a central tetrahedral oxyanion $[XO_4]^{m-}$ encapsulated in a cage made up of twelve MO_6 octahedra by sharing corners, which represent the most popular and largest subclasses of POMs with a variety of transformations in structure and composition of the incorporated hetero-atom in the interior or metal atoms on the surface [1]. While the surface oxygen atoms of the α-phosphomolybdate ion, $[\alpha\text{-}PMo_{12}O_{34}]^{3-}$, and α-phosphotungstate ion, $[\alpha\text{-}PW_{12}O_{34}]^{3-}$, are readily alkylated with alkylation agents such as methyl sulfate and trimetyloxonium tetrafluoborate [54,55], they are unable to be replaced directly by any an imdio delivery reagent so far. For example, reaction of $[\alpha\text{-}PMo_{12}O_{34}]^{3-}$ with *p*-methyphenyl isocyanate in hot pyridine failed to afford the expected imidosubstituted α-phosphotungstate, but a degradated decanuclear imido complex instead [96].We also tried the reaction of several Keggin-structure POMs containing molybdyl groups, including the α-phosphomolybdate, with aromatic primary amines in the presence of DCC. However, in most case, aromatic amines were oxidized instead of acting as imido delivery reagents due to the high oxidation nature of the Keggin POMs of interest.

The only successful case in the synthesis of Keggin-type imido derivatives was just reported last year by A. Proust and coworkers [97]. They obtained an imido complex, $[Bu_4N]_4[PW_{11}O_{39}(ReNC_6H_5)]$ (34), by the substitution reaction of the monovacant Keggin compound $[Bu_4N]_4[H_3PW_{11}O_{39}]$, a penta-dentate ligand, with the mononuclear complex $[Re(NPh)Cl_3(PPh_3)_2]$ containing a metal-imido functional group in acetonitrile [Eq(10)]. Compound 34 have clearly demonstrated by various techniques including ^1H and ^{14}N NMR, electrochemistry, and especially by ESI mass spectrometry. However, its X-ray structure has not yet been completed.

$$[H_3PW_{11}O_{39}]^{4-} + [Re(NPh)Cl_3(PPh_3)_2] \xrightarrow[\text{Refluxing, 1h}]{Et_3N,\ CH_3CN} [PW_{11}O_{39}(ReNC_6H_5)]^{4-} \quad (10)$$

In all, the functionalization of Keggin-type POMs with imido groups faces still formidable challenges and much more attentions should be paid into this interesting field in the future since the potential of Keggin POM-based molecular materials is expected to be greater than that of Lindqvist derivatives.

3. Reactive Chemistry of Organoimido Derivatives of POMs

While organoimido derivatives of POMs have been discovered for more than one decade [58] and a variety of synthetic approaches have been developed as mentioned above, the reactive chemistry of them has just been investigated very recently and only the organoimido derivatives of hexamolybdate have been involved [72, 73, 81, 98-100]. Both the inorganic cluster skeleton and the incorporated organic ligand in an organoimido POM can act as the active sites for their own enjoyed reactions. The Mo-N bond is also a possible reaction site, the correlative reactions such as hydrolysis, however, are trivial due to break of the Mo-N bond and leaving of the organic ligand from the cluster in most cases. So far, the reaction chemistry of organoimido derivatives of POMs has focused mainly on the incorporated organic ligands because it provides great potentials in controllable synthesis of organic-inorganic hybrids exploiting well-developed common organic synthesis techniques. In particular, for the imidosubstituted hexamolybdates, several important inorganic and organic reactions, which do not cleave the Mo-N bond, have been set up. These includes coordination [73], addition polymerization [72] and carbon-carbon coupling reactions [98], of which the carbon-carbon coupling has been proved to be a powerful tool for our molecule-module-assembly strategy in fabricating POM-based organic-inorganic hybrid materils [56].

3.1. Coordination

Compared to the corresponding parent cluster, the surface oxygen atoms on the cluster skeleton of an organically derivatived POM should be more reactive to coordinate with a metal ion due to the activation effect of incorporated organic ligands [53], nevertheless, such reactions have not been realized in the imidosubstituted hexamolybdates. On the other hand, if the incorporated imido groups bearing ligatable substituents such as hydroxyl, carboxyl and pryidyl, such derivatives can in principle behave as POM-functionalized organic ligands. According to this idea, we synthesized the compound 15 with a terpyridyl group (see figure 10) [73]. The coordination reaction of this hexamolybdate-functionalized ligand 15 with some bivalent transitional metal ions such as Zn^{2+} and Ru^{2+} has been confirmed by preliminary *in situ* UV studies. For example, on adding an excess of $Ru(tPy)Cl_3$ to a solution of functionalized ligand 15 in DMSO, the intense absorption at 392 nm of ligand 15 diminishes significantly, in the meanwhile, a new absorption band at 504 nm appears gradually. The new absorption is a characteristic M→L charge-transfer (MLCT) transition of the Ru-bisterpyridine complexes, which implies definitely that the new hexamolybdate-functionalized bisterpyridine-Ru complex 35 has formed by the complexation reaction shown in [Scheme 1]. Isolation of the complex 35 and its X-ray single crystal structure determination are currently in progress.

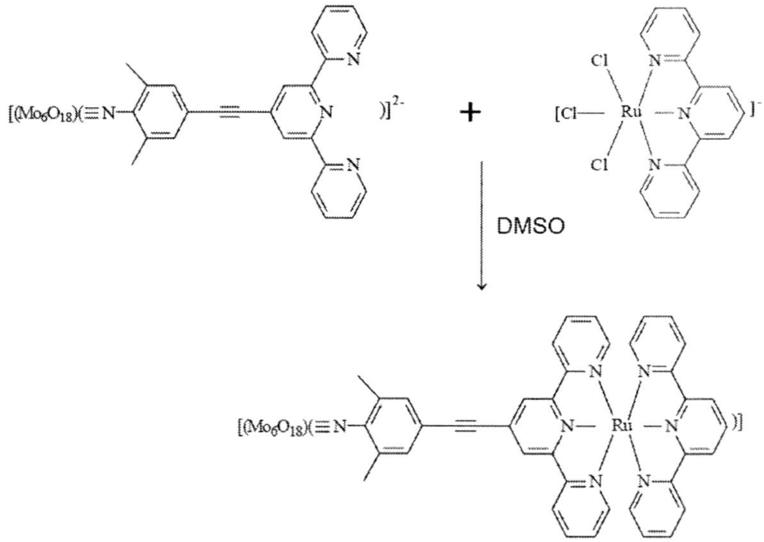

Scheme 1. Coordination reaction of compound 15 with $Ru(tPy)Cl_3$.

3.2. Addition Polymerization

The addition polymerization of organoimido derivatives of POMs bearing an unsaturated functional group such as the C=C double bond was just reported at the beginning of this new centenary [72]. In this report, E. A Maatta and co-workers discovered that, the *p*-styrenylimido hexamolybdate **9** and 4-methylstyrene could undergo conventional free radical-induced copolymerization initiated by AIBN [2,2'-azobis(2-methylpropionitrile)] in 1,2-dichloroethane and afforded a copolymer 36 with organoimido-POMs as pendants [Scheme 2]. The formation of copolymer 36 was confirmed by its IR spectrum, where the band of C=C stretching disappeared, but the characteristic doublet pattern of the terminal Mo-O and Mo-N vibrations in an monoimido hexamolybdate were still kept, compared to the precursor **9**. By comparison of the relative intensities of ^{1}HNMR due to the aryl groups, the α-CH_2 groups of the $[n\text{-}NBu_4]^+$ cations, and the styrenic CH_3 group, copolymer 36 was suggested to have a composition of nearly three 4-methyl styrene units per unit of **9**. This work demonstrates that the styrylimido ligand has similar chemical reactivity to that of the corresponding free stryrene and can be employed as a reactive site capable of delivering its metal complex into a conventional polymeric environment. Such a methodology should have broadly applicable potentials in the synthesis of novel inorganic/organic hybrid polymeric materials which incorporate covalently attached organoimido-POM complexes.

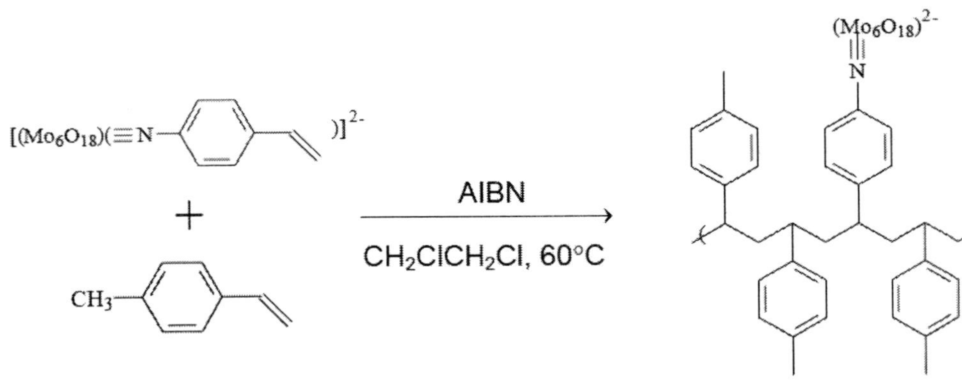

Scheme 2. Free radical-induced copolymerization Between compound **9** and 4-methylstyrene.

3.3. Carbon-Carbon Coupling and Its Application in POM-Based Organic-Inorganic Hybrids

It is well known that aryl halides and terminal acetylenes can undergo Pd catalyzed carbon-carbon coupling reactions [101-103]. The arylimido derivatives of POMs with a remote iodo or ethynyl groups, such as compounds 13 and 14, are in fact, POM-functionalized aryl halides and terminal acetylenes. If the iodo or ethynyl groups on functionalized POM clusters can undergo common Pd-catalyzed coupling reactions, and the POM clusters survive and the imido linkage does not break under the coupling reaction conditions, such organically-functionalized clusters would be invaluable building blocks for the rational synthesis of novel molecular POM–organic hybrids.

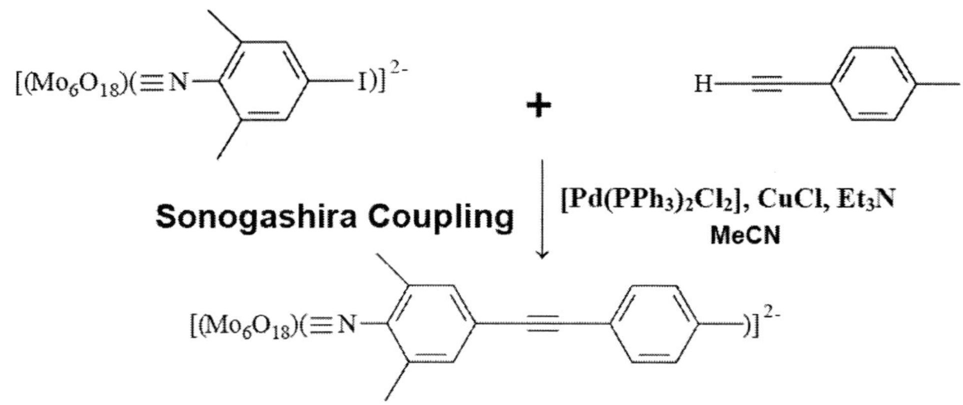

Scheme 3. Sonogashira coupling of 13 with 4-methyl-1-ethynylbenzene in acetonitrile at room temperature.

The Sonogashira coupling reaction of 13 with 4-methyl-1-ethynylbenzene [Scheme 3] or 3,5-di-tertbutyl-1-ethynylbenzene has been carried out in acetonitrile at room temperature under the protection of N_2 gas [98]. Compared to classic coupling reactions in organic synthesis [101-103], this reaction is extremely fast and finishes within only 20 minutes as monitored by thin-layer chromatography. As confirmed by the single X-ray diffraction study (figure 24), the hexamolybdate cluster survived indeed in the coupling product 37 under the reaction conditions, and obviously it activates the coupling reaction. The enhanced reactivity of the iodo function in 13 is presumably due to the electron-withdrawing nature of the hexamolybdate cluster. Prolonged reaction time and excess base were found to be harmful to the desired coupling products 37 and 38, as they slowly decomposed to the parent hexamolybdate cluster under the reaction conditions. Nevertheless, pure products can be usually obtained in more than 80% excellent yields.

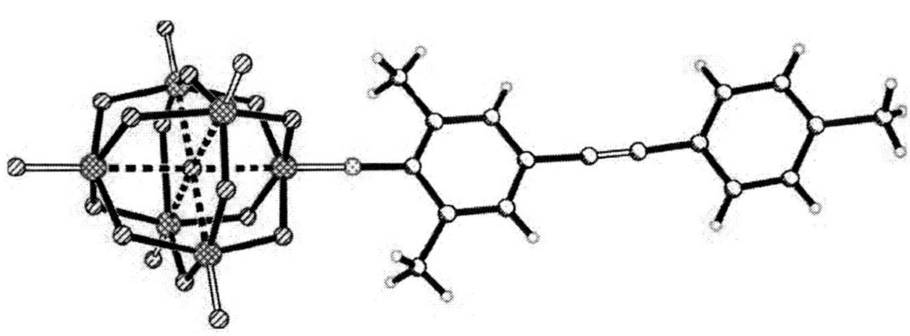

Figure 24. Ball-and-stick structure of the coupling product 37.

Hybrid Molecular Nanodumbbells

With our success in which iodo- or ethynylphenylimido-functionalized POM clusters can undergo Pd-catalyzed carbon-carbon coupling reactions, a great number of POM-based hybrids with various linking geometries, including covalently linked POM networks, can be rationally designed and easily prepared. For example, hybrid molecular nanodumbbells 39 and 40 with two hexamolybdate clusters linked by a rigid conjugated organic rod with different lengths in nanosize, shown in Scheme 4, have been synthesized by the Sonogashira coupling reaction between 13 and 14 in good yields [99].

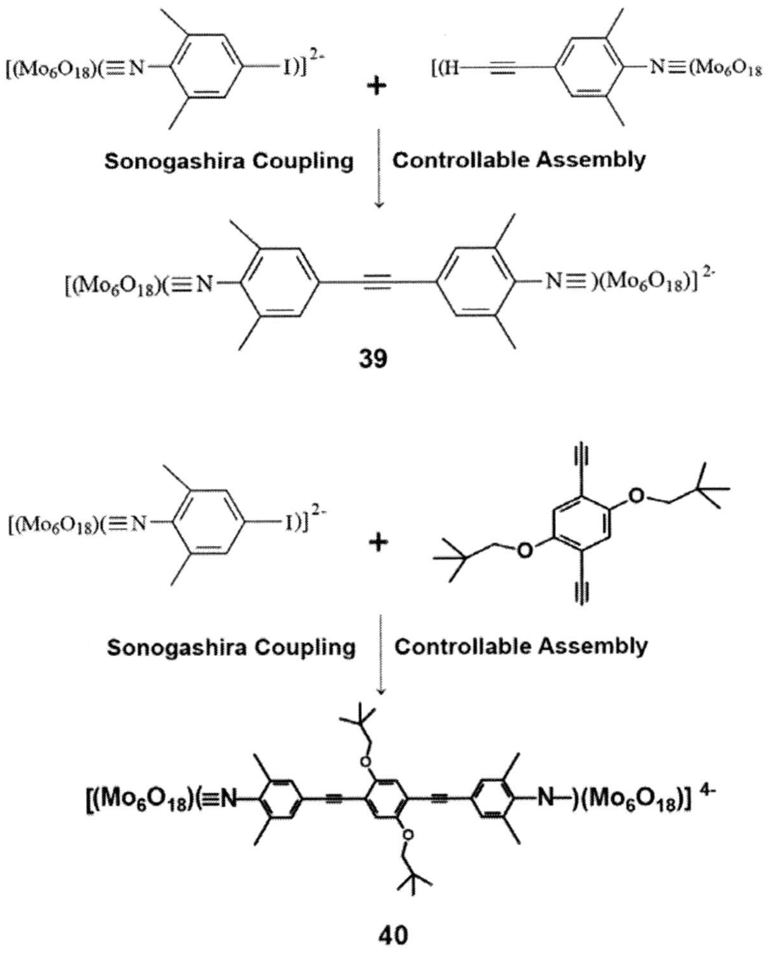

Scheme 4. Controllable assembly of hybrid nanodumbbells.

The X-ray crystal structure of nanodumbbell 39 has been determined (figure 25). Cyclic-voltammetry studies reveal a reversible one-electron reduction wave at -891 mV (vs Ag/Ag^{+}) for both nanodumbbells 39 and 40, which is very close to the reduction potential of other momosunstituted imido derivatives such as 13 (-909 mV), 14 (-893 mV), 37 (-893 mV), and 38 (-895 mV) [56]. UV/Vis absorption spectroscopy studies show that the lowest energy electronic transition is significantly bathochromically shifted from compounds 13 (362 nm) and 14 (368 nm), to hybrid dumbbells 39 (426 nm) and 40 (438 nm), thus indicating that the Mo≡N π bond is delocalized within the organic conjugated π system. Compared to the original oligo(phenylene ethynylene) ligands, which are strongly fluorescent molecules [104], both of two hybrid dumbbells show no fluorescence under excitation from 200 nm to 500 nm, which implies clearly that the conjugatedly-attached hexamolybdate cluster acts as an efficient fluorescence quencher in the dumbbells. These results suggest that strong electronic interactions through the conjugated organic bridge exist between the two clusters in a nanodumbbell.

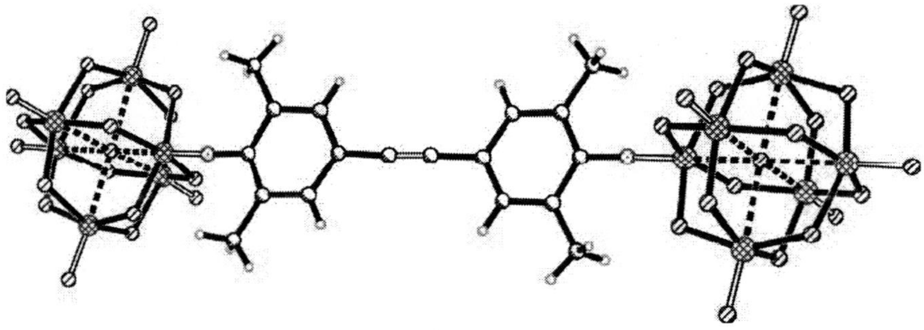

Figure 25. Ball-and-stick structure of the hybrid nanodumbbell 39.

Molecular Charge Transfer Hybrids

Charge-transfer hybrids based on organic donors and inorganic acceptors have being focused by chemists for several decades [100, 107, 108], due to their potential applications in solar energy conversion and storage such as photovoltaic cells, conducting materials including novel molecular conductors and superconductors, and nonlinear optical devices. POMs are one of the most important electron acceptors [50, 105], which have been widely exploited in preparing charge-transfer salts with organic cation donors, or charge-transfer supramolecular aggregates with neutral organic π donors through cocrystallization [106]. Efforts have recently been made to develop intramolecular charge transfer hybrids based on POMs[100]. Compound 11 is one of such examples, in which a ferrocenyl donor is bound covalently to a hexamolybdate acceptor and a charge-transfer transition from the Fe^{2+} center to the cluster is observed at 536nm [70]. However, this hybrid was obtained in only 19% yield through a direct functionalization reaction on the hexamolybdate cluster. Additionally, the short separation distance between the ferrocenyl donor and the cluster acceptor also renders the charge recombination facile as a coherent superexchange mechanism dominates [109]. These problems, however, are easy to be overcome with iodophenylimido-functionalized POM cluster building blocks through the Pd-catalyzed carbon-carbon coupling reactions.

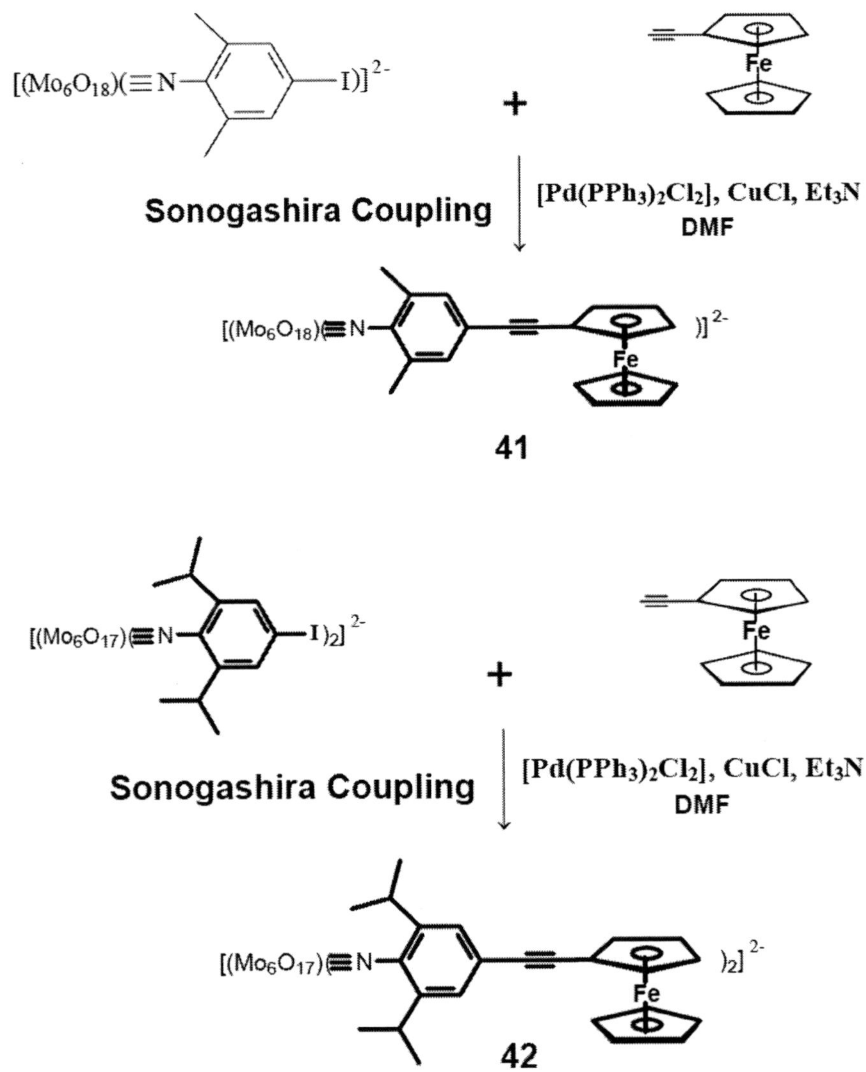

Scheme 5. Synthesis of two novel charge-transfer hybrids based on hexamolybdate cluster.

For example, two novel charge-transfer hybrids 41 and 42, shown in Scheme 5, in which one or two ferrocenyl groups are bonded indirectly to a hexamolybdate cluster through an extended π-conjugated bridge, have been prepared in over 65% good yields from the Sonogashira coupling reactions of 13 and 21 with ethynylferrocene [100], respectively. The X-ray molecular structure of 41 was shown in figure 26.The electronic spectra of these hybrids show a broad absorption tail extending beyond 550 nm, indicating the existence of charge-transfer transition from the ferrocenyl donor to the cluster acceptor. The observation of the clear charge-transfer transition indicates the contribution of charge-transfer resonance to the ground state in both 41 and 42 even though the donor-acceptor separation distance of 11.29 Å is rather long, signaling a through-bond charge-transfer nature made possible by the organic π-conjugated bridge. Cyclic voltammetry studies reveal a single-electron oxidation wave at 0.328 V and a single electron reduction wave at -0.831 V for 41 with one ferrocenyl unit. For the hybrid with two ferrocenyl units compound 42, a lower single electron reduction potential at -1.055 V and a two-electron oxidation wave at 0.331 V are observed. The reduction process in 41 and 42 is associated with the cluster skeleton. The observed cathodical shift of reduction potential from 41 to 42 is in agreement with the increasing number of the electron-donating imido ligand in 42. The oxidation wave in these two hybrids is attributed to the ferrocenyl unit. The almost same oxidation potential of them indicates negligible electronic interactions between the two ferrocenyls in 42.

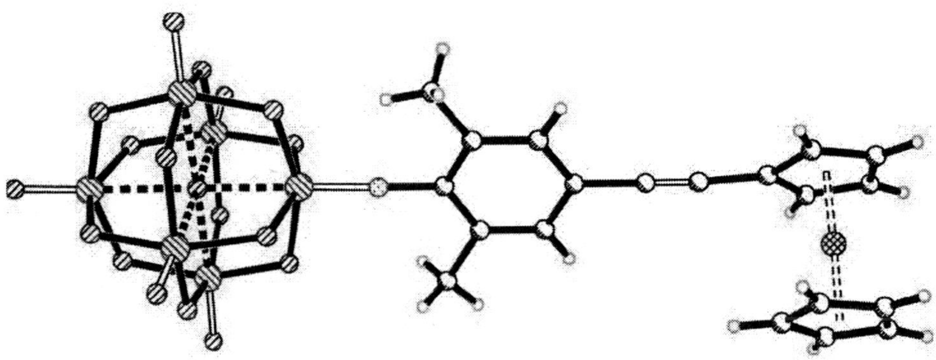

Figure 26. Ball-and-stick structure of the charge-transfer hybrid 41.

Hybrid Ploymers

Hybrid polymers containing POMs have been intensively studied for seeking new conducting materials. Conjugated polymeric hybrids containing covalently linked POM clusters, either in the main chain or in the side chains, at present, can be synthesized conveniently with the Pd-catalyzed coupling reactions. Starting from difunctionalized cluster 21 or 22, main-chain POM-containing polymers have been prepared as shown in Scheme 6. These polymers have an average molecular weight of about 160,000. Their electronic spectra are nearly identical to that of hybrid dumbbell 39, indicating that the organic π electrons delocalize only within the two imidobearing Mo atoms and the electronic interaction between the two orthogonally jointed organic π systems is limited. Cyclicvoltammetry studies reveal a reversible single-electron reduction wave at -1.2 V (vs Ag/Ag+) which is consistent with bis(arylimido)-substituted hexamolybdates. These polymers also show intense absorption in the visible range but with little fluorescence emissions, suggesting efficient fluorescence quenching of the embedded POM cluster on the organic phenylene acetylene units. Peng et al has studied their applications in photovoltaic cells and a power conversion efficiency of 0.15% was obtained, which is significantly higher than photovoltaic cells fabricated with other conjugated polymers in the same device configuration.

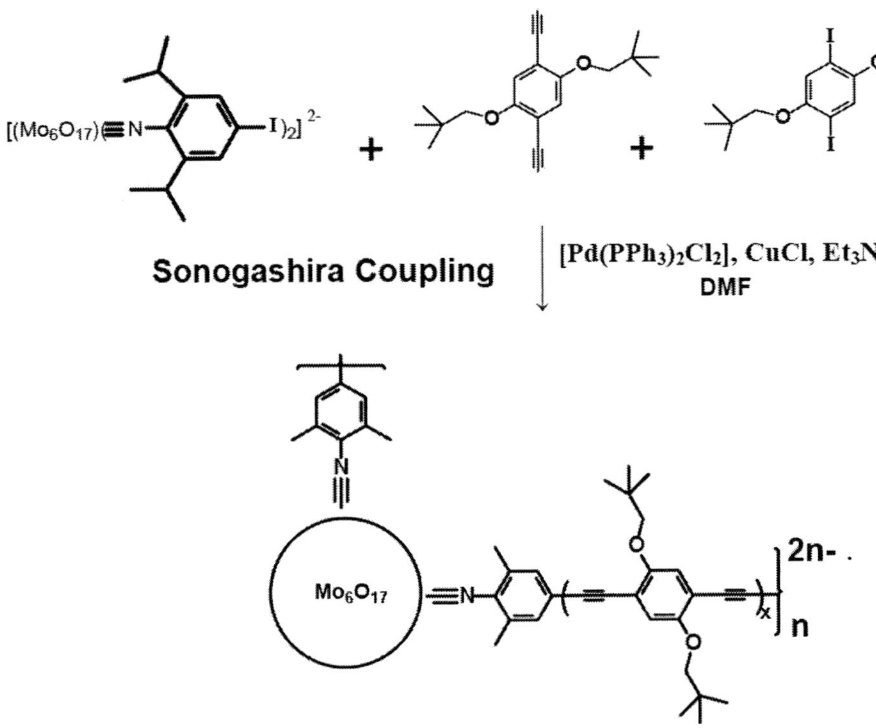

Scheme 6. Synthesis of polymeric hybrids based on hexamolybdate cluster.

Recently, the hexamolybdate cluster has been incorporated into poly(phenylene ethynylene) as conjugated side-chain pendants through the Pd-catalyzed coupling reactions in Peng's groups too. These polymeric hybrids have much more potential for applications in photovoltaic cells, owing to efficient fluorescence quenching in such materials to make free charge carriers reside in different structural units [110].

4. Conclusion and Outlook

The chemistry of organoimido derivatives of polyoxometalates, especially their reactive chemistry, is still in its infancy. Although classically organic synthetic techniques have given a fascinating promise in this field, to date, only a limited number of methods have been developed for the preparation of covalently bonded organoimido derivatives of POMs and their reactions have just stepped mainly on the Pd-catalyzed coupling reactions. Nonetheless, this new chemistry indeed opens a wide road to make novel molecular hybrids and polymeric hybrids based on POMs in a much more controlled manner. With the development of this chemistry, a large number of covalently bonded POM–organic hybrids, particularly those involving organic π-conjugated systems, including molecular, dendritic, and polymeric systems are sure to be prepared in the near future. In addition, efforts on exploring the electrical, optical and photophysical properties of this new class of hybrids are expected.

5. ACKNOWLEDGEMENTS

Y. G. Wei sincerely thanks Professor Zhonghua Peng for professional support to start the project when he was at UMKC. He also gratefully thanks Professors Hongyou Guo and Sijia Xue for helps in continuing this research in china. The contributions of undergraduate and graduate students: Ning Ge, Na Wang, Qiang Li, Yun Xia, Zicheng Xiao, Yi Zhu and Shaoyong Ke, are also gratefully acknowledged. This work is sponsored by NFSC-20201001, SRF for ROCS of SEM, MOST-TG2000077503 and the university funding from Hubei University of Technology.

6. REFERENCES

[1] Pope, M. T. *Heteropoly and Isopoly Oxometalates*. Berlin: Springer; 1983.

[2] Berzelius, J. J. Beitrag zur näheren Kenntniss des Molybdäns. Poggend. *Ann. Phys. Chem.* 1826, 6, 369–392.

[3] Pope, M. T.; Müller, A. Editors. Polyoxometalates: *From Platonic Solids to Anti-Retroviral Activity*. Dordrecht: Kluwer; 1994.

[4] Hill, C. L. Guest Editor. A Special Thematic Issue on Polyoxometalates. *Chem. Rev.* 1998, 98, 1-390.

[5] Pope, M. T.; Müller, A. Editors. Polyoxometalate Chemistry: *From Topology via Self-Assembly to Application.* Dordrecht: Kluwer; 2001.

[6] Zhang, S. W.; Wei, Y. G.; Yu, Q.; Shao, M. C.; Tang, Y. Q. Toward Quantum Line—Self-Assembly Process of Nanomolecular Building Blocks Leading to a Novel One-Dimensional Nanomolecular Polymer: Single-Crystal X-ray Structure of $\{[H_3O]^+_{12}\ \{(H_2O)MoO_{2.5}[Mo_{36}\ O_{108}(NO)_4(H_2O)_{16}]O_{2.5}Mo(H_2O)\}^{12-}\}$. *J. Am. Chem. Soc.* 1997, 119, 6440-6441.

[7] Jiang, C. C.; Wei, Y. G.; Liu, Q.; Zhang, S. W.; Shao, M. C.; Tang, Y. Q. Self-assembly of a Novel Nanoscale Giant Cluster: $[Mo_{176}O_{496}(OH)_{32}(H_2O)_{80}]$. *Chem. Commun.* 1998, 1937-1938.

[8] Liu, G.; Wei, Y. G.; Yu, Q.; Liu, Q.; Zhang, S. W. Polyoxometalate Chain-like Polymer: the Synthesis and Crystal Structure of a 1D Compound, $[Mo_{36}O_{108}(NO)_4(MoO)_2La_2(H_2O)_{28}]_n{\cdot}56nH_2O$. *Inorg. Chem. Commun.* 1999, 2, 434-437.

[9] Wei, Y. G..; Zhang, C.; Zhang, S. W.; Shao, M. C. A Spherical Nanascale Polyoxomolybdate Encapsulating Thirty Coordinated Butyrate: Synthesis and Structure Characterization of $(NH_4)_{42}[Mo_{132}O_{372}(CH_3CH_2CH_2COO)_{30}(H_2O)_{72}]$. Unpublished result.

[10] Müller, A.; Krickemeyer, E.; Meyer, J.; Bögge, H.; Peters, F.; Plass, W.; Diemann, E.; Dillinger, S.; Nonnenbruch, F.; Randerath, M.; Menke, C. $[Mo_{154}(NO)_{14}O_{420}(OH)_{28}(H_2O)_{70}]^{(25 \pm 5)-}$: A Water-Soluble Big Wheel with More than 700 Atoms and a Relative Molecular Mass of About 24000. *Angew. Chem. Int. Ed.* 1995, 34, 2122-2124.

[11] Müller, A.; Krickemeyer, E.; Bögge, H.; Schmidtmann, M.; Peters, F.; Menke, C. Meyer, J. An Unusual Polyoxomolybdate: Giant Wheels Linked to Chains. *Angew. Chem. Int. Ed.* 1997, 36, 484-486.

[12] Müller, A.; Krickemeyer, E.; Bögge, H.; Schmidtmann, M.; Beugholt, C.; Kögerler, P.; Lu, C. Z. Formation of a Ring-Shaped Reduced "Metal Oxide" with the Simple Composition $[(MoO_3)_{176}(H_2O)_{80}H_{32}]$. *Angew. Chem. Int. Ed.* 1998, 37, 1220-1223.

[13] Müller, A.; Shah, S. Q. N.; Bögge, H.; Schmidtmann, M. Molecular Growth from a Mo-176 to a Mo-248 Cluster. *Nature.* 1999, 397, 48-50.

[14] Müller, A.; Serain C. Soluble Molybdenum Blues-"des Pudels Kern". *Acc. Chem. Res.* 2000, 33, 2-10.

[15] Müller, A.; Krickemeyer, E.; Bögge, H.; Schmidtmann, M.; Peters, F. Organizational Forms of Matter: An Inorganic Super Fullerene and Keplerate Based on Molybdenum Oxide. *Angew. Chem. Int. Ed.* 1998, 37, 3359-3363.

[16] Müller, A.; Sarkar, S.; Shah, S. Q. N.; Bögge, H.; Schmidtmann, M.; Sarkar, S.; Kögerler, P.; Hauptfleisch, B.; Trautwein, A. X.; Schünemann, V. Archimedean Synthesis and Magic Numbers: "Sizing" Giant Molybdenum-Oxide-Based Molecular Spheres of the Keplerate Type. *Angew. Chem. Int. Ed.* 1999, 38, 3238-3241.

[17] Müller, A.; Polarz, S.; Das, S. K.; Krickemeyer, E.; Bögge, H.; Schmidtmann, M.; Hauptfleisch, B. "Open and Shut" for Guests in Molybdenum-Oxide-Based Giant Spheres, Baskets, and Rings Containing the Pentagon as a Common Structural Element. *Angew. Chem. Int. Ed.* 1999, 38, 3241-3245.

[18] Müller, A.; Koop, M.; Bögge, H.; Schmidtmann, M.; Peters, F.; Kogerler, P. Building blocks as disposition in solution: $[\{Mo^{VI}O_3(H_2O)\}_{10}\{V^{IV}O(H_2O)\}_{20}\{(Mo^{VI}/Mo^{VI}{}_5O_{21})(H_2O)_3\}_{10}(\{Mo^{VI}O_2(H_2O)_2\}_{5/2})_2(\{NaSO_4\}_5)_2]^{20-}$, a giant spherical cluster with unusual structural features of interest for supramolecular and magneto chemistry. *Chem. Commun.* 1999, 1885-1886.

[19] Müller, A.; Shah, S. Q. N.; Bögge, H.; Schmidtmann, M.; Kogerler, P.; Hauptfleisch, B.; Leiding, S.; Wittler, K. Thirty Electrons "Trapped" in a Spherical Matrix: A Molybdenum Oxide-Based Nanostructured Keplerate Reduced by 36 Electrons. *Angew. Chem. Int. Ed.* 2000, 39, 1614-1616.

[20] Müller, A.; Das, S. K.; Bögge, H.; Schmidtmann, M.; Botar, A.; Patrut, A. Generation of Cluster Capsules (I_h) from Decomposition Products of a Smaller Cluster (Keggin-T_d) while Surviving Ones Get Encapsulated: Species with Core-shell Topology formed by a Fundamental Symmetry-driven Reaction. *Chem. Commun.* 2001, 657-658.

[21] Müller, A.; Kogerler P.; Dress A. W. M. Giant Metal-oxide-based Spheres and Their Topology: from Pentagonal Building Blocks to Keplerates and Unusual Spin Systems. *Coord. Chem. Rev.* 2001, 222, 193-218.

[22] Müller, A.; Beckmann, E.; Bögge, H.; Schmidtmann, M.; Dress A. Inorganic Chemistry Goes Protein Size: A Mo_{368} Nano-Hedgehog Initiating Nanochemistry by Symmetry Breaking. *Angew. Chem. Int. Ed.* 2002, 41, 1162-1167.

[23] Müller, A.; Roy S. En Route from the Mystery of Molybdenum Blue via Related Manipulatable Building Blocks to Aspects of Materials Science. *Coord. Chem. Rev.* 2003, 245, 153-166.

[24] Müller, A.; Kuhlmann C.; Bögge, H.; Schmidtmann M.; Baumann M.; Krickemeyer E. $Mo^V{}_2O_4{}^{2+}$ directs the formation and subsequent linking of potential building blocks

under different boundary conditions: A related set of novel cyclic polyoxomolybdates. *Eur. J. Inorg. Chem.* 2001, (9), 2271-2277.

[25] Yamase T.; Prokop P. V. Photochemical Formation of Tire-shaped Molybdenum Blues: Topology of a Defect Anion, $[Mo_{142}O_{432}H_{28}(H_2O)_{58}]^{12-}$. *Angew. Chem. Int. Ed.* 2002, 41, 466-469.

[26] Yang, W. B.; Lu C. Z.; Lin X.; Zhuang H. H. A Novel Acetated 54-member Crown-shaped Polyoxomolybdate with Unprecedented Structural Features: $Na_{26}[\{Na(H_2O)_2\}_6(\{\mu_3\text{-}OH\}_4Mo^{V}_{20}Mo^{VI}_{34}O_{164}(\mu_2\text{-}CH_3COO)_4)]\cdot\approx 120H_2O$. *Chem. Commun.* 2000, 1623-1624.

[27] Yang, W. B.; Lu C. Z.; Lin X.; Zhuang H. H. Novel Acetate Polyoxomolybdate "Host" Accommodating a Zigzag-Chainlike “Guest” of Five Edge-Shared Sodium Cations: $Na_{21}\{[Na_5(H_2O)_{14}]\subset[Mo_{46}O_{134}(OH)_{10}(\mu\text{-}CH_3COO)_4]\}\cdot CH_3COONa\cdot\approx 95H_2O$. *Inorg. Chem.* 2002, 41, 452-454.

[28] Wassermann K.; Dickman M. H.; Pope M. T. Self-Assembly of Supramolecular Polyoxometalates: The Compact, Water-Soluble Heteropolytungstate Anion $[As_{12}^{III}Ce_{16}^{III}(H_2O)_{36}W_{148}O_{524}]^{76-}$. *Angew. Chem. Int. Ed.* 1997, 36, 1445-1448.

[29] Liu T. B. Supramolecular Structures of Polyoxomolybdate-Based Giant Molecules in Aqueous Solution. *J. Am. Chem. Soc.* 2002, *124,* 10942-10943.

[30] Liu T. B.; Diemann E.; Li H. L.; Dress A. W. M., Müller, A. Self-assembly in Aqueous Solution of Wheel-shaped Mo_{154} Oxide Clusters into Vesicles. *Nature.* 2003, 426, 59-62.

[31] Liu T. B. An Unusually Slow Self-Assembly of Inorganic Ions in Dilute Aqueous Solution. *J. Am. Chem. Soc.* 2003, 125, 312-313.

[32] Liu, G.; Cai, Y. G.; Liu, T. B. Automatic and Subsequent Dissolution and Precipitation Process in Inorganic Macroionic Solutions. *J. Am. Chem. Soc.* 2004, 126, 16690-16691.

[33] Bredas, J. L. Silbey, R. Editors. *Conjugated Polymers.* Dordrecht/ The Netherlands: Kluwer; 1991.

[34] Nalwa, H. S. Editor. Handbook of Organic Conductive Molecules and Polymers I – IV. Chichester/ England: Wiley; 1997.

[35] Pope, M.; Swenberg, C. E. Editors. *Electronic Processes in Organic Crystals and Polymers*. 2nd edition. Oxford: Oxford University Press; 1999.

[36] Heeger, A. J. Semiconducting and Metallic Polymers: The Fourth Generation of Polymeric Materials. *J. Phys. Chem. B.* 2001, *105*, 8475-8491.

[37] Kraft, A.; Grimsdale, A. C.; Holmes A. B. Electroluminescent Conjugated Polymers - Seeing Polymers in a New Light. *Angew. Chem. Int. Ed.* 1998, 37, 402-428.

[38] Kelly, S. M. *Flat Panel Displays: Advanced Organic Materials.* Cambridge: Royal Society of Chemistry; 2002.

[39] Katz, H. E.; Bao, Z. The Physical Chemistry of Organic Field-Effect Transistors. *J. Phys .Chem. B.* 2000, 104, 671-678.

[40] Katz, H. E.; Bao, Z.; Gilat, S. L. Synthetic Chemistry for Ultrapure, Processable, and High-Mobility Organic Transistor Semiconductors. *Acc. Chem. Res.* 2001, 34, 359-369.

[41] Katz, H. E. Recent Advances in Semiconductor Performance and Printing Processes for Organic Transistor-Based Electronics. *Chem. Mater.* 2004, *16*, 4748-4756.

[42] Newman, C. R.; Frisbie, C. D.; da Silva Filho, D. A.; Bredas, J.-L.; Ewbank, P. C.; Mann, K. R. Introduction to Organic Thin Film Transistors and Design of n-Channel Organic Semiconductors. *Chem. Mater.* 2004, *16*, 4436-4451.

[43] Hide, F.; Diaz-Garcia, M. A.; Schwartz, B. J.; Heeger, A. J. New Developments in the Photonic Applications of Conjugated Polymers. *Acc. Chem. Res.* 1997, 30, 430-436.

[44] McGehee, M. D.; Heeger, A. J. Semiconducting (Conjugated) Polymers as Materials for Solid-State Lasers. *Adv. Mater.* 2000, 12, 1655-1668.

[45] Dagani, R. Nobel Prize in chemistry. *Chem. Eng. News.* 2000, 78(42), 4-4.

[46] Shirakawa H. The Discovery of Polyacetylene Film: The Dawning of an Era of Conducting Polymers (Nobel Lecture). *Angew. Chem. Int. Ed.* 2001, 40, 2574-2580.

[47] MacDiarmid, A. G. "Synthetic Metals": A Novel Role for Organic Polymers (Nobel Lecture). *Angew. Chem. Int. Ed.* 2001, 40, 2581-2590.

[48] Heeger, A. J. Semiconducting and Metallic Polymers: The Fourth Generation of Polymeric Materials (Nobel Lecture). *Angew. Chem. Int. Ed.* 2001, 40, 2591-2611.

[49] Pope, M. T.; Müller, A. Polyoxometalate Chemistry: An Old Field with New Dimensions in Several Disciplines. *Angew. Chem. Int. Ed.* 1991, 30, 34-48.

[50] Yamase T. Photo- and Electrochromism of Polyoxometalates and Related Materials. *Chem. Rev.* 1998, 98, 307-326.

[51] Sanchez, C.; Soler-Illia, G. J. A. A.; Ribot, F.; Lalot, T.; Mayer, C. R.; Cabuil, V. Designed Hybrid Organic-Inorganic Nanocomposites from Functional Nanobuilding Blocks. *Chem. Mater.* 2001, *13*, 3061-3083.

[52] Katsoulis D. E. A Survey of Applications of Polyoxometalates. *Chem. Rev.* 1998, 98, 359-388.

[53] Gouzerh, P. Proust, A. Main-Group Element, Organic, and Organometallic Derivatives of Polyoxometalates. *Chem. Rev.* 1998, 98, 77-112.

[54] Knoth, W. H. Harlow, R. L. Derivatives of heteropolyanions. 3. O-alkylation of $Mo_{12}PO_{40}^{3-}$ and $W_{12}PO_{40}^{3-}$. *J. Am. Chem. Soc.* 1981, *103*, 4265-4266.

[55] Knoth, W. H. Farlee, R. D. Esters of phosphotungstic acid. Anhydrous phosphotungstic acid. *Inorg. Chem.* 1984, *23*, 4765-4766.

[56] Peng, Z. H. Rational Synthesis of Covalently Bonded Organic-Inorganic Hybrids. *Angew. Chem. Int. Ed.* 2004, 43, 930-935.

[57] Kang, H.; Zubieta, J. Co-ordination complexes of polyoxomolybdates with a hexanuclear core: synthesis and structural characterization of $(NBu^n{}_4)_2[Mo_6O_{18}(NNMePh)]$, *Chemical Communications*, 1988, (17), 1192-1193.

[58] Du, Y.; Rheingold, A. L.; Maatta, E. A. A Polyoxometalate Incorporating an Organoimido Ligand: Preparation and Structure of $[Mo_5O_{18}(MoNC_6H_4CH_3)]^{2-}$ *J. Am. Chem. Soc.* 1992, *114*, 345-346.

[59] Hur, N. H.; Klemperer, W. G.; Wang, R.-C. Tetrabutylammonium Hexamolybdate(VI) *Inorg. Synth.* 1990, 27, 77-78.

[60] Allcock, H. R.; Bissell, E. C.; Shawl, E. T. Crystal and molecular structure of a new hexamolybdate-cyclophosphazene complex. *Inorg. Chem.* 1973, 12, 2963-2967.

[61] Hsieh, T.- C.; Zubieta, J. A. Synthesis and Characterization of Oxomolybdate Clusters Containing Coordinatively Bound Organo-Diazenido Units: The Crystal and Molecular Structure of the Hexanuclear Diazenido-Oxomolybdate, $(NBu^n{}_4)_3[Mo_6O_{18}(N_2C_6H_5)]$ *Polyhedron.* 1986, 5, 1655-1657.

[62] Kwen, H.; Young, V. G., Jr.; Maatta, E. A. A Diazoalkane Derivative of a Polyoxometalate: Preparation and Structure of $[Mo_6O_{18}(NNC(C_6H_4OCH_3)CH_3)]^{2-}$ *Angew. Chem. Int. Ed.* 1999, 38, 1145-1146.

[63] Proust, A.; Thouvenot, R.; Chaussade, M.; Robert, F.; Gouzerh, P. Phenylimido Derivatives of $[Mo_6O_{19}]^{2-}$: Syntheses, X-ray Structures, Vibrational, Electrochemical, ^{95}Mo and ^{14}N NMR Studies. *Inorg. Chim. Acta* 1994, 224, 81-95.

[64] Errington, R. J.; Lax, C.; Richards, D. G.; Clegg, W.; Fraser, K. A. New Aspects of Non-Aqueous Polyoxometalate Chemistry: 3. Suface Oxide Reactivity. In: Pope, M. T.; Müller, A. Editors. *Polyoxometalates: From Platonic Solids to Anti-Retroviral Activity.* Dordrecht: Kluwer; 1994; p 112-113.

[65] Strong, J. B.; Ostrander, R.; Rheingold, A. L.; Maatta, E. A. Ensheathing a Polyoxometalate: Convenient Systematic Introduction of Organoimido Ligands at Terminal Oxo Sites in $[Mo_6O_{19}]^{2-}$. *J. Am. Chem. Soc.* 1994, 116, 3601-3602.

[66] Clegg, W.; Errington, R. J.; Fraser, K. A.; Holmes, S. A.; Schäfer, A. Functionalisation of $[Mo_6O_{19}]^{2-}$ with Aromatic Amine: Synthesis and Structure of a hexamolybdate Building Block with Linear Difunctionality. *Chem. Commun.* 1995, 455-456.

[67] Roesner, R. A.; McGrath, S. C.; Brockman, J. T.; Moll, J. D.; West, D. X.; Swearingen, J. K.; Castineiras, A. Mono- and di-functional aromatic amines with *p*-alkoxy substituents as novel arylimido ligands for the hexamolybdate ion. *Inorg. Chim. Acta.* 2003, 342, 37-47.

[68] Wei, Y. G.; Xu, B. B.; Barnes, C. L.; Peng, Z. H. An Efficient and Convenient Reaction Protocol to Organoimido Derivatives of Polyoxometalates. *J. Am. Chem. Soc.* 2001, *123*, 4083-4084.

[69] Stark, J. L.; Rheingold, A. L.; Maatta, E. A. Polyoxometalate clusters as building blocks: preparation and structure of bis(hexamolybdate) complexes covalently bridged by organodiimido ligands. *Chem. Commun.* 1995, 1165-1166.

[70] Stark, J. L.; Young, V. G., Jr.; Maatta, E. A Functionalized Polyoxometalate Bearing a Ferrocenylimido Ligand: Preparation and Structure of $[(FcN)Mo_6O_{18}]^{2-}$. A. *Angew. Chem. Int. Ed.* 1995, 34, 2547-2548.

[71] Strong, J. B.; Yap, G. P. A.; Ostrander, R.; Liable-Sands, L. M.; Rheingold, A. L.; Thouvenot, R.; Gouzerh, P.; Maatta, E. A. A New Class of Functionalized Polyoxometalates: Synthetic, Structural, Spectroscopic, and Electrochemical Studies of Organoimido Derivatives of $[Mo_6O_{19}]^{2-}$. *J. Am. Chem. Soc.* 2000, *122*, 639-649.

[72] Moore, A. R.; Kwen, H.; Beatty, A. M.; Maatta, E. A. Organoimido-polyoxometalates as polymer pendants. *Chem. Commun.* 2000, 1793-1794.

[73] Xu, B. B.; Peng, Z. H.; Wei, Y. G.; Powell, D. R. Polyoxometalates covalently bonded with terpyridine ligands. *Chem. Commun.* 2003, 2562-2563.

[74] Zhu, Y.; Wu, P. F.; Ge, N.; Wei, Y. G.; Wang, Y. The First Naphthylimido Derivatives of the hexamolybdate: Synthesis and Characterization of $(Bu_4N)_2[Mo_6O_{18}(NC_{10}H_7\text{-}1)]$. Submitted to publication.

[75] Wu, P. F.; Li, Q.; Ge, N.; Wei, Y. G.; Wang, Y.; Wang, P.; Guo, H. Y. An Easy Routine to Monofunctionalized Organoimido Derivatives of the Lindqvist Hexamolybdate. *Eur. J. Inorg. Chem.* 2004, 2819-2822.

[76] Li, Q.; Wu, P. F.; Wei, Y. G.; Xia, Y.; Wang, Y.; Guo, H. Y. Organic-Inorganic Hybrids: Preparation and Structural Characterization of $(Bu_4N)_2[Mo_6O_{17}(NAr)_2]$ and $(Bu_4N)_2[Mo_6O_{18}(NAr)]$ (Ar = o-$CH_3C_6H_4$). *Z. Anorg. Allg. Chem.* 2005, 631, in Press.

[77] Li, Q.; Wu, P. F.; Wei, Y. G.; Wang, Y.; Wang, P.; Guo, H. Y. An organic-inorganic hybrid of polyoxoanion cluster incorporating para-bromophenylimido ligand:

Preparation, spectroscopic studies and crystal structure of $(Bu_4N)_2[Mo_6O_{18}(NC_6H_4Br$-p)]. Submitted to publication.

[78] Xue, S. J.; Ke, S. Y.; Wei, Y. G. Synthesis, spectroscopic characterization and crystal structure of an arylimido derivative of with electron-withdrawing fluoro substituents. Unpublished result.

[79] Xu, L.; Lu, M.; Xu, B. B.; Wei, Y. G.; Peng, Z. H.; Powell, D. R. Towards Main-Chain-Polyoxometalate-Containing Hybrid Polymers: A Highly Efficient Approach to Bifunctionalized Organoimido Derivatives of Hexamolybdates. *Angew. Chem. Int. Ed.* 2002, 41, 4129-4132.

[80] Li, Q.; Wu, P. F.; Wei, Y. G.; Wang, Y.; Wang, P.; Guo, H. Y. Synthesis, structure and supramolecular assembly in the crystalline state of a bifunctionalized arylimido derivative of hexamolybdate. *Inorg.Chem. Commun.* 2004, 7, 524-527.

[81] Lu, M.; Xie, B.; Kang, J.; Chen, F.-C.; Yang, Y.; Peng, Z. H. Synthesis of Main-Chain Polyoxometalate-Containing Hybrid Polymers and Their Applications in Photovoltaic Cells. *Chem. Mater.* 2005, *17*, 402-408.

[82] Xia, Y.; Wu, P. F.; Wei, Y. G.; Wang, Y.; Guo, H. Y. A Hybrid Molecular Semiconductor: Synthesis and Structural Characterization of the Bifunctionalized Organoimido Derivative of Hexamolybdate, $(n\text{-}Bu_4N)_2[Mo_6O_{17}(\equiv NAr)_2]$ (Ar = o-$CH_3OC_6H_4$). Submitted to publication.

[83] Xia, Y.; Wu, P. F.; Wei, Y. G.; Guo, H. Y. Synthesis, Spectroscopic and Structure of an Unexpected Trans- Disubstituted Arylimido Derivative of the Hexamolybdate, $(n\text{-}Bu_4N)_2[Mo_6O_{17}(\equiv NAr)_2]$ (Ar = 2,6-$(CH_3)_2C_6H_4$). Submitted to publication.

[84] Strong, J. B.; Haggerty, B. S.; Rheingold, A. L.; Maatta, E. A. A Superoctahedral Complex Derived from a Polyoxometalate: the Hexakis(arylimido)hexamolybdate Anion $[Mo_6(NAr)_6O_{13}H]^-$. *Chem. Commun.* 1997, 1137-1138.

[85] Xia, Y.; Wu, P. F.; Wei, Y. G.; Guo, H. Y. The Unprecedent Bridging Oxo Replacement in a Trifunctionalized Organoimido Derivative of the Hexmolybate, $[Mo_6O_{16}(\equiv NAr)_2(\mu_2\text{-}NAr)]^{2-}$ (Ar = 2,6-$(CH_3)_2C_6H_4$). Submitted to publication.

[86] Johnson, A. W. *Ylides and Imines of Phosphorus*. Wiley: New York; 1993.

[87] Saunders, J. H.; Slocombe, R. J. The Chemistry of the Organic Isocyanates. *Chem. Rev.* 1948, *43*, 203-218.

[88] Patrick G. Section J– Aldehydes and Ketones: J6. Nucleophilic Addition– Nitrogen Nucleophiles. *Instant Notes in Organic Chemistry*. Reprint Edition; BIOS Scientific Publishers Limited: Beijing; 2002; p 183-184.

[89] Williams, A.; Ibrahim, I. T. Carbodiimide Chemistry: Recent Advances. *Chem. Rev.* 1981, *81*, 589-636.

[90] Kurzer, F.; Douraghi-Zadeh, K. Advances in the Chemistry of Carbodiimides. *Chem. Rev.* 1967, *67*, 107-152.

[91] Khorana, H. G. the Chemistry of Carbodiimides. *Chem. Rev.* 1953, *53*, 145-166.

[92] Nugent, W.; Mayer, J. E. *Metal-Ligand Multiple Bonds*. Wiley: New York; 1988; p122-125.

[93] Yan, L.-K.; Su, Z.-M.; Guan, W.; Zhang, M.; Chen, G.-H.; Xu, L.; Wang, E.-B. Why Does Disubstituted Hexamolybdate with Arylimido Prefer to Form an Orthogonal Derivative? Analysis of Stability, Bonding Character, and Electronic Properties on Molybdate Derivatives by Density Functional Theory (DFT) Study. *J. Phys. Chem. B.* 2004, *108*, 17337-17343.

[94] Mohs, T. R.; Yap, G. P. A.; Rheingold, A. L.; Maatta, E. A. An Organoimido Derivative of the Hexatungstate Cluster: Preparation and Structure of $[W_6O_{18}(NAr)]^{2-}$ (Ar = 2,6-(i-Pr)$_2$C$_6$H$_3$). *Inorg. Chem.* 1995, *34*, 9-10.

[95] Wei, Y. G.; Lu, M.; Cheung, C. F.- C.; Barnes, C. L.; Peng, Z. H. Functionalization of $[MoW_5O_{19}]^{2-}$ with Aromatic Amines: Synthesis of the First Arylimido Derivatives of Mixed-Metal Polyoxometalates. *Inorg. Chem.* 2001, *40*, 5489-5490.

[96] Proust, A.; Taunier, S.; Artero, V.; Robert, F.; Thouvenot, R.; Gouzerh, P. The Unexpected Reactivity of *p*-Tolylisocyanate towards the Keggin Anion α-$[PMo_{12}O_{40}]^{3-}$. *Chem. Commun.* 1996, 2195 – 2196.

[97] Dablemont, C.; Proust, A.; Thouvenot, R.; Afonso, C.; Fournier, F.; Tabet, J.-C. Functionalization of Polyoxometalates: From Lindqvist to Keggin Derivatives. 1. Synthesis, Solution Studies, and Spectroscopic and ESI Mass Spectrometry Characterization of the Rhenium Phenylimido Tungstophosphate $[PW_{11}O_{39}\{ReNC_6H_5\}]^{4-}$. *Inorg. Chem.* 2004, *43*, 3514-3520.

[98] Xu, B. B.; Wei, Y. G.; Barnes, C. L.; Peng, Z. H. Hybrid Molecular Materials Based on Covalently Linked Inorganic Polyoxometalates and Organic Conjugated Systems. *Angew. Chem. Int. Ed.* 2001, 40, 2290-2292.

[99] Lu, M.; Wei, Y. G.; Xu, B. B.; Cheung, C. F.- C.; Peng, Z. H.; Powell, D. R. Hybrid Molecular Dumbbells: Bridging Polyoxometalate Clusters with an Organic π-Conjugated Rod. *Angew. Chem. Int. Ed.* 2002, 41, 1566-1568.

[100] Kang, J.; Nelson, J. A.; Lu, M.; Xie, B.; Peng, Z. H.; Powell, D. R. Charge-Transfer Hybrids Containing Covalently Bonded Polyoxometalates and Ferrocenyl Units. *Inorg. Chem.* 2004, *43*, 6408-6413.

[101] Sonogashira, K.; Tohda, Y.; Hagihara, N. Convenient synthesis of acetylenes. Catalytic substitutions of acetylenic hydrogen with bromo alkenes, iodo arenes, and bromopyridines. *Tetrahedron Lett.* 1975, 50, 4467-4470.

[102] deSousa, P. T. The Palladium-catalysed Cross Coupling Reaction of Acetylenic Compounds with Aryl Halides and Related Compounds. *Quimica Nova.* 1996, 19, 377-382.

[103] Wang, Y. F.; Deng, W.; Liu, L.; Guo, Q. X. Recent Progress in Sonogashira Reaction. *Chin. J. Org. Chem.* 2005, 25, 8-24.

[104] Ziener, U.; Godt, A. Synthesis and Characterization of Monodisperse Oligo(phenyleneethynylene)s. *J. Org. Chem.* 1997, 62, 6137-6143.

[105] Sadakane, M.; Steckhan, E. Electrochemical Properties of Polyoxometalates as Electrocatalysts. *Chem. Rev.* 1998, 98, 219-238.

[106] Coronado, E.; Gómez-García, C. J. Polyoxometalate-Based Molecular Materials. *Chem. Rev.* 1998, 98, 273-296.

[107] Gust, D.; Moore, T. A.; Moore, A. L. Molecular Mimicry of Photosynthetic Energy and Electron Transfer. *Acc. Chem. Res.* 1993, 26, 198-205.

[108] Wasielewski, M. R. Photoinduced Electron Transfer in Supramolecular Systems for Artificial Photosynthesis. *Chem. Rev.* 1992, 92, 435-461.

[109] Weiss, E. A.; Ahrens, M. J.; Sinks, L. E.; Gusev, A. V.; Ratner, M. A.; Wasielewski, M. R. Making a Molecular Wire: Charge and Spin Transport through *para*-Phenylene Oligomers. *J. Am. Chem. Soc.* 2004, *126*, 5577-5584.

[110] Xu, B. B.; Lu, M.; Kang, J.; Wang, D. G.; Brown, J.; Peng, Z. H. Synthesis and Optical Properties of Conjugated Polymers Containing Polyoxometalate Clusters as Side-Chain Pendants. *Chem. Mater.* 2005, *17*, 2841-2851.

In: Advances in Organometallic Chemistry Research
Editor: Kenji Yamamoto, pp. 139-175
ISBN: 978-1-60021-779-1

Chapter 6

STRUCTURAL AND BIOLOGICAL CHEMISTRY OF ORGANOTIN(IV) COMPLEXES

Saqib Ali[1], **Saira Shahzadi***[1] and **Moazzam H. Bhatti***[2]*
[1] Department of Chemistry, Quaid-i-Azam University, Islamabad-45320 Pakistan
[2] Department of Chemistry, Allama Iqbal Open University, Islamabad, Pakistan

ABSTRACT

The amazing structural diversity in organotin compounds is discussed in the systems containing Sn-O and Sn-S donor ligands. It is demonstrated that there exist a fascinating range of structural diversity for organotin(IV) complexes, including difference in coordination number and molecular geometry. Difference in the structure is correlated with the nature of tin and ligand bonded R groups. Despite the large number of different structures found in organotin(IV) carboxylates, there is limited range of coordination geometries about the Sn atom. Thus, the four coordinated Sn atom in triorganotin(IV) complexes is invariably distorted tetrahedral, five coordinate Sn is distorted trigonal bipyramidal. A large range has been observed for diorganotin carboxylates structures, where five-, six- and seven- coordinate geometries have been reported. The Sn atom in mono-organotin has only been demonstrated to exist in distorted octahedral geometries (the single exception being a pentagonal bipyramidal geometry).

From the discussion of the structural chemistry of organotin(IV) complexes of S donor ligands, it is clear that there exists a rich diversity in Sn atom geometries and coordination modes of the sulfur donor ligands themselves. As in related carboxylates systems, the assignment of coordination numbers to the Sn centers in some compounds is controversial.

Nevertheless, some general trends are discussed with the most obvious being that as the number of organic substituents at the Sn atom increases the overall coordination number at the Sn atom decreases. This phenomenon is usually achieved by increased asymmetry in the mode of coordination of the sulfur donor ligands.

---

* Prof. Dr. Saqib Ali : drsa54@yahoo.com
* Dr. Saira Shahzadi : sairashahzadi@yahoo.com
* Dr. Moazzam H. Bhatti: moazzamhussain_b@yahoo.com

A theoretical explanation has been proposed, based on valence bond theory, which suggests, in effect that as the number of tin-bound organic substituents increases dative resonance forms of the sulfur donor ligands predominate which leads to an increased asymmetry in the mode of coordination of the sulfur donor ligand in accordance with the experimental observations.

Knowledge of the biological chemistry of the tin is needed for the rational development of more effective compounds and newer uses, for example as antitumour agents. This chapter is particularly concerned with the molecular interactions of tin compounds with enzymes and biological membranes and with the metabolic processing of tin compounds. It deals with biochemical activities of organotin compounds leading from complex formation, through effects on individual enzymes and biological membranes to their inhibition of ATP synthesis and finally to their actions on whole cell systems in the nervous system, in, immune suppression and as antitumour agents.

INTRODUCTION

The increasing interest in organotin(IV) carboxylates of substituted benzoic acids in recent years has, to a large extent, been prompted by their new structural diversity [1] and broad therapeutic activity [2]. Information on the structure of organotin(IV) carboxylates continues and at the same time some new applications of such a high importance are being discovered which are relevant to ecological medicinal applications. The increasing interest in the chemistry of organotin (IV) compounds has led to the extended studies on their reactions with different biomolecules e.g., carbohydrates [3-5], nucleic acid derivatives [6-8], amino acids [9-11] and peptides [12-14].

In general triorganotin(IV) compounds display a large array of biological activity than their diorganotin and monoorganotin analogues. This has been attributed to their ability to bind proteins [15-17]. Further more many organotin(IV) carboxylates have been found to posses anticancer activity in a variety of tumor cells and the structure of these organotin(IV) compounds have been characterized both in solid and solution forms [18]. Dialkytin(IV) compounds have selective effect on lymphocytes [19] which can be used in cancer chemotherapy. Among diorganotin(IV) compounds tetra-organodistannoxanes are an important class of compounds owing to applications in catalysts [20] and their interesting biological activity [21,22].

The aim of this chapter is to summarize the each category of organotin carboxylates in order to illustrate the diverse modes of coordination/association of the carboxylate ligand/dithiocarbamate ligand as determined by crystallographic methods and relate it to a common structural motif and biological chemistry of organotin(IV) complexes.

RESULTS AND DISCUSSION

We will discuss the structural chemistry of organotin(IV) complexes under the two categories:

1) Oxygen donor ligands

2) Sulfur donor ligands

OXYGEN DONOR LIGANDS

Oxygen donor ligands coordinates to the organotin moiety in monodentate or bidentate fashion in different stochiometry giving the diversity of structures. We will discuss them as:

1. 4-Coordinated

When the oxygen donor ligands acts as monodentate then the geometry around the tin atom is distorted tetrahedral as reported in the literature [23-25]. Representative structure of this class is given in figure 1, which shows that, the Sn atom exists in a distorted tetrahedral geometry defined by three *ipso*-C atoms of the phenyl groups and O1 atom [Sn1-O1 = 2.0557(14) Å] of the (E)-3-(3-fluorophenyl)-2-(4-chlorophenyl)-2-propenoate. The major distortion from the ideal tetrahedral geometry is found in the O1-Sn1-C1 angle of 96.54(7)°. The relatively close interaction between O2 and Sn1 [Sn1...O2 = 2.8309(5) Å], does not disrupt the O1-Sn1-C7 and O1-Sn1-C13 angles significantly, which are 108.91(7)° and 110.10(7)°, respectively. However it is noteworthy that C7-Sn1-C13 angle of 119.06(8)° in the next major distortion from the ideal geometry. The monodentate mode of coordination of (E)-3-(3-fluorophenyl)-2-(4-chlorophenyl)-2-propenoate is reflected in the disprate O1-C19 and O2-C19 bond distances of 1.313(2) and 1.229(2) Å, respectively, with the longer separation being associated with the stronger Sn1-O1 interaction [23].

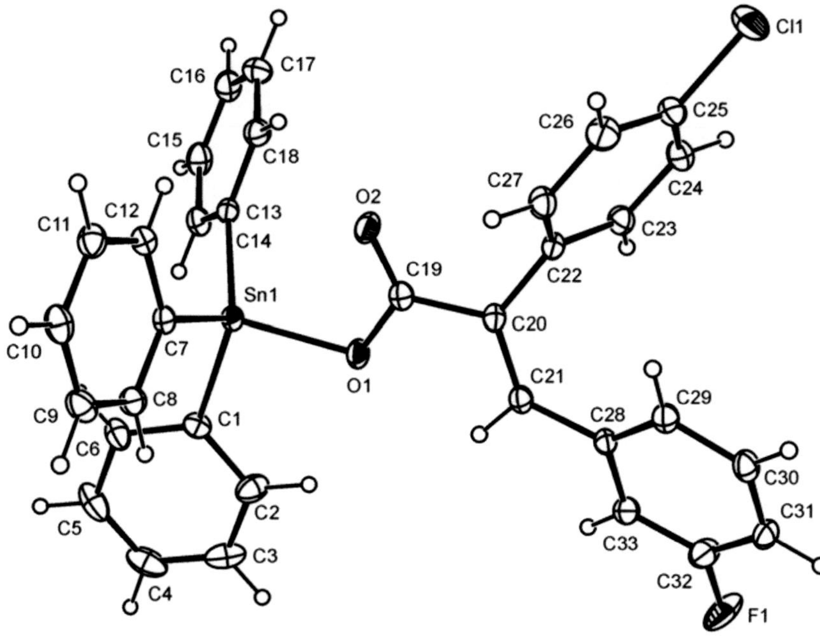

Figure 1. The structure of triphenyltin(IV)[3-(3-fluoro-phenyl)-2(4-chlorophenyl)-2 propenoate] [23].

Another example of this class is given in figure 2 in which also the distorted tetrahedral geometry is observed with space group P21/c [24].

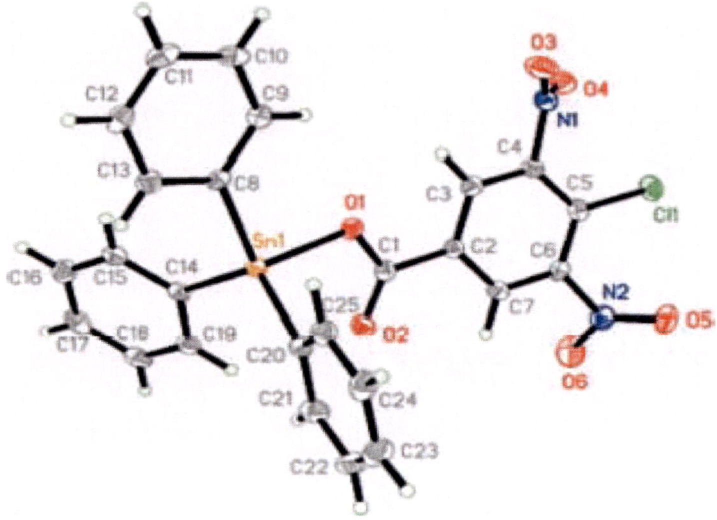

Figure 2. The structure of (4-chloro-3,5-dinitrobenzoato)triphenyltin(IV) [24].

2. 5-Coordinated

Among the organotin(IV) complexes, trimethyltin derivatives forms the polymeric structure with trigonal bipyramidal geometry [26-29]. The representative structure of this class is given in figure 3. The structure (figure 3) shows two independent monomers in the asymmetric unit, forming polymeric chains lying parallel to the *b* axis that are separated by one half unit cell length along *a* axis. In each polymeric chain, both O atoms of the acid moiety are bonded to Sn atoms with significantly different Sn-O distances, *viz*, Sn1-O1 2.129(2) and Sn1-O2 2.578(3) Å for one, Sn2-O5 2.159(2) and Sn2-O6 2.508(3) Å for the other polymer [28]. The shorter Sn-O distances are indicative of covalent bonds and longer distances clearly represent dative bonds. It is interesting to note that the shorter and longer Sn-O distances in two polymers are significantly different from each other as well. In both polymer chains (figure 4), the O-Sn-O angles are close to linear [170.46(9) and 171.92(9)°], the C-Sn-C angles lie between 111.22(17) and 123.91(16)°, deviating from ideal value of 120°.

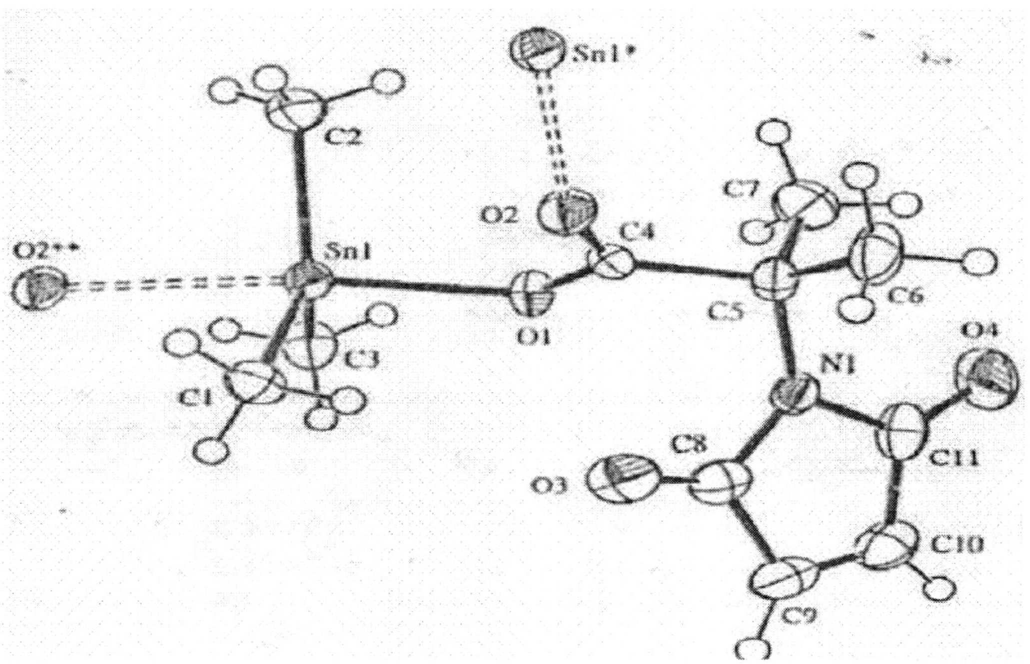

Figure 3. The structure of polymeric[2-(N-maleoyl-2-methyl propionato]trimethyltin(IV) [28].

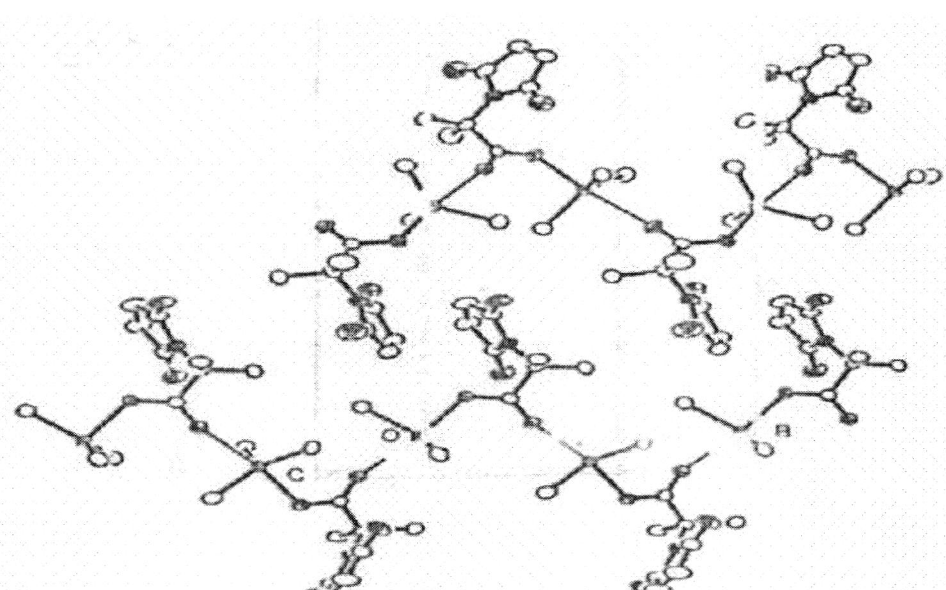

Figure 4. Two polymeric chains of polymeric[2-(N-maleoyl-2-methyl propionato]trimethyltin(IV) [28].

Literature [30-33] have shown that coordination of different alkyl groups with O-donor ligands does not change the coordination geometry of triorganotin complexes and they show the polymeric structure with the trigonal bipyramidal geometry. Examples are given in figure 5 [30] and figures 6 and 7 [31].

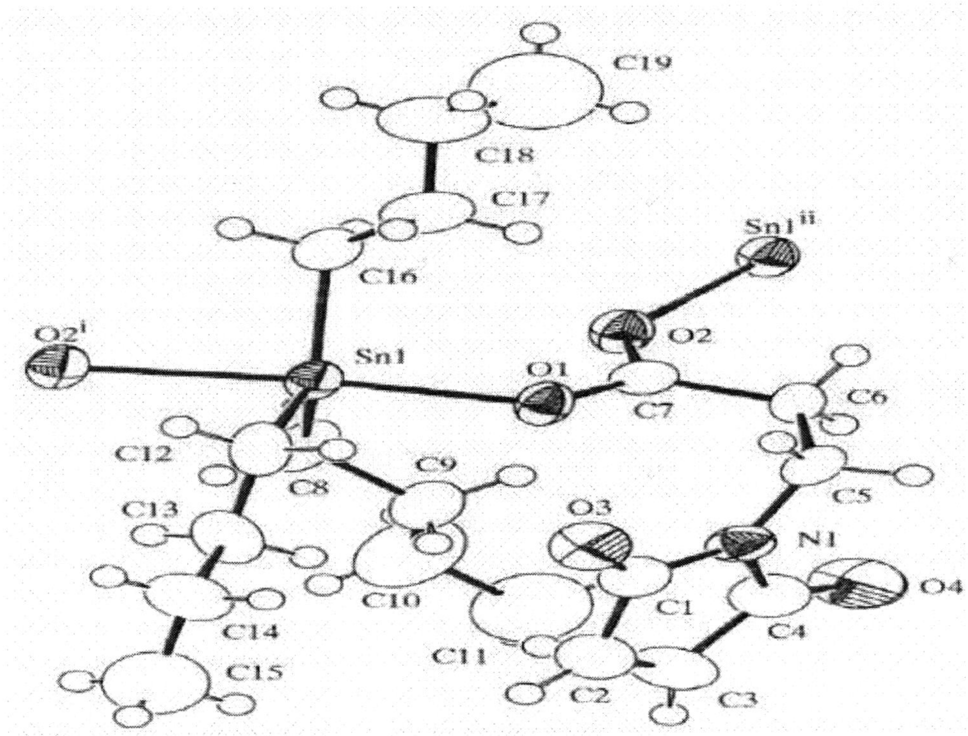

Figure 5. The structure of tri-*n*-butyl(N-maleoyl-β-alaninato)tin [30].

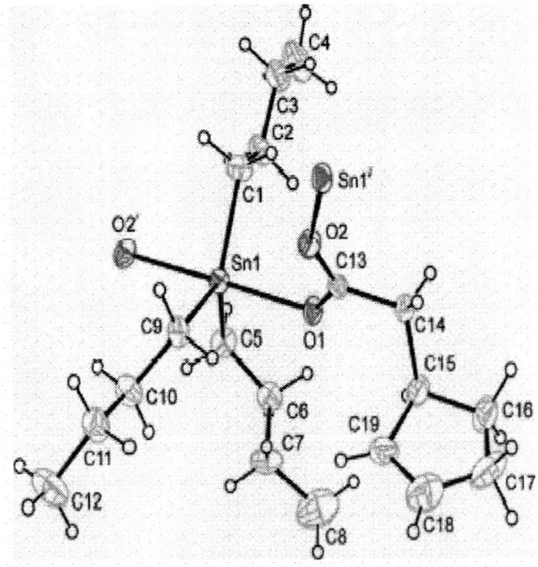

Figure 6. The structure of *catena*-poly[[tri-*n*-butyltin(IV)]-μ-cyclopent-2-enylacetato- O:O] [31].

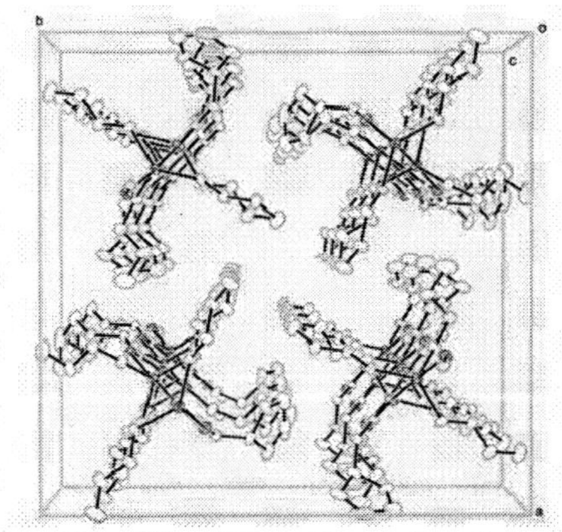

Figure 7. The structure of catena-poly[[tri-n-butyltin(IV)]-μ-cyclopent-2- enylacetato-O:O] [31].

The tetrameric structure figure 8 is also reported in the literature [34] which is formed with tributyltin moiety. The structure forms an isolated cyclotetrameric 16-membered ring that is located about crystallographic centre of inversion so that two independent *n*-Bu_3SnL units comprise the asymmetric unit. This aggregation arises as the carboxylate ligands are bidentate bridging, albeit forming asymmetric bond distances of approximately 2.2 and 2.6 Å. The individual tin atom coordination geometry is based on the *trans*-R_3SnO_2 trigonal bipyramid with the above mentioned disparity in the Sn-O bond distances not withstanding. The axial angle of 171.19(10)° for the Sn1 atom is more significantly distorted from the ideal angle compared with Sn2 atom, for which the equivalent angle is 177.62(11)°. There are three $R_3Sn(O_2CR`)$ structures reported in the literature [34-36] that adopt tetrameric motif. The rationalization of the appearance of these cyclotetrameric structures, which may be thought of as being intermediate between monomeric and polymeric extremes described above, remains elusive.

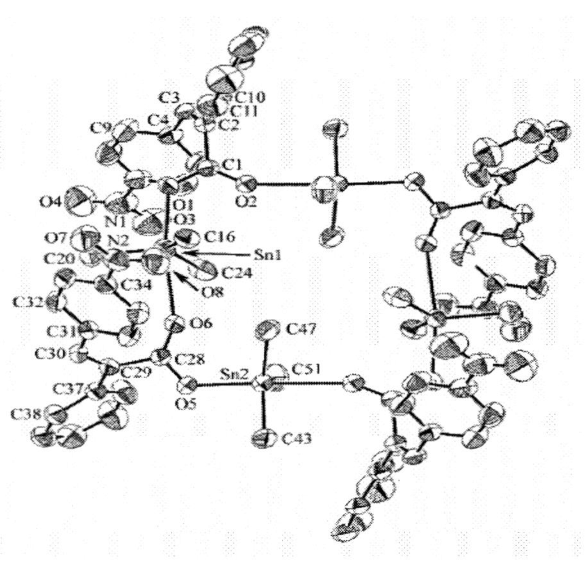

Figure 8. The structure of *catena*-poly[[tri-*n*-butyltin(IV)]-μ-cyclopent-2- enylacetato-O:O] [34].

3. 6-Coordinated

In organotin complexes, distorted octahedral geometry results by asymmetrically coordination of the carboxylate ligands with Sn atom. Representative structure of this class is given in figure 9 [37]. The structure is composed of discrete monomeric molecules in which six-coordinated Sn atoms are bonded to two *n*-butyl groups and two 2,3-bis(4-chlorophenyl)propenoates ligands. The geometry around the Sn atom is highly distorted octahedral and can be best described as a skew-trapezoidal planar geometry with two additional axial ligands. The Sn atom lies 0.190(2) Å out of the plane formed by asymmetrically bonded O atoms of the carboxylate ligands, while the two *n*-butyl groups lie above and below this plane. The mean Sn-C distance of 2.117(13) Å and a C-Sn-C angle of 139.18(18)° is observed. The carboxylate ligands are asymmetrically coordinated to the Sn atoms, with Sn-O covalent bonds [mean = 2.098(6) Å] that are significantly shorter than the dative bonds [mean = 2.602(12) Å].

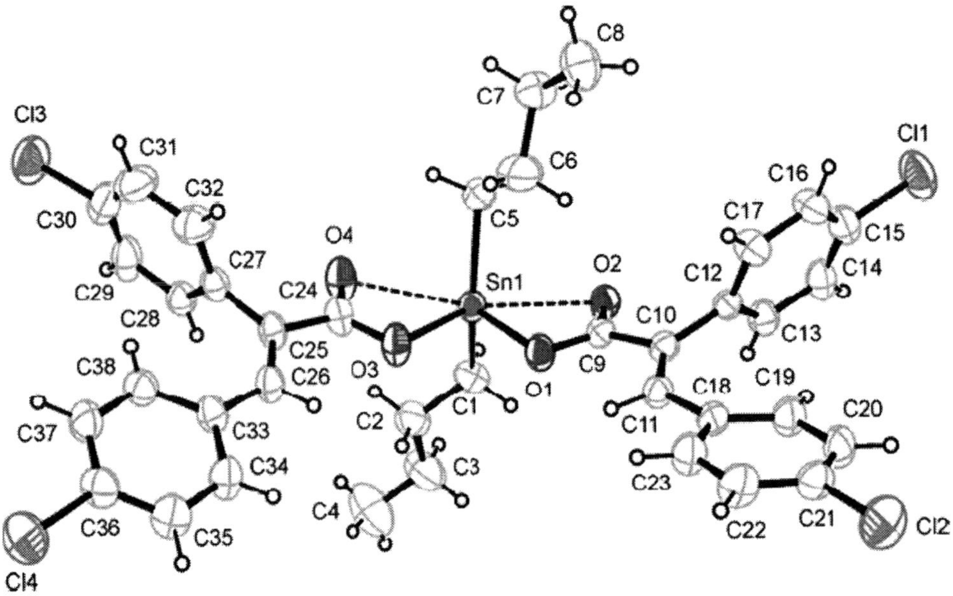

Figure 9. The structure of bis[2,3-bis(4-chlorophenyl)propenoato-O,O/] di-*n*-butyltin(IV) [37].

Distorted octahedral geometry was observed by coordination of different R groups with O-donor ligands. Below is the example of this class with dimethyltin moiety figure 10 [38].

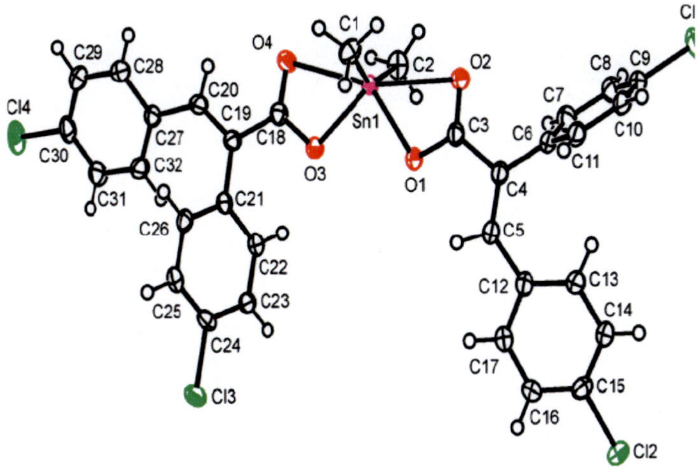

Figure 10. The structure of bis[2,3-bis(4-chlorophenyl)propenoato-O,O/]-dimethyltin(IV) [38].

Distannoxanes

Within the diorganotin complexes, distannoxanes are reported in the literature with trigonal bipyramidal geometry [39-42]. These distannoxanes observed with different alkyl groups e.g., ethyl [39] (figure 11), methyl [40] (figure 12) and butyl group [41,42] (figures 13 and 14). Figure 11 shows that complex is composed of two independent centrosymmetric dimmers lying about inversion centres. In each dimmer, the central Sn_2O_2 core is fused with two four membered (Sn_2O_2) and two six membered Sn_2O_2C rings. The endocyclic Sn atoms are six coordinate in a skew trapezoidal bipyramidal arrangement. The exocyclic Sn atoms are five coordinate and show distorted trigonal bipyramidal geometry. The cyclopentenylacetate ligand shows different modes of coordination with tin. In both dimmers, the Sn-C distances lie in a very narrow range [2.118(4)-2.134(4) Å], while the Sn-O distances range between 2.042(2) and 2.314(3) Å for strong bonds and between 2.683(3) and 2.658(3) Å for relatively weaker Sn-O bonds.

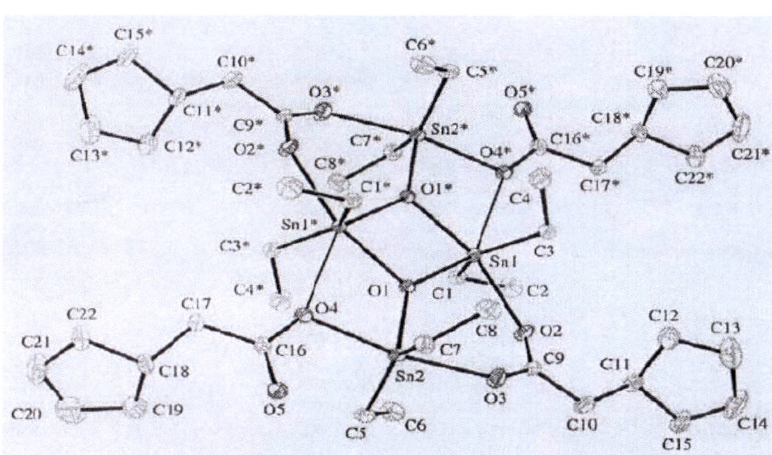

Figure 11. The structure of bis(cyclopentenylacetato)tetraethyldistannoxane dimmer [39].

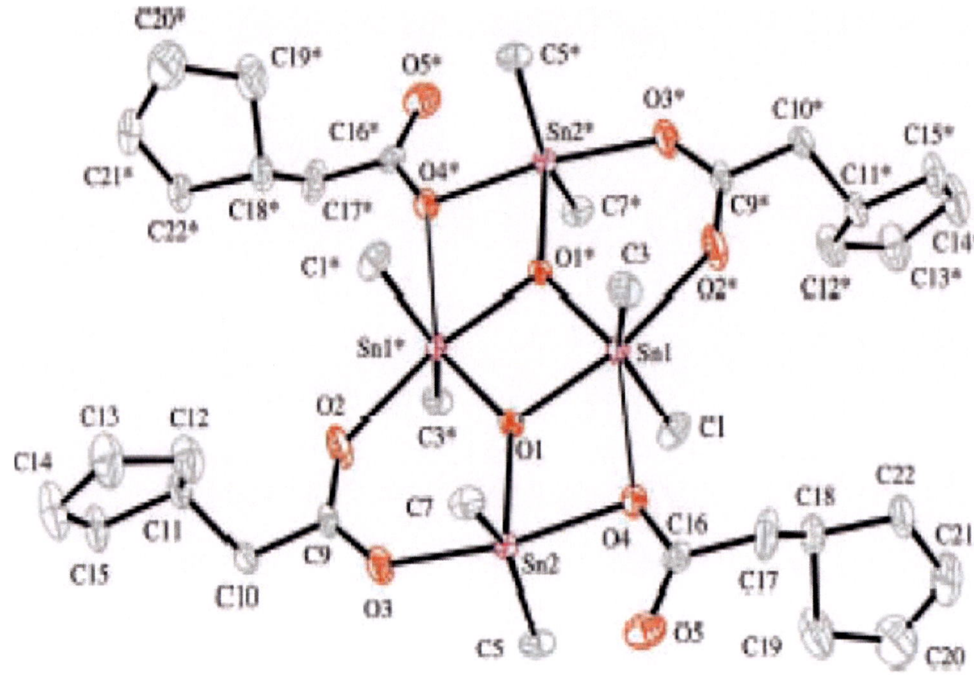

Figure 12. The structure of bis(cyclopentenylacetato)tetramethyldistannoxane dimmer [40].

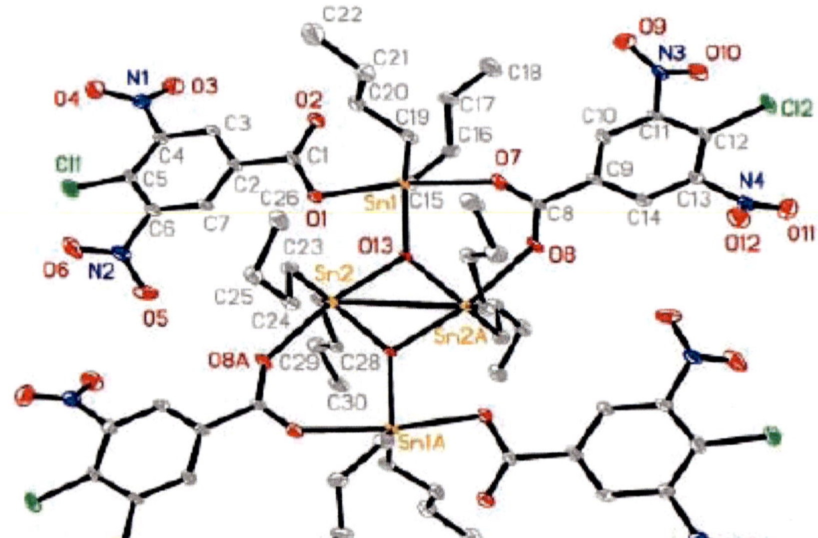

Figure 13. The structure of octa-*n*-butylbis(μ2-4-chloro-3,5-dinitrobenzoato O:O/)bis(4-chloro-3,5-dinitrobenzoato-O)di-μ3-oxo-tetratin(IV) [41].

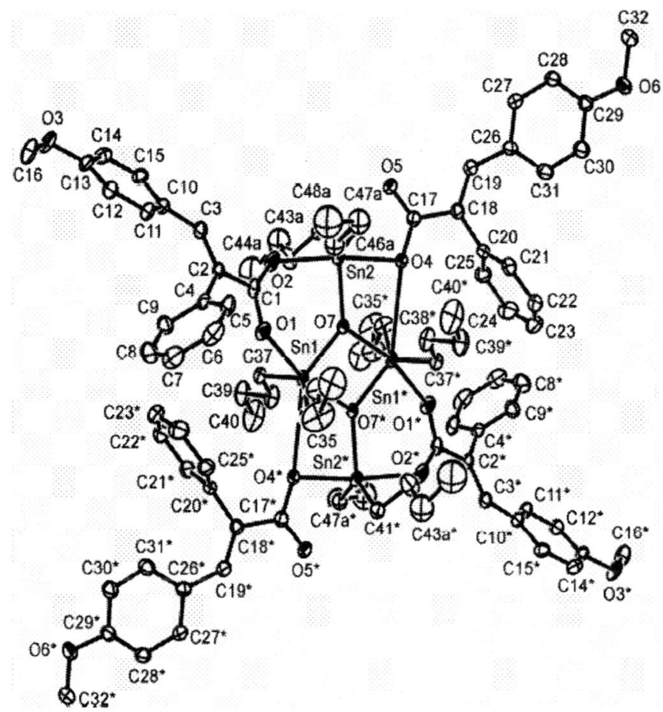

Figure 14. The structure of tetrabutylbis[3-(4-methoxyphenyl)-2-phenyl-2 propenoato]distannoxane dimmer [42].

SULFUR DONOR LIGANDS

It was demonstrated that seemingly closely related compounds may adopt different structures even though the only difference between the compounds was in the nature of the tin or carboxylate-bound R groups.

1. 4- Coordinated

Representative compound of this class with distorted tetrahedral geometry is shown in figure 15 [43]. The mean Sn-C [2.135 Å] and Sn-S bond lengths [2.148 Å] are comparable with those in related structures such as Ph_3SnSPh [44]. The distortion in tetrahedral geometry is evidence by the spread of angles about the central metal. Tetragonal geometry is also observed in organotin complexes with S donor ligands. Structure is given in figure 16 [45]. As shown in the figure 16, the two carbamates are monodentate, coordinating to the Sn atom through S1 and S3.The bond distances of S atoms 2 and 4 with Sn are 2.919(14) and 2.536(11) Å, which are two long to be strong covalent bonds. The C15-Sn-C19 linkage is not linear, having an angle of 135.3(16)°, which is much larger than the value expected for a regular tetrahedron. The coordination geometry is best described as distorted tetragonal. Another important distortion is caused by the asymmetric Sn-S bond lengths. Therefore, the angle S1-Sn-S3 of 83.03(14)°, is not consistent with true tetrahedral geometry, but instead with distorted tetragonal geometry. As a result, the Sn atom exists in distorted tetragonal geometry in which basal plane is defined by four S atoms, and the axial positions are occupied by two butyl substituents.

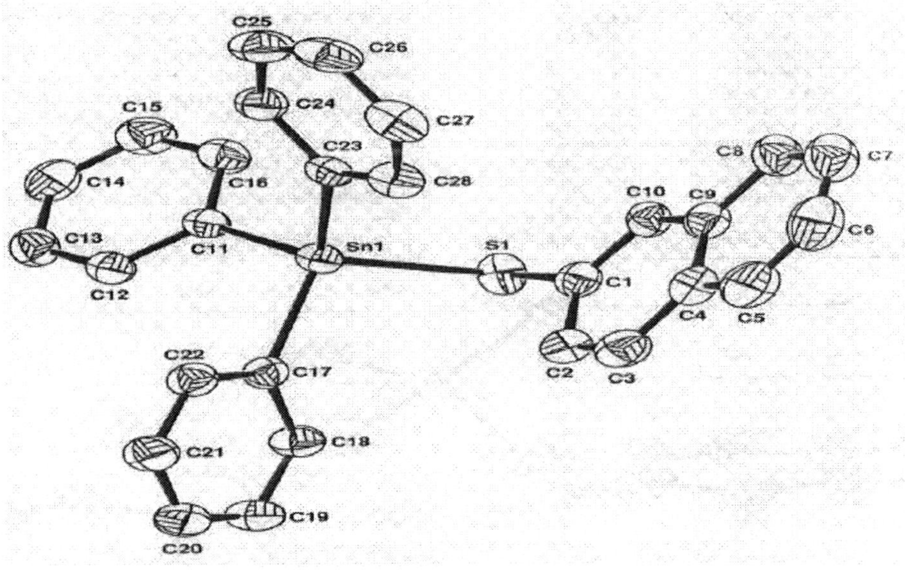

Figure 15. The structure of (naphthalenethiolato-S)triphenyltin(IV) [43].

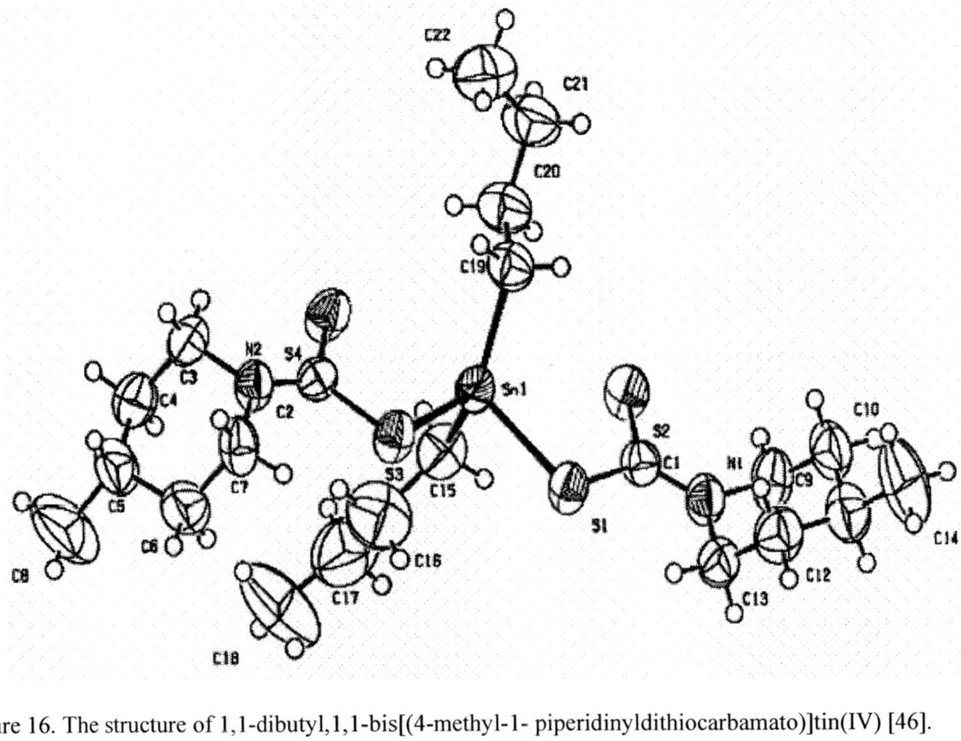

Figure 16. The structure of 1,1-dibutyl,1,1-bis[(4-methyl-1- piperidinyldithiocarbamato)]tin(IV) [46].

2. 5- Coordinated

Figures 17 and 18 show the molecular structure of this class of complexes [47]. The intermolecular bond distances and angles are given in tables 1 and 2. From both, figures and tables it can be seen that the coordination geometry about tin atom is penta-coordinated distorted trigonal bipyramidal with two R groups attach to tin and S(1)/(S2) occupying equatorial positions while Cl(1) and S(2)/S(1) occupy the apical position. In this way the ligand behaves as a bidentate species and chelates the tin atom by means of sulfur atoms [48]. Due to being apart of chelate the angle S(1)-Sn-S(2) is not 90^0 but only $68.25(2)^0$, so the S(2) cannot occupy exactly the corresponding trans apical position of Cl(1) and the angle between the apical groups is 154.19(3) . The sum of the equatorial angles is $358.37(10)^0$ showing some distortion from ideal bond angle of 360^0. In complexes, the one of Sn-S bond length (Sn-S(1) 2.466(7) Å) is shorter than the other (Sn-S(2) 2.739(9) Å) suggesting the unsymmetrical coordination of the ligand. Further the shorter Sn-S bond lengths are very close to the sum of the covalent radii of tin and sulfur and the longer Sn-S distances are significantly less than the sum of van der waals radii (4.0 Å) [49]. The Sn-C distances are (Sn-C(1) 2.109(3) and Sn-C(2) 2.112(3) Å) The Sn-Cl bond lengths (Sn-Cl(1) 2.485(8) Å) lies in the range of the normal covalent radii, 2.37–2.60 Å [50].

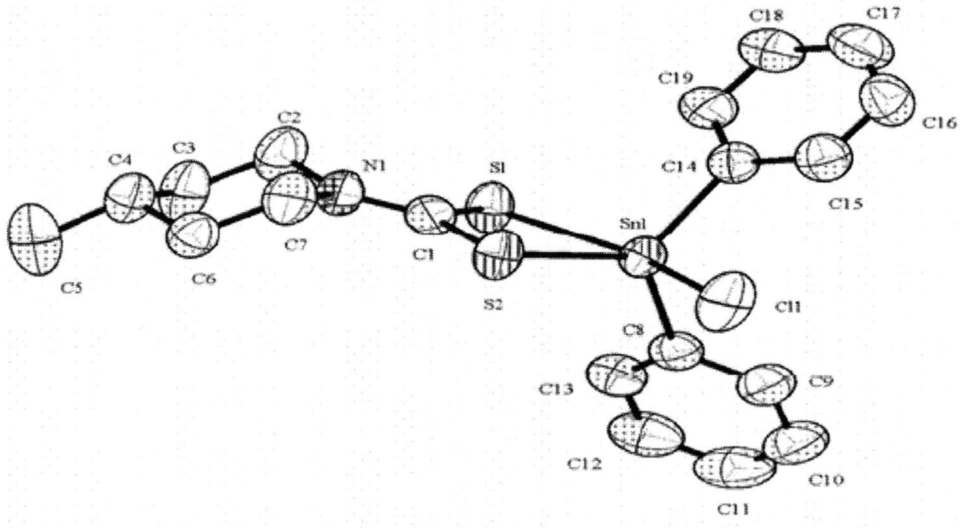

Figure 17. The structure of chlorodiphenyltin(4-methyl-1-piperidine dithiocarboxylato) [47].

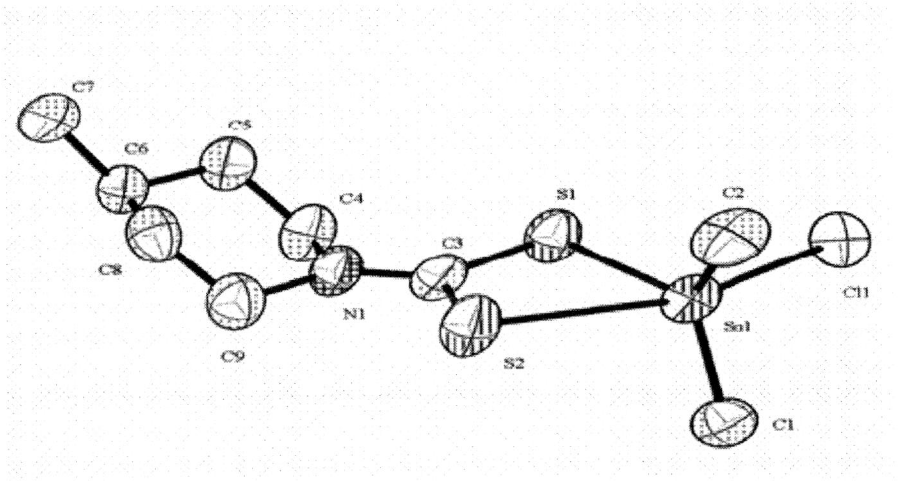

Figure 18. The structure of chlorodimethyltin(4-methyl-1-piperidine dithiocarboxylato)[47].

Table 1. Selected bond lengths (Å) and bond angles (°) for chlorodiphenyltin(4-methyl-1-piperidinedithiocarboxylato) and chlorodimethyltin(4-methyl-1-piperidine dithiocarboxylato) [47]

chlorodiphenyltin(4-methyl-1-piperidinedithiocarboxylato)		chlorodimethyltin(4-methyl-1-piperidine dithiocarboxylato)	
Bond lengths			
Sn(1)–C(1)	2.109 (3)	Sn(1)–C(8)	2.129 (2)
Sn(1)–C(2)	2.112 (3)	Sn(1)–C(14)	2.137 (2)
Sn(1)–S(1)	2.466 (7)	Sn(1)–S(2)	2.461 (6)
Sn(1)–Cl(1)	2.485 (8)	Sn(1)–Cl(1)	2.467 (6)
Sn(1)–S(2)	2.739 (9)	Sn(1)–S(1)	2.657 (6)
S(1)–C(3)	1.743 (3)	S(1)–C(1)	1.720 (2)
S(2)–C(3)	1.697 (3)	S(2)–C(1)	1.745 (2)
N(1)–C(3)	1.317 (3)	N(1)–C(1)	1.302 (3)
N(1)–C(9)	1.469 (4)	N(1)–C(2)	1.474 (3)
N(1)–C(4)	1.471 (3)	N(1)–C(7)	1.481 (3)
		Bond angles	
C(1)–Sn(1)–C(2)	126.88 (14)	C(8)–Sn(1)–C(14)	114.63 (8)
C(1)–Sn(1)–S(1)	112.96 (9)	C(8)–Sn(1)–S(2)	122.21 (6)
C(2)–Sn(1)–S(1)	118.53 (10)	C(14)–Sn(1)–S(2)	122.75 (6)
C(1)–Sn(1)–S(2)	94.76 (9)	C(8)–Sn(1)–C(11)	96.76 (6)
C(2)–Sn(1)–S(2)	92.50 (11)	C(14)–Sn(1)–C(11)	96.60 (6)
S(1)–Sn(1)–S(2)	68.25 (2)	S(2)–Sn(1)–S(1)	69.98 (18)
Cl(1)-Sn(1)-S(2)	154.19(3)	Cl(1)-Sn(1)-S(1)	153.55(2)
C(3)–S(1)–Sn(1)	91.23 (9)	C(1)–S(1)–Sn(1)	83.94 (7)
N(1)–C(3)–S(2)	123.6 (2)	N(1)–C(1)–S(1)	123.48 (17)
N(1)–C(3)–S(1)	119.6 (2)	N(1)–C(1)–S(2)	120.41 (17)
N(1)–C(4)–C(5)	110.2 (2)	N(1)–C(2)–C(3)	110.7 (2)

The crystal structure (figure 19) shows the tin atom is coordinated to the two sulfur atoms of the ligand and three carbon atoms of the phenyl groups [52]. The Sn-S bonds are (Sn(1)-S(1) 2.485(6)Å and Sn(1)-S(2) 2.926(6) Å). Thus shorter bond length is closed to the sum of the covalent radii of tin and sulfur (2.42 Å) but longer Sn-S bond length is much smaller than the van der Waals radii (4.0 Å). In this way, the ligand behaves as a bidentate species more correctly anisobidentate ligand and chelates the tin atom by means of sulfur atoms giving *cis*-trigonal bipyramid geometry around the tin atom. The *cis*-trigonal bipyramid environment around tin is defined with C(13) and S(2) in the apical position and S(1), C(1) and C(7) in the equatorial position. The sum of the equatorial angle, C(1)-Sn(1)-C(7) 111.10(8)°, C(1)-Sn(1)-S(1) 111.89(6)°, C(7)-Sn(1)-S(1) 126.78(6)°, is 349.8° instead of the ideal 360° thus bipyramid is distorted. Due to being apart of chelate, the angle S(1)-Sn(1)-S(2) is not 90° but only 65.35(19)° so the S(2) cannot occupy exactly the corresponding *trans* apical position of C(13) and the angle between the apical groups is 156.67(6)°. The longer S-C bond (S(1)-C(19) 1.785(2) Å) is associated with shorter Sn-S bond and the shorter C-S bond (S(2)-C(19) 1.687(2) Å) is associated with the longer Sn-S bond. The Sn-C bond length, Sn(1)-C(13) 2.170(2) Å, is very similar to equatorial ones, Sn(1)-C(1) 2.144(2)Å, Sn(1)-C(7) 2.145(2) Å. Finally the S-C bond lengths are characteristic of the dithiocarbamate group that these distances are both intermediate between the values expected for single and double bond [51].

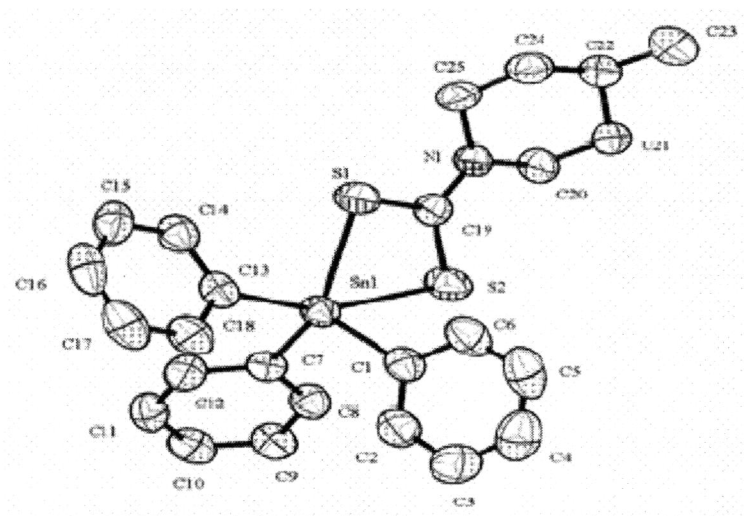

Figure 19. The structure of (4-Methylpiperidine-dithiocarbamato- S,S')triphenyltin(IV) [51].

Crystal structure of *catena*-poly[[trimethyltin(IV)]-μ-4-methyl-4H-1,2,4-triazole-3-thiolato-S:N] [52] shown in figures 20-22, demonstrate that it consist of long chain polymer and Sn atom is bonded to three methyl groups in equatorial positions. The axial positions are occupied by S and N atoms of 4-methyl-4H-1,2,4-triazole-3-thiolato anion, with almost linear S-Sn-N angle; the Sn atom has a distorted trigonal bipyramidal geometry.

Table 2. Selected bond lengths (Å) and bond angles (°) for (4-methylpiperidine-dithiocarbamato-S,S')triphenyltin(IV) [51]

Bond Distances (Å)		Bond Angles (°)	
Sn(1)–C(1)	2.144 (2)	C(1)–Sn(1)–C(7)	111.10 (8)
Sn(1)–C(7)	2.145 (2)	C(1)–Sn(1)–C(13)	109.66 (8)
Sn(1)–C(13)	2.170 (2)	C(7)–Sn(1)–C(13)	101.27 (8)
Sn(1)–S(1)	2.485 (6)	C(1)–Sn(1)–S(1)	111.89 (6)
Sn(1)–S(2)	2.926 (6)	C(7)–Sn(1)–S(1)	126.78 (6)
S(1)–C(19)	1.785 (2)	C(13)–Sn(1)–S(1)	92.39 (6)
S(2)–C(19)	1.687 (2)	C(1)–Sn(1)–S(2)	86.35 (6)
N(1)–C(19)	1.318 (3)	C(7)–Sn(1)–S(2)	87.84 (6)
N(1)–C(25)	1.471 (3)	C(13)–Sn(1)–S(2)	156.67 (6)
N(1)–C(20)	1.474 (3)	S(1)–Sn(1)–S(2)	65.35 (19)

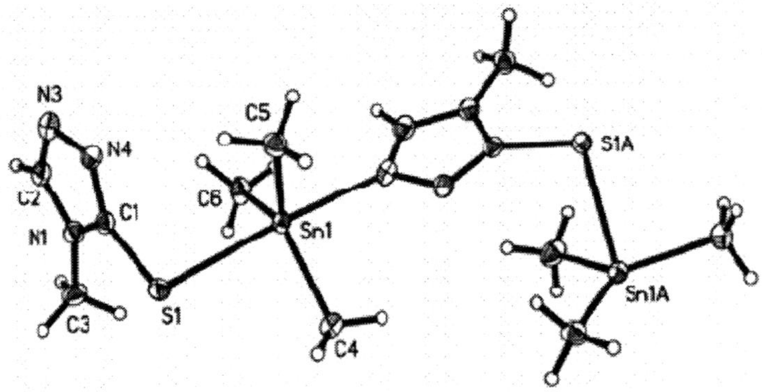

Figure 20. The structure of *catena*-poly[[trimethyltin(IV)]-μ-4-methyl-4H-1,2,4 triazole-3-thiolato-S:N] [52].

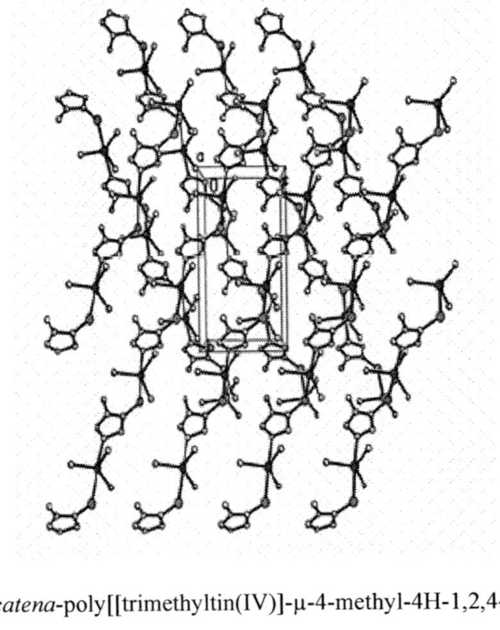

Figure 21. The structure of *catena*-poly[[trimethyltin(IV)]-μ-4-methyl-4H-1,2,4- triazole-3-thiolato-S:N] [52].

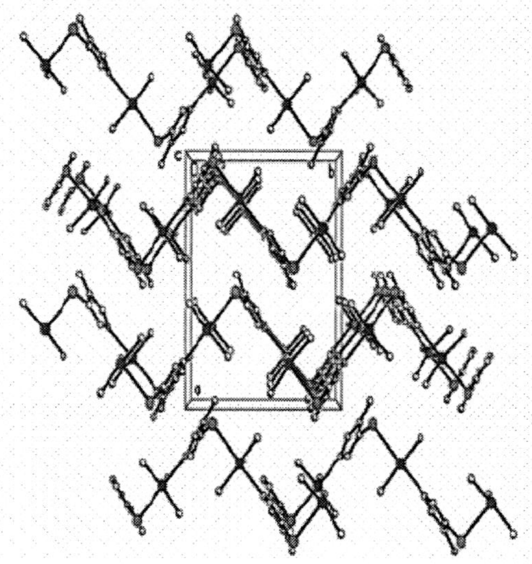

Figure 22. The structure of *catena*-poly[[trimethyltin(IV)]-μ-4-methyl-4H-1,2,4- triazole-3-thiolato-S:N] [52].

3. 6- Coordinated

When the dithiocarbamate acts as bidentate and coordinate to Sn atom through two sulfur atoms then the hexa-coordinated geometry results. The representative structure is given in figure 23 [53]. As shown in the figure 23, there are two independent complexes in the asymmetric unit, with closely comparable geometries. The coordination geometry of Sn is highly distorted octahedral. The S atoms lie in the equatorial plane, with S-Sn-S angles distorted significantly from 90°. The methyl groups define C-Sn-C angles of 138.5(2) and 134.0(2)° in two independent molecules, intermediate between a *cis* and *trans* configuration [53].

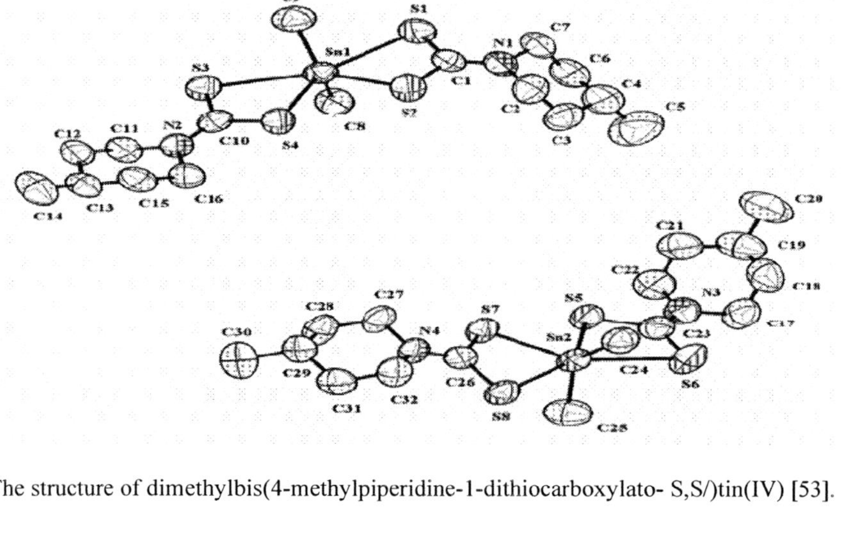

Figure 23. The structure of dimethylbis(4-methylpiperidine-1-dithiocarboxylato- S,S/)tin(IV) [53].

BIOLOGICAL CHEMISTRY OF ORGANOTIN(IV) COMPLEXES

Albert [54] has listed the principles of selective toxicity as accumulation, cytology and biochemistry, and these can be resolved into:

i) Uptake by the organism
ii) Transport and distribution within the organism
iii) Excretion
iv) Presence of sensitive organs, tissues, cells and sub-cellular organelles
v) Affinity for and quantity of receptor molecules
vi) Sequestration in harmless complexes or deposits
vii) Metabolic conversion to less or more toxic compounds

All of these apply to individual organism, but the toxicology and biological chemistry must take into account the global or environmental distribution of tin compounds. This can be divided into several processes:

viii. Manufacture and release into the environment
ix. Mobility e.g., volatility in air, solubility in water
x. Trapping in and leaching from non-leaving materials, such as clay or humic acids
xi. Distribution and concentration via food chains
xii. Environmental, including geochemical and biological processing

TYPES OF TOXICITY

Tetraorganotin Complexes

It is clear that the toxicity of the tetraorganotins is due to their rapid conversions to triorganotin derivatives.

Monoorganotin Complexes

The monoorganotin derivatives have been little studied but appear to be relatively nontoxic. Whether this is due to their intrinsic lack of biological activity or due to lack of absorption and formation of insoluble complexes with anions is not known.

Diorganotin Complexes

Early work demonstrated that this group of compounds are irritant when externally applied and cause tissue necrosis in a similar way to organoarsenicals when systematically administered. Poisoning can be treated by 2,3-dimerceptopropanol and their *in vitro* effects can be explained on the basis of high affinity for dithiols.

An important observation was that di-*n*-octyltin dichloride, when administered to rats in their food, produced thymus atrophy [55]. This has now been confirmed in several laboratories [56-58]. The route of administration to rats is not important, since thymus atropy with little or no loss of body weight is obtained after i,v. injection [59]. It is stated that monooctyltin and trioctyltin are inactive [59], where as tributyltin when fed at 150 ppm for 2 weeks is effective [60]. Since a rather greater effect is produced by 50 ppm for two weeks of dibutyltin [59] and in the absence of information about possible metabolism of tri- and dibutyltin, it is not possible to state that tributyltin is itself active in causing thymus atrophy. Changes are also also found in other lymphoid tissues such as the spleen, and suppression of many immunological processes has been demonstrated [58,60,61]. Lipophilic properties may be an important factor in increasing toxicity. For example, bis-(β-carbobutoxyethyl)- and bis-(β-carbomethoxyethyl)tin dichlorides do not cause a decrease in thymus weight. The explanation given for this is that both compounds are hydrolyzed *in vivo* to bis-(β-carboxyethyl)tin dichloride which is itself inert [62]. *In vitro*, the former compounds are active on lymphocyte oxidation of glucose and viability, whereas the latter hydrolysis product is not [62]. The hydrolysis product with two carboxyl groups will be freely water soluble and much less lipophilic than the parent compounds or other thymotoxic trialkyltins.

Triorganotin Complexes

The toxicity of the triorganotins may be classified into three types: (1) their irritant properties, (2) increase in nervous system water content (intramyelinic vacuoles), and (3) neuronal necrosis in selected areas of the central nervous system.

Irritancy

When applied to exposed surfaces such as skin, the stomach, or the bronchiolar epithelium, inflammation and damage is seen [63-65]. The more lipophilic compounds are the most active. In any consideration of mechanism it is important to remember that the particular anion associated with the triorganotin will greatly influence its physical properties and rate of penetration through membrane. Another factor is that in terms of dose, the local concentration applied to the skin, stomach, or as a particle in the bronchus can be very high. The generalized illness produced in animals by some of higher molecular weight compounds such as triphenyltin, when administered systematically, has not been characterized.

Intramyelinic Vacuoles

Since the original observation [66] that triethyltin causes an increase in water containing sodium and chloride in the brain and spinal cord of rats, the observation has been repeatedly confirmed and extended by others. The increased water is seen as spaces by light microscopy and has been shown to be located at the splitting of the intraperiod lines, and the changes are now called intramyelinic vacuoles [67]. Although not so extensive, intramyelinic vacuoles can be seen in the peripheral as well as the central nervous system [68,69]. The lesions can be produced in many mammalians species and are therefore a general response of mammalian nervous system. There is no evidence that the increase in water originated by movement from swollen cells in the central nervous system, and it seems likely that the formation of intramyelinic vacuoles is the primary event [70]. There is no morphological evidence for changes in the oligodendroglia or Schwann cells. However, it has been shown that the synthesis of myelin is prevented by triethyltin [71,72]. Triethylin combines with a component of myelin [73] but there is no evidence to associate this with the development of the lesion. The structure activity relationships are puzzling. Trimethyltin is inactive, and in decreasing order of activity are triethyltin, tri-*n*-propytlin, and tri-*n*-butyltin [66] and the lesion once produced by feeding triethyltin in the diet is completely reversible with no residual pathology [66].

Neuronal Necreosis

Although difference in the symptoms seen in the rats after administration of trimethyltin and triethyltin have been known for many years [67,74], the morphological correlate of the toxicity of trimethyltin was not discovered until 1979 [75]. The LD_{50} to rats is 12 to 13 mg trimethyltin chloride per kilogram, and the signs of poisoning include tremors, hyperexcitability, aggressive behavior, weight loss and convulsions. Electroencephalorgaphic changes have been detected [76]. After single or repeated smaller doses bilateral and symmetrical neuronal alterations in the hippocampus are seen with fewer changes in the pyriform cortex, amygdaloid nucleus, and neocortex.

In surviving rats, signs of poisoning gradually disappeared and the animals appeared normal, although there brains were severely damaged with cell loss in the hippocampi and each pyriform cortex.

The average concentrations of trimethyltin in the whole brain associated with neuronal damage is approximately 10 nmol/g wet weight of brain from several species [77]. Trimethyltin will undoubtedly circulate *in vivo* as the chloride and hydroxide, both of which will penetrate lipoprotein memberanes and rapidly enter tissues [78].

There appears to be no concentration of trimethyltin in any area in the brain of rats (figure 24) or marmosets [79].

	Intramyelinic vacuoles	Neuronal necrosis in hippocampus
Me_3Sn^+	-	+
Me_3EtSn^+	+	+
$Me_3Et_2Sn^+$	+	+
Et_3Sn^+	+	-
Me_3Sn^+/Et_3Sn^+	+	+

Figure 24. The ability of various triorganotins to cause neuronal necrosis in the hippocampus and intramyelinic vacuoles in the rat [80,81].

The structure activity relationships among triorganotins for the productions of neuronal necrosis and intramyelinic have been recently extended and are shown in figure 25. This should provide the basis for testing the validity of any hyprothesis of the mechanisms of toxicity.

Brain Region	mg blood/q tissue	Trimethyltin (nmol/q wet wt)
Cortex	14.8±0.9	7.2±1.0
Amygdala	12.6±0.9	8.7±0.7
Hippocampus	13.0±0.5	9.1±0.5
Thalamus	11.3±0.5	9.3±0.5
Cerebellum	19.0±0.3	9.0±0.4
Cerebrum	13.5±0.3	10.8±1.6

Figure 25. Distribution of trimethyltin in different areas of the rat brain [82]. The amount of blood in various areas were determined as previously described [83]. The concentration of trimethyltin in various areas was corrected for blood content using the determined concentration of 552±12 nmol trimethyltin per millilitier of blood.

Recently, it has been found that in addition to trimethyltin, tributyltin impaired the sensory organs when the compound was given intraperitoneally [84,85]. In particular, tributyltin produced damage and neuronal destruction in the pyriform and olfacatory cortices after intraperitoneal exposure (figure 26).

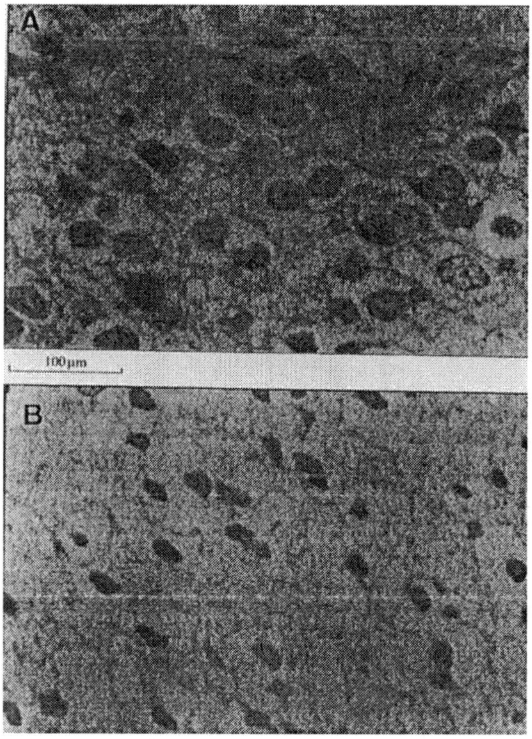

Figure 26. Olfactory cortex, (A) control, (B) tributyltin-treated, 3 days. Extensive destruction of the cortical neurons was observed.

Dibutyltin, dioctyltin and tributyl compounds cause severe thymus atrophy (table 3) [86,87]. This atropy is reversible (figure 27) [88], however, and its degree correlates with the concentration of organotin in thymus (figure 28) [89,90]. At histopathological examination, there is marked depletion of lymphocytes in the thymic cortex rather than in the thymic medulla (figure 29) [91].

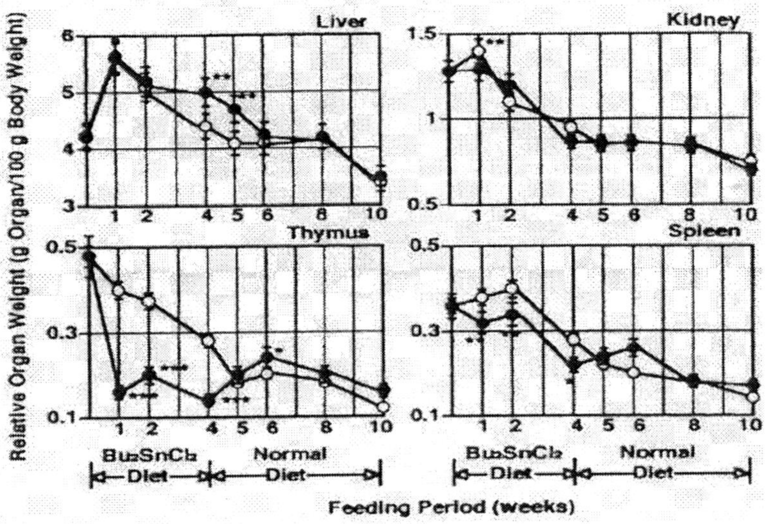

Figure 27. Relative organ weights of rat fed 0 or 100 ppm dibutyltin dichloride for 4 weeks and after that period a normal diet for 6 weeks. Vartical bars denote the SE of the mean for ten determination; those marked with asterisks differ significantly (Student's t-test) from the corresponding control values: *$p<0.05$, **$p<0.01$, ***$p<0.001$. (O) control (0 ppm) group, (•) 100 ppm group.

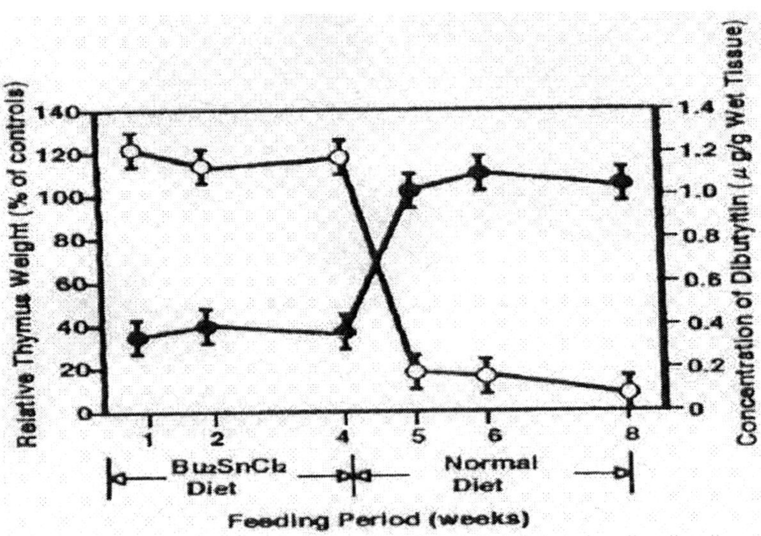

Figure 28. Relationship between thymus atropy and the concentration of dibutyltin in the thymus of rats fed 100 ppm Bu_2SnCl_2 for 4 weeks and after that period a normal diet for 6 weeks. (O) Concentration of Bu_2SnCl_2 in the thymus (μg g^{-1} tissue). (•) relative thymus weights are given as percentages of control values. Vertical bars denote the SE of the mean for ten determinations.

Table 3. Relative organ weights of rat fed various organotin compounds (percentage of controls)

	Body weight	Thymus	Adernal	Liver	Spleen	Kidney	*N*
Bu_2SnCl_2	97	48***	125	93	98	96	6
Bu_3SnCl	90	58***	132	93	75*	86*	7
$BuSnCl_3$	95	106	116	91	89	106	6
Bu_4Sn	101	88	119	89	79*	96	6
Pr_2SnCl_2	84	89	98	82**	88	97	7
Me_2SnCl_2	96	92	97	85*	112	99	6
$MeSnCl_3$	105	85	103	101	92	100	6
Oct_2SnCl_2	101	39***	100	104	116	100	7
Ph_3SnCl	91	100	113	89*	87	99	6
$SnCl_4$	88	83	110	92	73**	103	6

Note: Wister-derived weanling rats (males, 40-45 g) fed independently 100 ppm organotin for 10 days. All values are the mean of 6-7 animals; those marked with asterisks after significantly (Student's t-test) from the corresponding control values: *$p<0.05$, **$p<0.02$, ***$p<0.01$, ****$p<0.001$.

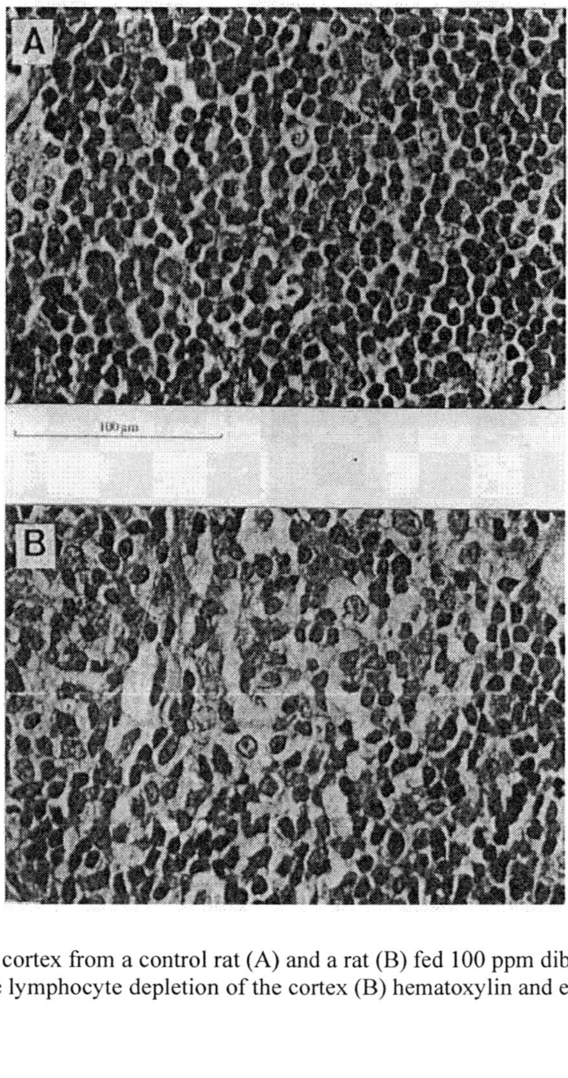

Figure 29. Part of thymic cortex from a control rat (A) and a rat (B) fed 100 ppm dibutyltin dichloride for 2 weeks. Note the complete lymphocyte depletion of the cortex (B) hematoxylin and eosin.

Carcinogenicity Studies

Very few experimental studies keyed to carcinogenicity have been concluded with organotins. Triphenyltin hydroxide was administered to rats and mice in their feed over a 78-week period followed by an additional 26-week observation period. No increase in tumor formation over that of the control subjects was seen [92].

Tin in Cancerous Tisúes

Considerable interest has been expressed in the role, if any, of trace metals in carcinogenicity. A number of trace metals catalyze or stimulate tumorigenicity arising from the presence of nonmetallic carcinogen [93-95].

Pfeilsticker, by noting the tumor/normal tissue ratio for 13 trace metals, clearly shows that the greatest variation lies with tin [96].

Tumor Affinity for Tin

Atsushi et al., in 1973, noted that tin has affinity for tumors [97]. This propensity was confirmed by various investigators who prepared imaging agents using tin-labeled technetium complexes as imaging agents for tumor localization [98].

Tin as Antitumor Agent

Early studies showed that tetraphenyltin, hexaphenylditin, triphenyltin bromide and triphenylpropyltin are inactive towards transplanted mouse cancers [99]. In addition, tin (II) chloride was found to be inactive against spontaneous tumors in mice and rats [100]. However, in 1972, Brown [101] demonstrated that tumour growth in rodents was retarted by repeated injections of triphenyltin acetate, and interest in tin compounds as anti-tumour agents has grown steadily throughout the 1970's and 1980's. Of the 11,000 or so metal and metalloid compounds tested up to August 1980, by the U.S. National Cancer Institute, 1434 were tin derivatives and, of these, 170 showed antitumour activity [102-104].

A series of diorganotin dihalide and di-pseudohalide complexes, $R_2SnX_2.L_2$ (where R = Me, Et, Pr, Bu or Ph; X = F, Cl, Br, I or NCS; and L = N- or O- donor ligands), which were modeled on the active platinum complexes has been screened against P388 lymphocytic leukaemia in mice [105-107]. The octahedral diorganotin complexes studied contain cis-leaving group such as halogen or isothiocyanate, an essential requirement for activity in platinum analogues [108], and certain of these were found to inhibit the cancer (table 4).

Table 4. The activity of diorganotin complexes of bidentate N- donor ligands towards P388 lymphocytic leukaemia in mice [107]

Complex[a]	Dose (mg/kg^{-1})	Average T/C % [b]
Et_2SnCl_2.phen	200	c
	100	152
	50	148
	25	139
	12.5	131
Et_2SnI_2.phen	200	159
	100	149
	50	141
	25	125
	12.5	108
$Et_2Sn(NCS)_2$.bipy	100	c
	50	c
	25	152
	12.5	157
	6.25	145
Bu_2SnCl_2.AMP	100	c
	50	139
	25	116
	12.5	116

[a]phen = 1,10-phenanthroline; bipy = 2,2`-bipyridyl; AMP = 2-aminomethylpyridine

[b]Median survival time of the treated mice group divided by that of the control group, percent. A compound is termed active if it has T/C % ≥ 120.

[c]Toxic dose.

Although the T/C % values obtained for the diorganotin complexes (table 4) were lower than those found for the platinum derivatives (200-300) against some tumour system [109], there is no evidence to suggest that organotin compounds display a high nephrotoxicity [109], as do many of platinum complexes. A number of organotin complexes which were found to possess reproduceable activity towards P388 were subsequently screened against other tumour systems, including B 16 melanocarcinoma, $CD8F_1$ mammary tumour , CX-1 colon xenograft, colon 38, L 1210 lymphoid leukaemia, LX-1 lung xenograft, Lewis lung carcinoma and MX-1 breast xenograft, but found to be inactive in each case [107]. Interestingly, the inorganic tin(IV) complexes $SnCl_4$.phen and $SnCl_4$.bipy show mild activity towards P388.

Several other diorganotin compounds have been found to exhibit significant activity against P388 lymphocytic leukaemia, and these are summarized in table 5.

Certain organotin compounds have shown activity towards other tumours. It has been claimed that dialkyltin derivatives of fluorouracil will reduce solid tumour size upon injection [108]. However, no regression of chemo-induced or transplanted solid tumour was evident when the metalloporphyrin dichlorotin (IV)-tetraammonium-meso-tetra-p-sulphophenylporphyrin, $Cl_2Sn\{TPP(SO_3^-NH_4^+)_4\}$, was administered by intraperitonal injection [109]. Interestingly, the diethyltin dichloride complexes of the N-donor ligands

3,4,7,8-tetramethyl-1,10-phenanthroline and 2-(2-pyridyl)benzimidazole showed good activity against a murine renal cell carcinoma [110], whereas they showed only marginal activity in routine testing against P388 leukaemia [107]. Similarly, dibutyltin dichloride and dioctyltin dichloride, which are both inactive towards P388 [107], exert a selective cytotoxic action or T-lymphocytes and may, therefore, hold potential as anti-T-cell tumour agents [111].

Schwarz et al., [112] showed in elaborate isolation studies that lack of dietary tin leads to severe growth abnormalities in rats. It has been concluded that tin is a vital trace nutrient in rats [113] and, possibly in humans [114]. In addition, unlike other trace metals, tin levels in human and rodent tumours are reported to be significantly lower than those of surrounding non-neo-plastic tissue [115] and, although the tin content is low, tin displays a strong affinity for the tumour site [116].

Table 5. Diorganotin compounds showing activity towards P388 lymphocytic leukaemia in mice [107]

Compound	T/C %[a]/Dose (mg/kg^{-1})	Referecnce
Et_2SnO	125/50, 154/25, 137/12.5, 127/6.25	[115]
$ClPh_2SnOH$	188/25, 163/12.5, 135/6.25	[115]
$Bu_2Sn(adenine)_2$	131/12.5, 123/6.25	[63]
$Ph_2Sn(adenine)_2$	169/100, 135/50, 145/25, 123/12.5	[116]
Me_2Sn(glycylglycine)	139/25, 126/12.5	[116]
Bu_2Sn(glycylglycine)	150/3.12	[116]
Ph_2Sn(glycylglycine)	141/13.2, 128/1.56	[116]
Me_2Sn(penicillamine)	148/400	[117]
Bu_2Sn(penicillamine)	120/3.12	[117]
Ph_2Sn(cysteine)	181/50	[118]
$Ph_2Sn\{Ph_2P(S)S\}_2$	142/12.5, 130/6.25	[108]
$Bu_2Sn(OC_6H_4CHNNCSSMe)$	124/6.25	[108]
$Bu_2Sn(OC_6H_4CHNC_6H_4F)_2$	122/12.5	[108]

[a]T/C is the ratio of survival times (in days) of treated and untreated mice. A compound is considered to be active at T/C % $\geq$ 120.

Recent studies involving continuous dosing of mammary cancer prone mice with the dibutyltin dichloride complexes of 2,2`-bipyradyl or 1,10-phenanthroline [117], a bibutyltin derivative of histidine [117], or tributyltin fluoride [118], each administered at 1 ppm or 10 ppm levels in drinking water, have shown a significant reduction in tumour growth rates. It was suggested that the tin intake, whatever its form, is biochemically converted to anticarcinogens through the body with lumph fluids [118]. The anticarcinogenic organotin compound must be continuously replenished by dietary tin intake [113].

Antitumour activity of organotin compounds in vivo towards various tumour system [119] (table 6). As a result, n-Bu_2SnCl_2 was found to exhibit reproducible activity in vivo towards Ehrlich ascites tumour, IMC carcinoma, P388 lymphocytic leukemia and sarcoma 180 systems, in decending order of activity. In particular, the compound showed the highest activity against the Ehrlich ascites tumour system and gave T/C value (life span) in the range of 98 to 186, with the dosage ranging from 0.1 to 3.0 mgkg^{-1}, the maximum activity in the present case being observed at a single dose of 3 mg kg^{-1}. Generally, a total of one to two injections of high dose level (2-3 $mgkg^{-1}$), at which the compound does not show any toxicity, were more effective against any tumour than five low dose injections (0.1-0.3 $mgkg^{-1}$) [120,121].

Figure 29 indicates the antotumour effect of dibutyltin on Ehrlich ascites tumour cells [119]. The survival period of control group was about 22 days. An increase in survival was observed upon administration of dibutyltin. In particular, a total of one to two injections of high dose levels was more effective against the tumour cellsthan five- low dose injections [122].

Table 7 present the effect of dibutyltin on tumour injection-promotion system in mouse skin carcinogenesis [113]. When the optimum dose of dibutyltin was presented at the first phase TPA promotion stage 1 (stage I promotion) of the two-stage promotion system, the most effective inhibitory effect was obtained. In particular, when 5 μg of dibutyltin was given at the TPA promotion stage, the occurrence of papillomas was completely suppressed (figure 30) [122].

Table 6. Screening data for antitumour activity of dibutyltin compound [112]

Tumour	Dose[a]		Life span[b]	
	2 mg/kg^{-1} i.p	Injection time	Survival time (days)	T/C %
Sarcoma 180	0	5	15.1±0.4	100
	0.1	5	17.6±0.3	117
	0.3	5	17.8±0.4	118
	1.0	5	15.8±0.5	105
	2.0	2	15.0±0.3	100
	3.0	4	13.2±0.3	100
	3.0	1	13.0±0.3	87
IMC-carcinoma	0	5	16.5±0.3	86
	0.1	5	18.6±0.4	100
	0.3	5	19.1±0.4	113
	1.0	5	19.8±0.5	116
	2.0	2	20.8±0.3	120
	3.0	4	21.3±0.7	126
	3.0	1	22.0±0.7	129
Lymphocyte leukemia P-388	0	5	10.4±0.3	127
	0.1	5	10.6±0.2	133
	0.3	5	12.1±0.5	100
	1.0	5	11.8±0.4	102
	2.0	2	12.3±0.5	116
	2.0	4	12.2±0.5	114
	3.0	1	12.1±0.3	118

[a] A total of one to five injections were given at daily intervals in each experiment.

[b] The increase in survival of treated animals over controls is expressed as T/C (%). The values of survival time are the mean ± SE of 10 animals per group.

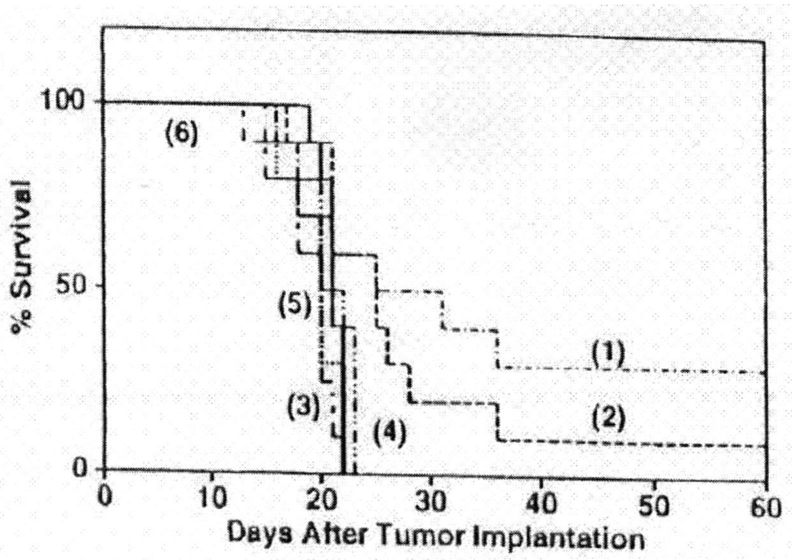

Figure 30. Antitumour effect of Bu_2SnCl_2 on Ehrlich ascites tumour cells. Bu_2SnCl_2 was injected i.p. into dd-Y mice daily from days 1 to 5 after Ehrlich tumour cells (5×10^6) were inoculated i.p. into the mice. Survival times are the means of 10 animals per group. (1) 3 mg/kg^{-1}×1 or 2(-.-), (2) 2 mg/kg^{-1}×2 or 4(---), (3) 1 mg/kg^{-1}×5 (--.-), (4) 0.3 mg/kg^{-1} ×1 (-..-), (5) 0.1 mg/kg^{-1}×5(……) (6) controls (____).

Table 7. Effect of Bu_2SnCl_2 on tumour-promotion system in mouse skin carcinogenesis

Initiation	Promotion				Tumour responce	
	Stage I		Stage II		Papilloma (number/ mouse)	5 of positive controls in (Exp. 10)
1 Acetone	1 wk TPA	4 times	Mezerein	28 times	0	0
2 DMBA	1 wk Acetone	4 times	Mezerein	28 times	0	0
3 DMBA	1 wk TPA	4 times	Acetone	28 times	2.4±0.1	25
4 DMBA	1 wk TPA	4 times	Mezerein	28 times	9.6±0.3	100
5 DMBA + Bu_2Sn (5 μg)	1 wk TPA+ Bu_2Sn (5 μg)	4 times	Mezerein+ Bu_2Sn (5 μg)	28 times	0	0
6 DMBA+ Bu_2Sn (50 μg)	1 wk TPA+ Bu_2Sn (50 μg)	4 times	Mezerein+ Bu_2Sn (50 μg)	28 times	16.9±0.7	175
7 DMBA+ Bu_2Sn (0.5 μg)	1 wk TPA	4 times	Mezerein	28 times	20.2±0.8	210.4
8 DMBA+ Bu_2Sn (5 μg)	1 wk TPA	4 times	Mezerein	28 times	5.6±0.2	58.3
9 DMBA+ Bu_2Sn (50 μg)	1 wk TPA	4 times	Mezerein	28 times	19.2±0.8	200

Initiation was with 100 nmol of DMBA (in Exp. 1, only acetone); 1 week later, twice-weekly applications of 5 μg of TPA or acetone were begun (stage I). Starting on the third week of promotion, the mice received twice-weekly applications of 2.5 μg of mezerein or acetone (Stage II) for the rest of the 14-week experiment. Bu_2SnCl_2 was applied 30 min before treatment with the initiator or promoters. At the end of the experiment period, >95 % of the mice were alive. Tumour response is expressed as the mean ± SE of the papilloma number per mouse as a percentage of the positive control (in Exp. 4). Each value is the result of 10 animals per group [113].

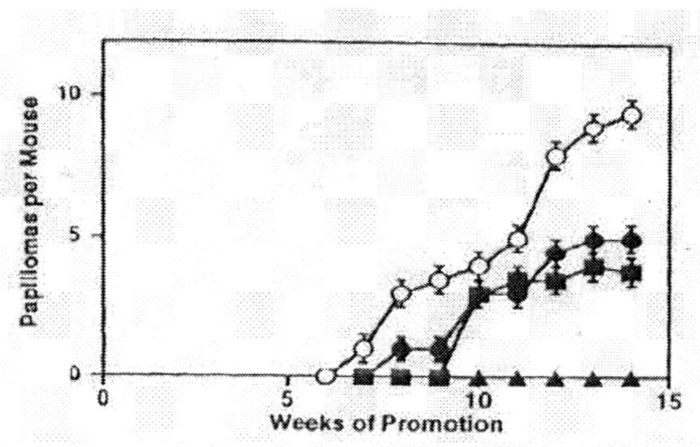

Figure 31. Inhibitory effect of Bu_2SnCl_2 on two-stage mouse skin carcinogenesis system of initiation and promotion. The mice were initiated with 100 nM of DMBA and promoted with 5 μg of mezerein. Bu_2SnCl_2 (5 μg) was applied 30 min before treatment with the initiator and promoters. Each point is expressed as papilloma number per mouse and vertical bars denote the SE of the means for 10 animals per group. (O) controls (•) DMBA-Bu_2SnCl_2 (Initiation stage). (▲) TPA-Bu_2SnCl_2 (promotion stage I). (■) mezerein-Bu_2SnCl_2 (promotion stage II).

Figure 31 demonstrates the effect of dibutyltin on two stage in vitro transformation of BALB/c 3T3 cells. The cells were initiated with MCA and promoted with TPA. Dibutyltin was added at approximately the same time as MCA or TPA. Line A represents the entire negative control, while line B represents the case of TPA treatment alone, namely promotion treatment only. Under these conditions the cells do not transform. However, when the cells were initiated with MCA and promoted with TPA as in line D, the cells were transformed completely. This represents position control. On the other hand, when 300 nM of dibutyltin existed at the MCA initiation stage (line E), the transformation was completely suppressed. Conversely, (line F), when the same amounts 300 nM of dibutyltin existed at the TPA promotion stage, the transformation was not suppressed. When a low concentration (3×10^8) of dibutyltin was presented at the MCA initiation stage and TPA promotion stage, respectively (lines G and H), no suppression of transformation was observed. These results verify that dibutyltin is a potent inhibitor of the MCA initiation stage in the two stage cell transformation system (figure 31) [121,122].

Figure 32. Effect of Bu_2SnCl_2 on two-stage in vitro transformation BALB/c 3T3 cells. The cells (1×10^4) were initiated with MCA (0.1 μg ml^{-1}) and promoted with (0.1 μg ml^{-1}). Bu_2SnCl_2 was added simentaneously with either MCA or TPA. A: whole negative controls. B: MCA alone (promotion blank). C: TPA alone (initiation blank). D: MCA-TPA (positive controls), E: [MCA + Bu_2Sn (3×10^7 M)]-TPA, F: MCA-[TPA + Bu_2Sn (3×10^7 M)], G: [MCA + Bu_2Sn (3×10^8 M)]-TPA, H: H: MCA-[TPA + Bu_2Sn (3×10^8 M)].

In order to make the best use of the antiproliferative activity of tin, a wide variety of tin compounds has been synthesized and tested for possible anticancer agents, stimulating the basic structure, such as octahedral diorganotin dihalide complexes. Table 8 shows the number of metal and metalloid compounds tested by the National Cancer Institute (NCI). Among the compounds tested, the most numerous are tin (2008) with Pt (1538) next in descending order.

Table 8. Metal and metalloid compounds tested by the National Cancer Institute

										II A Al 90	IV A	V A	VI A
Sc 8	Ti 87	V 86	Cr 253	Mn 273	Fe 883	Co 844	Ni 734	Cu 1275	Zn 861	Ga 59	Ge 226	As 1254	Se 734
Y 10	Zr 59	Nb 28	Mo 170	Tc 0	Ru 145	Rh 257	Pd 435	Ag 145	Cd 124	In 18	Sn 2008	Sb 353	Te 51
La 27	Hf 15	Ta 15	W 118	Re 41	Os 25	Ir 41	Pt 1538	Au 121	Hg 823	Tl 37	Pb 288	Bi 81	Po 0
Ce 30	Pr 23	Nd 27	Pm 0	Sm 20	Eu 79	Gd 18	Tb 40	Dy 22	Ho 12	Er 18	Tm 12	Yb 13	Lu 11

Although the T/C % values obtained for the diorganotin complexes (table 4) were lower than those found for the platinum derivatives (200-300 mg) against the same tumour system [123], there is no evidence to suggest that organotin compounds display a high nephrotoxicity [124], as do many of the platinum complexes [123]. A number of organotin complexes which were found to possess reproducible activity towards P388 were subsequently screened against tumour systems, including B16 melanocarcinoma, CD8F1 mammary tumour, CX-1 colon xenograft, colon 38, L 1210 lymphoid leukemia, LX-1 lung xenograft, Lewis lung carcinoma and MX-1 breast xenograft, but were found to be inactive in each case [107]. Interestingly, the inorganic tin(IV) complexes, $SnCl_4$.phen an $SnCl_4$.bipy, show the mild activity towards P388 [125].

Antitumour Active Organotin Compounds

Many organotin compounds have been tested by the National Cancer Institute [126]. Percentage of active compounds against P-388 lymphocytic leukemia in mice is indeed very high upto 50 % for complexed diorganotin dihalides and 48% for uncomplexed R_2SnX_2 compounds. The percentage of active tri- and monoorganotin halides against P-388 lymphocytic leukemia in mice is 9% [127] and that of tetraorganotin compounds only 2% [and 7% for inorganic tin compounds]. The percentage of active diorganotin dihalides against L1210 leukemia is, however, only 1% (and zero for all the other types of tin compounds except tetraorganotins (0.4%). For instance, diorganotin derivatives of 5-fluorouracil are anticarcinogenic and cause shrinkage of solid tumors upon direct injection, [128], but 5-fluorouracil is already an antitumor agent. Crow and Smith [129] reported several examples of marginal activity against ascetic P-388 lymhocytic leukemia in mice of diorganotin compounds. Diethyltin dichloride shows a T/C value (ratio of medium survival times of treated mice and untreated animals) of 125%. Values of 150% have been obtained for the ortho-phenanthrolin complexes of diethyltin iodide and thiocyanide [130]. Diphenyltin chloride hydroxide shows a value of 163% [127]. Tin prophyrins also possess antitumour activity. [130].

Diorganotion(IV) complexes with adenine and glycylglycine have reproducible (but again marginal) antileukemic effects, 21 with T/C values reaching 169% against P-388 lymphocytic leukemia in mice.

Heimann, Atassi,et al.reported comparable T/C values for two complexes of diorganotin dichlorides, NSC-314263 and NSC-328378, against P-388 leukemia, whereas against renal adenocarcinoma values of 218 and 173%, respectively, have been obtained.

Contact Inhibition

The study of the biology of tumor cells was basically initiated by Beebe and Ewing who in 1906 attempted to transplant canine lymphosarcoma in the blood of infected and unifected dogs. Many thousands of articals have appeared over the intervening years providing both data and speculation in to the moephological and physiological distinctions between normal and neoplastic cells. One very basic difference is that cancer cells growing *in vivo* or *in vitro* lack contact inhibition. In normal cell growth, when the cell wall contacts a neighboring cell, growth ceases. This inhibition is totally or partially lacking in neoplastic cells [131]. Contact inhibition occurs because neighboring cells are in continuous chemical contact, and the static cell transmits a chemical message to the growing cell signaling it to stop growing, at least in that direction. Cancer cells do not grow at the more rapid rate that normal cells- they just don,t know when to stop growing. The key mutation in the cancerous cell has been said to be the one that disables the cells mechanisms for turning off growth [132]. A being tumor can be distinguished from a malignant tumor in that the former grows in such a way that the neoplastic cells adhere to and overgrow each other, pushing aside normal tissue, whereas the malignant cells adhere to, overgrow, and destroy normal surrounding tissue. (The invasive wedges look like the claws of the crab and thus the name: cancer:, given long ago by the father of medicine, Hippoctares). The mechanism of contact inhibition, or postconfluence inhibition when in vitro, is unknown. The inhibitor mechanism at the ultracelluear level involves micrefilaments essential to cell membrane movement [132]. Tissue homeostatic factors probably move from cell via gap junctions. It has been speculated that tumor promoters may act by inhibition of intercellular communication through these junctions [133].

Structure-Toxicity Relationship

In the formula R_nSnX_{4-n}, R is generally a hydrocarbon and X a halide or monovalent anionic ligand, although the oxides of triorganotin compounds $(R_3Sn)_2O$ are also important. As a rule X has relatively little effect on toxicity, unless it also has some toxic action, because the organotin compounds are hydrolyzed in aqueous media. Exceptions showing reduced toxicity occur when the X radical forms a very stable link or links with the tin atom, particularly bifunctional radicals which form bidentate complexes with the tin atom, for example dithiols with dialkyltins and radicals containing a nitrogenous ligand which can form five-coordinate complexs with triorganotins [134-136]. Radicals forming five-coordinate polymers have less effect, due to dissociation on dilution. The reduction in toxicity may reflect not only perturbation of equilibria between free organotin cation or hydroxide and binding sites but also inhibition of mobility, uptake and the halide-hydroxide exchange catalyzed by triorganotins. The effect of the X groups on toxicity are greater in dialkyltins than in trialkyltins, and are particularly important in the balance between toxicity and antitumour activity. The difference may result from the dependence of the toxic action of dialkyltins on the formation of bifunctional links with dithiol compounds comparable to the links with bifunctional X radicals, in contrast to the trialkyltins which form weaker bifunctional or single links with hydrophobic bonding to the carbon skeleton (which seldom occurs with the X radicals) contributing to the high affinity of some binding sites.

The number of R groups has a crucial effect on toxicity. The general rule is that for all organism the triorganotin compounds are the most toxic, followed by the diorganotins, while the mono-organotin compounds are of low toxicity. The toxic effects of tetraorganotins are delayed and appear to be due to their metabolic conversion to the triorganotin compounds.

The effect of the R group depends on the class of organism. Triethyltin is the most toxic to mammals, both trimethyltin and tripropyltin are also toxic, but tributyltin is markedly less toxic and tri-*n*-octyltin has very low toxicity. Tributyltin and tripropyltin were the most toxic to gram-positive bacteria and fungi, but gram-negative bacteria were much less sensitive, triethyltin and tripropyl being the most toxic [137].

Conclusion

Organotin carboxylates have a rich diversity of structural variations. It was found that organotin carboxylates exhibit the limited range of coordination geometries about the Sn atoms. Thus the 4-coordinate Sn is invariably distorted tetrahedral, 5-coordinate Sn is distorted trigonal bipyramidal, 6-coordinate Sn is distorted octahedral. For the diorganotin complexes 4, 5 and 6-coordinate geometries have been reported. Whereas, for triorganotin complexes, both 4 and 5-coordinate geometries are reported.

The effects of organotins have been described and vary from general damage to skin or gastric mucosa, to selective damage, to specific cell types, particularly in the nervous system. Some tin compounds are being gradually but surely eliminated as environmental contaminants, while other are expected to be developed as new medicines. In the coming years, tin studies may be able to contribute valuable suggestions in the areas of thymic immunity and immune supervisory control, and in carcinogenesis and aging.

References

[1] Cameron, T.S; Bakshi, P.K; Thangarasa, R; Grindley, T.B. *Can. J. Chem.* 1992, 70, 1623-1674.

[2] Jancso, A; Nagy, L; Moldrheim, E; Sletten, E. *J. Chem. Soc. Dalton Trans.* 1999, 10, 1587-1594.

[3] Barbieri, R; Silvertri, A; Piro, V. *J. Chem. Soc. Dalton Trans.* 1990, 12, 3605-3613.

[4] Li, Q; Yang, P; Wang, H; Guo, M. *J. Inorg. Biochem.* 1996, 64, 181-195.

[5] Cashion, J.D; Bomazetis, G; James, B.D. *J. Organomet. Chem.* 1980, 185, 433-441.

[6] Hynes, M.J; Dowd, M.O. *J. Chem. Soc. Dalton Trans.* 1987, 3, 553-563.

[7] Arena, G; Gianguzza, A; Pellerito, L; Musumeci, S; Purello, R; Rizzarelli, E. *J. Chem. Soc. Dalton Trans.* 1990, 8, 2603-2605.

[8] Buzas, N.N; Gajda, T; Kuzmann, E; Vertes, A; Burger, K. *Main Group Met. Chem.* 1995, 11, 641-649.

[9] Arena, G; Cali, R; Contino, A; Musumeci, A; Musumeci, S; Purello, R. *Inorg. Chim. Acta.* 1995, 237, 187-242.

[10] Harrison, P.G; Sharpe, N.W. *Appl. Organomet. Chem.* 1989, 3, 141-153.

[11] Preut, H; Vornefeld, M; Huber, F. *Acta Crystallogr. Sect. C.* 1991, 47, 264-266.

[12] Surdy, P; Rubini, P; Buzas, N; Henry, B; Pellerito, L; Gajda, T. *Inorg. Chem.* 1999, 38, 346-352.

[13] Davies, A.G; Smith, P.J. *Adv. Inorg. Chem. Radiochem.* 1980, 23, 1-13.

[14] Aldridge, W.N. in: Zuckerman, J.J. (Ed.), *Organotin Compounds: New Chemistry and Applications,* Adv. Chem. Ser., Vol. 168, American Chemical Society: Washington, 1976.

[15] Gielen, M; Tiekink, E.R.T. in: Gielen, M; Tiekink, E.R.T. (Eds.), Metallotherapeutic Drug and Metal-based Diagnostic Agents: Tin Compounds and Their Therapeutic Potential, Wiley, New York, (Chapter 22), 2005.

[16] Ogwuru, N; Khoo, L.E; Eng, G. *Appl. Organomet. Chem.* 1998, 12, 409-413.

[17] Goh, N.K; Chu, C.K; Khoo, L.E; Whalen, D; Eng, G; Smith, F.E; Hynes, R.C. *Appl. Organomet. Chem.* 1998, 12, 457-482.

[18] Seinen, W; Willems, M.I. *Toxicol. Appl. Pharmacol.* 1976, 35, 63-75.

[19] Miller, K; Scott, M.P; Foster, J.R. *Clin. Immunol. Immunopathol.* 1983, 30, 62-65.

[20] Ho, B.Y.K; Zuckerman, J.*J. Inorg. Chem.* 1973, 12, 1552-1561.

[21] Gielen, M; Lelieveld, P; de Vos, D; Willem, R. in: Keppler B.K. (Ed.), *Metal Complexes in Cancer Chemotherapy,* VCH, Weinheim, 1993.

[22] Shahzadi, S; Shahid, K; Ali, S; Mazhar, M; Badshah, A; Ahmed, E; Malik, A. *Turk. J. Chem.* 2005, 29, 273-287.

[23] Sadiq-ur-Rehman; Shahid, K; Ali, S; Bhatti, M.H.; Parvez, M. *J. Organomet. Chem.,* 2006, 690, 1396-1408.

[24] Aziz-ur-Rehman; Ali, S; Helliwell, M; Shahzadi, S. *Acta Cryst.,* 2006, E62, m1656-m1657.

[25] Aziz-ur-Rehman; Ali, S; Helliwell, M; Shahzadi, S. *Acta Cryst.,* 2006, E62, m1778-m1779.

[26] Parvez, M; Ali, S; Bhatti, M.H; Mazhar, Choudhary, M.A. *Acta Cryst.,* 1999, C55, 1429-1431.

[27] Tahir, M.N; Ulku, D; Danish, M; Ali, S; Badshah, A; Mazhar, M. *Acta Cryst.,* 1997, C53, 183-185.

[28] Parvez, M; Ahmad, S; Ali, S; Bhatti, M.H; Mazhar, M. *Acta Cryst.,* 2004, E60, m554-m556.

[29] Aziz-ur-Rehman; Ali, S; Haq, M.N; Shahzadi, S; Wurst, K. *Acta Cryst.,* 2006, E62, m451-m453.

[30] Aziz-ur-Rehman; Ali, S; Helliwell, M; Shahzadi, S. *Acta Cryst.,* 2006, E62, m1778-m1779.

[31] Parvez, M; Ali, S; Bhatti, M.H; Khokhar, M.N; Mazhar, M; Qureshi, S.I. *Acta Cryst.*, 1999, C55, 1427-1429.
[32] Ahmad, A; Khan, A; Ali, S; Parvez, M. *Acta Cryst.*, 2006, E62, m1192-m1194.
[33] Danish, M; Ali, S; Mazhar, M; Badshah, A; Masood, T; Tiekink, E.R.T. *Main Group Met. Chem.*, 1995, 18, 27-34.
[34] Sadiq-ur-Rehman; Ali, S; Badshah, A; Malik, A; Ahmed, E; Jin, G-X; Tiekink, E.R.T. *Appl. Organomet. Chem.*, 2004, 18, 401-408.
[35] Ng, S.W; Kumar Das, V.G; Pelizzi, G; Vitali, F. *Heteroatom. Chem.*, 1990, 1, 433-438.
[36] Gielen, M; El-Khloufi, A; Biesemans, M; Kayser, F; Willem, R. *Organometallics*, 1994, 13, 2849-2854.
[37] Sadiq-ur-Rehman; Ly, H.V; Ali, S; Badshah, A; Parvez, M. *Acta Cryst.*, 2004, E60, m1144-m1146.
[38] Sadiq-ur-Rehman; On-On Leung, B; Ali, S; Badshah, A; Parvez, M. *Acta Cryst.*, 2004, E60, m984-m986.
[39] Ahmad, A; Khan, A; Ali, S; Parvez, M. *Acta Cryst.*, 2006, E62, m1088-m1090.
[40] Ahmad, A; Khan, A; Ali, S; Parvez, M. *Acta Cryst.*, 2006, E62, m1167-m1169.
[41] Aziz-ur-Rehman; Shahzadi, S; Ali, S; Helliwell, M. *Acta Cryst.*, 2006, E62, m1734-m1736.
[42] Parvez, M; Sadiq-ur-Rehman; Shahid, K; Ali, S; Mazhar, M. *Acta Cryst.*, 2004, E60, m1465-m1467.
[43] Kalsoom, A; Mazhar, M; Ali, S; Mahon, M.F; Molloy, K.C; Chaudry, M.I. *Appl. Organomet. Chem.*, 1997, 11, 47-55.
[44] Speziali, N.L; Guimaraes, B.G; Silva, R.M; Duarte P.H; Aguiar, S.R. *Acta Cryst.*, 1994, C50, 1059-1062.
[45] Clarke, P.L; Cradwick M.E; Wardell, J.L. *J. Organomet. Chem.*, 1973, 63, 279-285.
[46] Zia-ur-Rehman; Shahzadi, S; Ali, S; Badshah, A; Jin, G-X. *J. Iran. Chem. Soc.*, 2006, 3(2), 157-160 .
[47] Shahzadi, S; Ali, S; Bhatti, M.H; Fettouhi, M; Athar, M. *J. Organomet. Chem.*, 2006, 691, 1797-1802.
[48] Ali, S; Ahmad, S.U; Sadiq-ur-Rehman; Shahzadi, S; Parvez, M; Mazhar, M. *Appl. Organomet. Chem.*, 2005, 19, 200.
[49] Ma, C; Zhang, J; Zhang, R. *Can. J. Chem.*, 2003, 81, 1070-1075.
[50] Eng, G; Song, X; Duong, Q; Strickman, D; Glass, J; May, L. *Appl. Organomet. Chem.*, 2003, 17, 218-225.
[51] Shahzadi, S; Ali, S; Fettouhi, M. (unpublished results).
[52] Aziz-ur-Rehman; Ali, S; Shahzadi, S; Helliwell, M. *Acta Cryst.*, 2006, E62, m2328-m2329.
[53] Shahzadi, S; Ali, S; Fettouhi, M. *Acta Cryst.*, 2006, E62, m1178-m1180.
[54] Albert, A. *Selective Toxicity*, 7th Edn, Chapman and Hall, London and New York, 1985.
[55] Seinen, W; Willems, M.I. *Toxicol. Appl. Pharmacol.*, 1976, 35, 63-69.
[56] Hennighausen, G; Lange, P. *Arch. Toxicol., Suppl.*, 1979, 315-319.
[57] Funahashi, N; Iwasaki, I; Idc, G. *Acta Pathol. Jpn.*, 1980, 30, 955-966.
[58] Seinen, W; Vos, J.G; Brands, R; Hooykass, H. *Immunopharmacology*, 1, 1979, 343-355.
[59] Seinen, W; Vos, J.G; van Spange, I; Snock, M; Brands, R; Hooykass, H. *Toxicol. Appl. Pharmacol.*, 1977, 42, 197-212.

[60] Seinen, W; Penninks, A. *Ann. NY. Acad. Sci.,* 1979, 320, 499-517.
[61] Seinen, W; Vos, J.G; van Krieken, R; Penninks, A; Brands, R; Hooykaas, H. *Toxicol. Appl. Pharmacol.,* 1997, 42, 213-224.
[62] Pennincks, A.W; Sienen, W. *Food Chem. Toxicol.,* 1982, 20, 909-910.
[63] Verschoyle, R.D; Little, R.A. *J. Appl. Toxicol.,* 1981, 1, 247-256.
[64] Kao, J; Hall, J; Holland, J. M. *Toxicol. Appl. Pharmacol.,* 1983, 68, 206-217.
[65] Middleton, M.C; Pratt, J. *Invest. Dermatol.,* 1977, 68, 379-386.
[66] Magee, P.N; Stoner, H.B; Barnes, J.M. *J. Pathol. Bacteriol.,* 1957, 53, 107-109.
[67] Barnes, J.M; Stoner, H.B. *Br. J. Ind. Med.,* 1958, 15, 15-22.
[68] Graham, D. I; Gonatas, N.K. *Lab. Invest.,* 1973, 29, 628-632.
[69] Scheinberg, L.C; Taylor, J.M; Herzog. I; Mandell, S. *J. Neuropathol. Exp. Neurol.,* 1966, 25, 202-213.
[70] Jacobs, J.M; Cremer, J.E; Cavanagh, J.B. *Neuropathol. Appl. Neurobiol.,* 1977, 3, 169-181.
[71] Blaker, W.D; Krigman, M.R; Thomas, D.J; Mushak, P; Morell, P.J. *Neurochem.,* 1981, 36, 143-153.
[72] Eto, Y; Suzuki, K; Suzuki, K.J. *Lipid Res.,* 1971, 12, 570-579.
[73] Lock, E.A; Aldridge, W.N. *J. Neurochem.,* 1975, 25, 871-876.
[74] Lock, E.A; Scales, M.D; Little, R.A. *Toxicol. Appl. Pharmacol.,* 60, 121-130, 1981.
[75] Tobias, R.S. *Organomet. Chem. Rev.,* 1966, 1, 93-129.
[76] Ray, D.E. *J. Appl, Toxicol.,* 1981, 1, 145-151.
[77] Brown, A.W; Aldridge, W.N; Street, B.W; Verschoyle, R.D. *Am. J. Pathol.,* 1979, 97, 59-82.
[78] Doctor, S.V; Sultatos, L.G; Murphy, S.D. *Toxicol. Lett.,* 1983, 17, 43-49.
[79] Brown, A.W; Verschoyle, R.D; Street, B.W; Aldridge, W.N; Grindley, H. *J, Appl. Toxicol.,* 1984, 4, 12-21.
[80] Seawright, A.A; Brown, A.W; Ng, J.C; Hrdlicka, J. Experimental pathology of Short-chain alkyllead compounds, in *Biological Effects of Organolead Compounds*, Grandjean, P. Ed., CRC Press, Boca Raton, Fla., 177, 1984.
[81] Krnjevic, K. *Physiol. Rev.,* 1974, 54, 418-540.
[82] Ticku, M.K; Olsen, R.W. *Biochim. Biophys. Acta.,* 1977, 464, 519-529.
[83] Doctor, S.V; Sultatos, L.G; Murphy, S.D. *Toxicol. Lett.,* 1983, 17, 43-49.
[84] Hirano, Y; Sato, H; Arakawa, Y. *Jpn. J. Hyg.,* 1995, 50, 195-199.
[85] Hirano, Y; Sato, H; Arakawa, Y. *Jpn. J. Hyg.,* 1998, 50, 196-201.
[86] Arakawa, Y. *Trace Nutrients Res.,* 1994, 11, 1-12.
[87] Yamazaki, N; Arakawa, Y; Nagahashi, M. *Jpn. J. Hyg.,* 1980, 35, 206-218.
[88] Arakawa, Y. *Proc.* 5th *Symposium on Roles of metal in Biological Reactions, Biology and Medicines*, June 6-7, Kanazawa, 82, 1990.
[89] Arakawa, Y; Shinkawa, K; Morimoto, A. *Trace Nutrients Res.,* 1989, 6, 43-54.
[90] Arakawa, Y. Sugiyama *Chemical and Industrial Laboratory Annual Report.* 151, 1990.
[91] Stoner, H. B. *Br. J. Ind. Med.*, 23, 1966, 222-229.
[92] Anon., Bioassay of Triphyenyltin Hydroxide for Possible Carcinogenicity, *Natl, Cancer Inst. Rep.Ser.* NCI-CG-TR-139, 1978.
[93] Dixon, J. R., Lowe, D.B., Richards, D. E., Cralley, J. J., and Strokinger, H. E., *Cancer Res.*, 30, 1970, 1060-1063.
[94] Schwartz, M. K., *Cancer Res.*, 35, 1975, 3481-3487,

[95] Sunderman, F.W. *Human Pathol.*, 1973, 4, 549-552.
[96] Pfeilsticker, K., *Z. Klin. Chem.*, 1965, 3, 145-150.
[97] Glenn, J.J; Haynie, T.P; Konikowski, T. Tumor localization with Technetium-99m, in *Tumor Localization with Radioactive Agents*, Rep. int. Atomic Energy Agency, Vienna September, 1967.
[98] Knapp, F. F. *Nuclear Medicine Technology Prog.* Oak Ridge Natl. Lab. W-7405- ENG-26, May, 1979.
[99] Krause, *E. Bar.*, 1926, 62, 135-141.
[100] Kanisawa, M; Schroeder, H.A. *Cancer.*, 1967, 27, 1192-1195.
[101] N .M. Brown, Ph.D Thesis, Clemson University, 1972.
[102] Sadler, P.J. *Chem. Br.*, 1982, 18, 182-188.
[103] Crowe, A.J; Smith, P.J. *Chem. Ind. (London),* 1980, 200-202.
[104] Crowe, A.J; Smith, P.J; Atassi, G. *Chem.-Biol. Interactions.*, 1980, 32, 171-178.
[105] Crowe, A.J; Smith, P.J; Atassi, G. *Inorg. Chem. Acta.*, 1984, 93, 179-184.
[106] Saxena, A; Tandon, J.P. *Cancer Lett.*, 1983, 19, 73-76.
[107] Denechaud, M; Level, M; Dabadie, M; Ducassou, D; Pommier, J.C. *Bull. Cancer (Paris),* 1981, 68, 40-48.
[108] Vandendris, M; Dumont, P; Heimann, R; Atassi, G. *Cancer Chemother. Pharmacol.*, 1983, 11, 182-187.
[109] Seinen, W. Ph.D Thesis, University of Utrechat, 1978.
[110] Schwarz, K; Milne, D.B; Vinyard, E. *Biochem. Biophys. Res. Commun.*, 1970, 40, 22-29.
[111] Schwarz, K. in *'Newer Trace Elements in Nutrition'*, ed. Meritz, W; Cornatzer, W.E. Marcel Dekker, New York, 1971, p.313.
[112] Cardarelli, N.F; Cardarelli, B.M; Marioneaux, M. *J. Nutrition, Growth and Cance.r,* 1983, 1, 181-185.
[113] Mulay, I.L; Roy, R; Knox, B.E; Suhr, N.H; Delaney, W.E. *J. Natl. Cancer Ints.*, 1971, 47, 1.
[114] Ando, A; Hisada, K; Ando, I. *J. Radioisotopes.*, 1973, 22, 7-18.
[115] Cardarelli, N.F; Cardarelli, B.M; Libby E.P; Dobbins, E. *Austral. J. Exp. Biol. Med.*, 1984, 62, 209-221.
[116] Cardarelli, N.F; Quitter, B.M; Allen, A.S; Dobbins, E; Libby, E.P; Hager, P; Sherman, L.R. *Austral. J. Exp. Biol. Med.*, 1984, 62, 199-213.
[117] Arakawa, Y. *Chemistry of Tin,* Ed., Smith, P.J.Blackie Academic and Professional, 2nd Edition, 1998.
[118] Arakawa, Y; Abe, T; Yu, T.H. *J. Pharmacobio. Dyn.*, 1987,10, S-2.
[119] Arakawa, Y. *In Chemistry and Technology of Silicon and Tin,* eds. Kumar Das, V.G; Gielen, M. Oxford University Press, Oxford, Chapter 23, 319, 1992.
[120] Arakawa, Y. *Biomed. Res. Trace Element.*, 1995, 6, 57-69.
[121] Rosenburg, B. *Naturwissenchaften.*, 1973, 60, 399-405.
[122] Anon. Occupational Exposure to Organotin Compounds, U.S. DHEW (NIOSH) publ. No. 77, 1976.
[123] Cusack, P.A. Ph. D thesis, University of London, 1982.
[124] Narayanam, V.L. Strategy for the discovery and development of novel anticancer agents, in *Structure activity relationships of Antitumour Agents*, Reinhoudt, D.N;

Connors, T.A; Pinedo, H.M; Van de Poll, K.W. Eds., Martinus Nijhoff, The Huge, 5, 1983.

[125] Schwarz, M.K. *Cancer Res.,* 1975, 35, 3481-3492.

[126] Ozaki, S; Sakai, H. *Chem. Ind.*, 1983, 8, 301-304.

[127] Crow, A.J; Smith, P.J; Atassi, G. *Chem. Biol. Interactions.*, 1980, 32, 171-182.

[128] Deuechaud, M; Laval, M; Dabadie, M; Ducassou, D; Pommier, J.C. *Bull. Cancer (Paris),* 1981, 68, 408-412.

[129] Abercrombie, M; Ambrose, E.J. *Exp. Cell. Res.,* 1958, 15, 332-345.

[130] Abercrombie, M; Heaysman, J.E.M; Pegrum, S.M. *Exp. Cell. Res.,* 1970, 60, 437- 444.

[131] Gotlieb, A.I; Heggeness, M.H; Ash, J.F; Singer, S.J. *J. Cell. Physiol.,* 1979, 100, 563-578.

[132] Harris, A. *Contact inhibition of cell locomotion, in cell Communication,* Cox, R.P. Ed., John Wiley and Sons, New York, 420, 1981.

[133] Trosko, J. E; Yotti, L. P; Dawson, B; Chang, C.C. *In vitro assay for tumor promoters, in Short Term Tests For Chemical Carcinogenesis,* Stich, H.F; San, R. H.S. Ed., Springer-Verlag, New York, 420,1981.

[134] Blunden, S.J; Smith, P.J; Sugavanam, B. *Pestic. Sci.,* 1984, 15, 253-282.

[135] Molloy, K.C; Purcell, T.G; Cunningham, D; McCardle, P; Higgins, T. *Appl. Organomet. Chem.,* 1987, 1, 119-131.

[136] Blunden, S.J; Patel, B.N; Smith, P.J; Sugavanam, B. *Appl. Organomet. Chem.,* 1987, 1, 241-250.

[137] Sijperstein, A.K; Luijten, J.G. A; van der Kerk, G.J.M. *Fungicides: An Advanced Treatise,* Vol. 2, ed., D. C. Torgeson, Academic Press, New York, 331, 1969.

In: Advances in Organometallic Chemistry Research
Editor: Kenji Yamamoto, pp. 177-200
ISBN: 978-1-60021-779-1

Chapter 7

THE ROLE OF IONIC LIQUIDS IN PALLADIUM CATALYZED C-C BOND FORMING REACTIONS

Anna M. Trzeciak *** and Józef J. Ziółkowski*** *

Uniwersity of Wrocław, Faculty of Chemistry,
14 F. Joliot-Curie St., 50-383 Wrocław, Poland

1. ABSTRACT

Green chemistry is a strategy of changes in chemical processes aimed at ensuring a safer, cleaner environment in the 21st century. In designing reactions following the green chemistry principles, chemists pay special attention to selecting substrates and chemicals that are as harmless to the environment and to human health as possible. One way to achieve this goal is by replacing traditional toxic solvents with ionic liquids. The use of ionic liquids in C–C bond forming reactions like Heck, Sonogashira, Suzuki, and carbonylation offers many spectacular advantages including facilitation of catalyst separation from the organic products. It is important and of great topical interest because many products of C–C coupling reactions are used as pharmaceuticals or agricultural chemicals, which can contain only very limited amounts of metals. This article presents catalytic systems containing palladium catalyst precursors, both soluble and heterogenized complexes as well as palladium nanocolloids, applied in C–C bond forming processes performed in ionic liquids. The applicability of ionic liquids and the effect of their molecular structure on the reaction course are discussed. Special attention is given to the reactions of ionic liquids with palladium precursors leading to the formation of new species and modification of the catalytic properties of the catalytic system. As an example, reactions of imidazolium halides with palladium precursors are discussed in detail. The article, based on literature data and results obtained in our lab, presents the state of knowledge in the field and formulates some expectations for future design of catalytically active and environmentally friendly palladium-based systems for C–C bond forming reactions.

---

* ania@wchuwr.pl
* jjz@wchuwr.pl

2. INTRODUCTION

The term "green chemistry" was first used by P. Anastas in the Program of the US Environmental Protection Agency (EPA). The term was proposed as a designation of a new approach to the problems of environmental protection and a new point of view on the contemporary role of chemistry [1]. The principles of green chemistry were formulated by P. Anastas and J. Warner [2] and subsequently supplemented by N. Winterton [3] with emphasis on the particular importance of sustainable development of chemistry and the design of environmentally friendly chemical processes. In consequence of such a new way of understanding the present-day role and function of chemistry, the whole chemical process, from the selection of raw materials (substrates), if possible non-toxic, through the reaction media (solvents), to the final products, should be carefully examined from the point of view of the green chemistry principles. Thinking about chemical reactions in terms of the rules of green chemistry is very inspiring and prompts not only the design of completely new processes but also the modification of existing ones to eliminate the use of toxic compounds (solvents, substrates). Visible effects of such thinking include applications of ultrasonic and microwave energy to chemical reactions as well as a search for new, non-toxic solvents, possibly leading to total elimination of environmentally unfriendly ones.

Catalysts and catalytic processes play a particularly important role in green chemistry. They are considered the pillars or even the foundation of green chemistry.

In this chapter we will discuss only a selected number of processes catalyzed by transition metal compounds or organometallic complexes and performed in ionic liquids. The application of catalysts usually brings about significant energy savings since most catalytic reactions occur in milder conditions compared with non-catalytic processes. Very often the application of catalysts makes it possible to obtain the required products with high selectivity and yield following the green chemistry principles called "atoms economy" and to obtain high yields of much-desired products with no or low amounts of wastes.

One extremely important area of organometallic catalyst application is asymmetric synthesis in the pharmaceutical industry, used in the production of certain medicines. Very often only one optical isomer in a mixture exhibits biological activity, and therefore it is important to produce it with high yield.

Proper selection of the solvent is a very important factor in designing new chemical processes. The replacement of volatile organic compounds (VOC), which are toxic or flammable, with environmentally friendly solvents like water or ionic liquids is usually a very reasonable solution. Undoubtedly water is the most environmentally acceptable solvent; however, it is very often not easy to carry out catalytic organic reactions in water [4 - 7]. Many organic substrates and organometallic catalysts are not soluble in water; besides, water may often cause partial decomposition of the catalyst. These difficulties may be partially omitted or even totally eliminated by application of surfactants or phase-transfer agents.

The use of water-soluble ligands and synthesis of water-soluble organometallic catalyst precursors also present a reasonable, if not cheap, solution [6, 7]. Moreover, it is not quite safe for the environment because of the presence of toxic phosphorus compounds.

Recently ionic liquids have been very intensively studied as new "green" solvents, fairly safe and convenient media for organic reactions catalyzed by metal complexes [8 - 15]. One may observe an increasing interest in the application of ionic liquids not only in laboratory-scale organic synthesis but also on an industrial scale [8 - 15]. The rising interest in wide use of ionic liquids stimulates complementary studies into their interesting properties, including anti-microbial [16] and antielectrostatic ones [17].

3. Ionic Liquids as Catalytic Reaction Media

The first compound demonstrating ionic liquid properties, $[HNEt_3]NO_3$ (m.p. 12°C), was obtained as far back as 1914 [18]. However, only in recent years have a large number of ionic liquids, mainly pyridinium and imidazolium salts with melting points below 100°C, been synthesized and commercialized as a result of very intensive studies (Scheme 1). It is estimated that ionic liquids composed with different cations and anions, both modified with substituents, can represent even up to 10^{18} compounds. Unfortunately it is very difficult to design ionic liquids with certain desired properties, mainly because there is no simple correlation between structure and properties (like m.p.) [8 - 15].

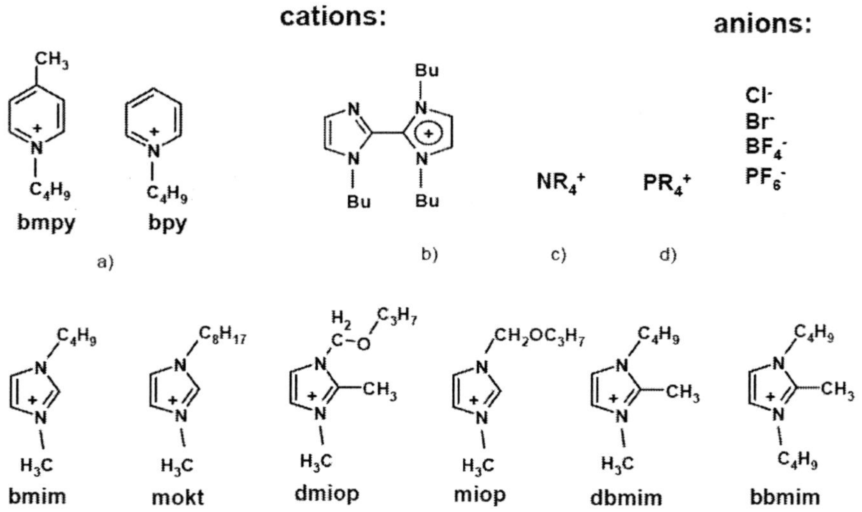

Scheme 1. Ionic liquids a) pyridinium salts, b) bis-imidazolium salts, c) tetraalkilammonium salts, d) phosphonium salts, e) imidazolium salts.

Regarding ionic liquids as catalytic reaction media, the following properties should be listed:

- High range of temperature stability, up to 200°;
- Insignificant vapor pressure;
- Good solvents for organic, inorganic, and organometallic compounds as well as for many gases like H_2, O_2, CO;
- Polarity and lipophility depending on the kinds of cations and anions forming a given ionic liquid;
- Possibility to use ionic liquids in biphasic systems like organic solvent/ionic liquid or water/ionic liquid.

It is worth noting the possibility of the formation of a two-phase system in which one phase contains organic reagents and the other one an ionic liquid with a dissolved catalyst. The efficiency of a catalytic reaction may significantly increase when an initially two-phase system turns into one phase. When the reaction is completed, after cooling down the reaction mixture, the catalyst can be easily separated from the organic products. Such a procedure makes it possible to recycle the catalyst and, in the case of high solubility of the catalyst in the ionic liquid and low leaching, to synthesize the organic products free of metal traces. The extraction of organic products can be performed with an organic solvent or with supercritical CO_2. The last procedure is rather expensive and technically complicated but at the same time very promising and practicable on a large scale [19].

Unlike many other solvents (mainly organic), ionic liquids cannot be considered inert components of the reaction mixture, as they may react, mainly with metal complexes used as catalyst precursors, to form new compounds demonstrating different catalytic activity. Therefore, when selecting the ionic liquid solvent and the potential modifying ligand, at least two parameters should be taken into account:

- the general coordination ability of the ionic liquid depending on the kind of cation and anion; and
- the particular coordinating ability of imidazolium ionic liquids to form N-heterocyclic metal carbene complexes as a result of deprotonation of the imidazolium cation.

Metal carbene complexes with N-heterocyclic ligands may be formed in reaction of a metal complex used as a catalyst precursor and imidazolium halides [20, 21]. This reaction proceeds easily when one ligand coordinated to the metal in the catalyst precursor undergoes protonation. The origin of carbene complex formation can be explained as a result of proton abstraction in the C2 position and the formation of metal–carbene bonding in the M-C2 position. An example of the formation of a metal-N-heterocyclic carbene complex, Pd(bmim-y)$_2$I$_2$, is presented in Scheme 2 as a result of Pd(OAc)$_2$ reaction with [bmim]I [22].

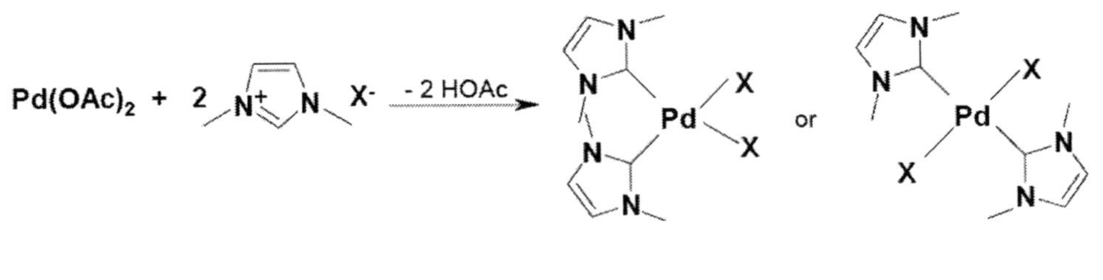

Scheme 2. Reaction of Pd(II) N-heterocyclic carbenes formation.

Palladium carbene complexes have found many applications in organic reactions, such as hydrosililation of alkenes, alkynes, and ketons, hydrogenation and hydroformylation of alkenes, Heck, Suzuki, and metathesis reactions [23 - 27].

4. APPLICATION OF IONIC LIQUIDS AS REACTION MEDIA IN C–C BOND FORMING REACTIONS

In modern organic synthesis there are a number of reactions that lead to the formation of new C–C bonds. There are first of all such reactions as alkoxycarbonylation, Sonogashira, Suzuki, and Heck reactions, very frequently used not only in advanced organic synthesis on a laboratory scale but also industrially in the production of drugs and medicines. Most of these reactions are catalytic processes in which palladium complexes (such as Pd(II) compounds or Pd(0) colloids) are used as catalyst precursors. During these reactions aryl halides (common substrates) are transformed to the products: the corresponding esters (in alkoxycarbonylation), diphenylacetylenes (in the Sonogashira reaction), biphenyls (in the Suzuki reaction), and cinnamates or phenylcinnamates (in the Heck reaction) (Scheme 3).

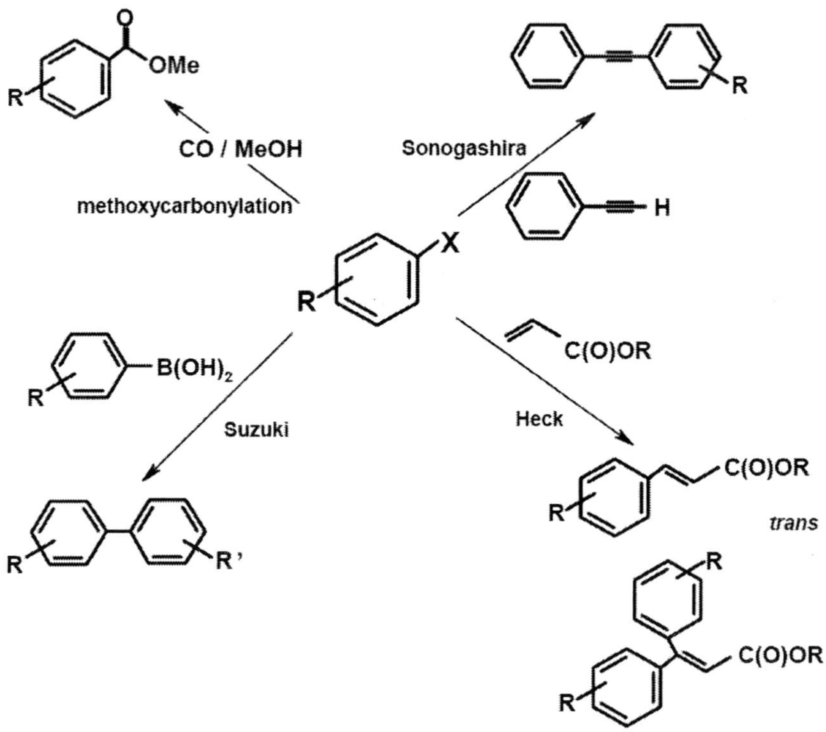

Scheme 3. C-C bond forming reactions catalyzed by palladium compounds.

One of the most important advantages of the application of ionic liquids as reaction media for these processes is the possibility to obtain reaction products practically free from palladium traces. Although the same substrates (aryl halides) are used in all the reactions, their mechanisms are quite different except for the first step, i.e. oxidative addition of aryl halide to the Pd(0) reactive form of the catalysts (Scheme 4).

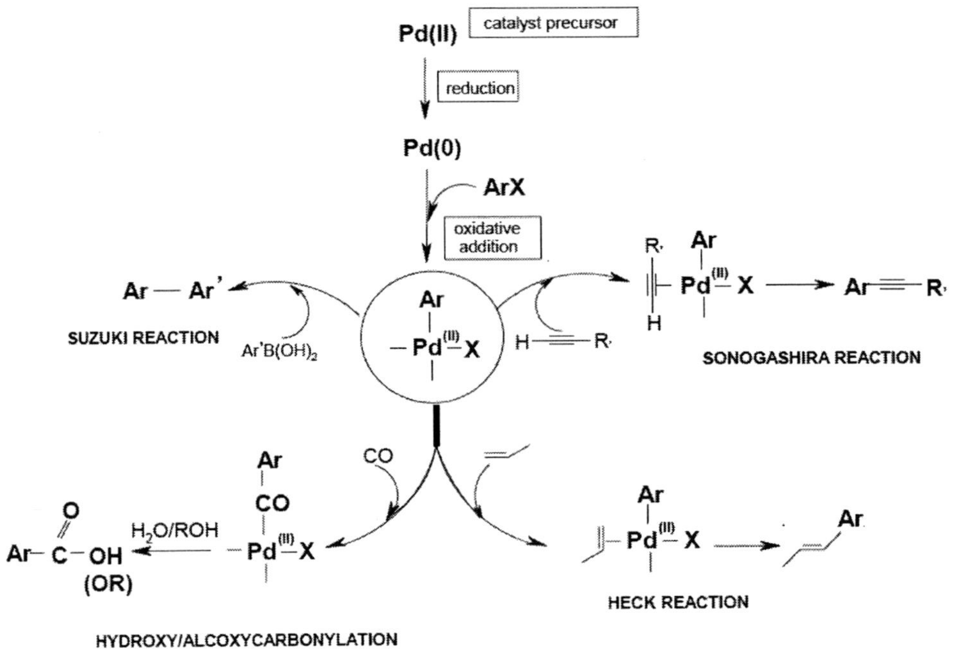

Scheme 4. Mechanisms of palladium catalyzed C-C bond forming reactions [73].

In this process, Pd(0) in the form of a monomolecular complex or Pd(0) as a colloid reacts with an aryl halide (ArX) to form a corresponding Pd(II) intermediate complex. Interaction of this intermediate with the other substrate leads to the formation of final products.

4.1. Hydroxy-/Alkoxycarbonylation Reaction of Aryl Halides

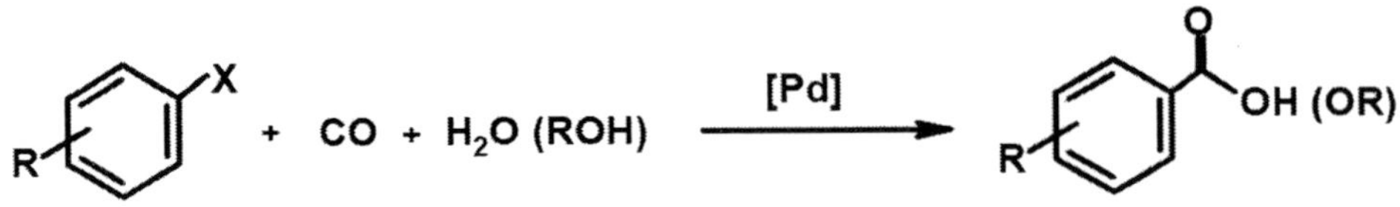

Scheme 5. Hydroxy-/alkoxycarbonylation reaction.

The first paper demonstrating the application of an ionic liquid in alkoxycarbonylation was published in 2001 [28]. The authors presented the results of mono- and di-alkoxy- and aminocarbonylation of bromo- and iodobenzene catalyzed with $Pd(OAc)_2$ with the addition of a fourfold excess of PPh_3 dissolved in $[bmim]BF_4$ and $[bmim]PF_6$ (Scheme 6). The same reactions carried out in MeOH or EtOH were less effective. However, in recycling procedures in the ionic liquid, a yield decrease was observed in successive runs [28]. When the addition of free phosphine versus palladium increased to 20-fold, the yield of the product increased to 99% in the first entry and remained relatively high (74–78%) in the fourth entry [28].

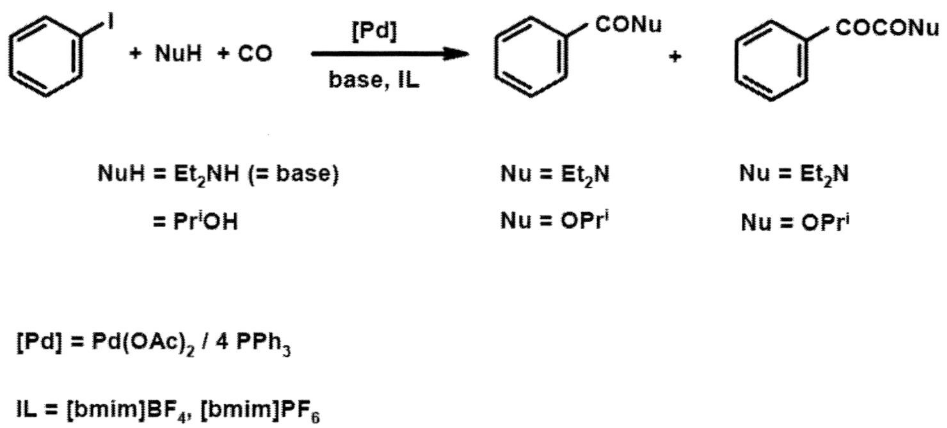

Scheme 6. General scheme of alcoxy- and amidocarbonylation of iodobenzene [28].

A year later the butoxycarbonylation reactions of 4-bromoacetophenone catalyzed by a Pd(II) carbene complex with a benzotiazole ligand in ionic liquids ([bmim]BF_4, [bmim]Cl, ([bmim]Br, Aliquat® (tricaprilmethylammonium chloride) or in ammonium salts [Bu_4N]Cl, [Bu_4N]Br, [Bu_4N]I) were studied (Scheme 7) [29]. The highest yield of ester (78%) was obtained in the presence of [Bu_4N]Br, whereas in imidazolium salts used as the reaction medium the yield did not exceed 16%. Also N-butylpyridinium tosilate was found a very poor solvent as less than 5% of the product was obtained.

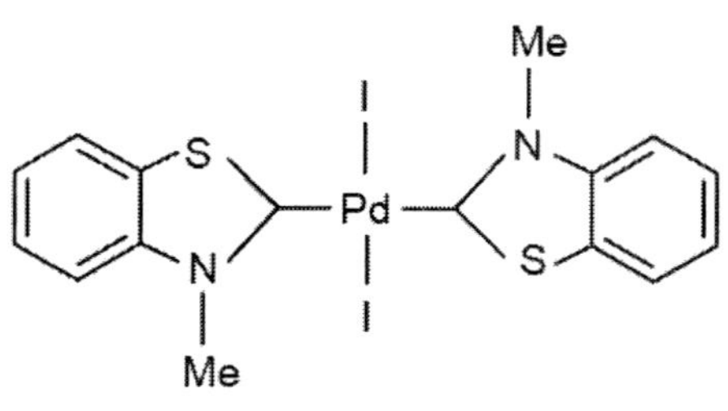

Scheme 7. Pd(II) carbene complex with benzotiazole ligands [29].

It was also found that, in the presence of a 20-fold excess of PPh_3, $Pd(OAc)_2$ effectively catalyzed hydroxycarbonylation of iodobenzene, bromobenzene, and benzylchloride derivatives in organic solvents as well as in ionic liquids such as [bmim]PF_6, [bmim]BF_4, or Aliquat 336® ([$CH_3(C_8H_{17})_3N$]Cl) [29]. In comparison to the reactions carried out in organic solvents like benzene, dichlorometane, THF, or DMF, the application of ionic liquids caused an increase in benzoic acid yield to 83% in [bmim]PF_6, whereas in [bmim]BF_4 and Aliquat 336® the obtained yields of the product were 55% and 50% respectively. Taking advantage of the fact that [bmim]PF_6 is not miscible with water, a method of separation of the reaction products from the catalyst was developed, and benzoic acid, amine hydroiodide, and amine were extracted with water, whereas the catalyst remained in the ionic liquid and was used for subsequent experiments. In five successive runs, only a little drop in product yield was observed. The same catalytic system in [bmim]PF_6 with NaOH added as a base was used for hydroxycarbonylation of benzyl chloride derivatives, and quite good yields of phenylacetic acid derivatives as reaction products were obtained [30].

Methoxycarbonylation of iodobenzene, which produces methylbenzoate, was carried out in ionic liquids using $PdCl_2$(cod) as a catalyst precursor [31]. This Pd(II) complex reacts with imidazolium or pyridinium halides forming $[IL]_2[PdX_4]$ type complexes considered as another form of catalyst precursor [31] (Scheme 8).

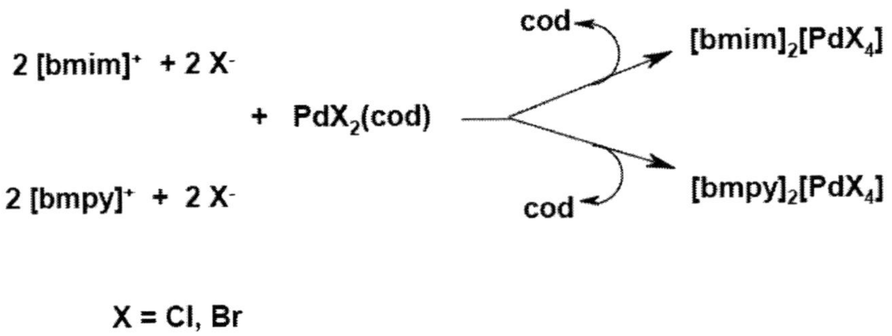

Scheme 8. Synthesis of Pd(II) anionic complexes of $[IL]_2[PdX_4]$ type (IL = ionic liquid cation) [31].

These compounds were isolated and their structural determination showed square planar arrangement of the $[PdX_4]^{2-}$ anion interacting with imidazolium or pyridinium cations (Scheme 8). When used as starting palladium catalyst precursors, $[IL][PdX_4]$ complexes exhibited activity comparable to $PdCl_2$(cod) under the same reaction conditions [31] (Scheme 9).

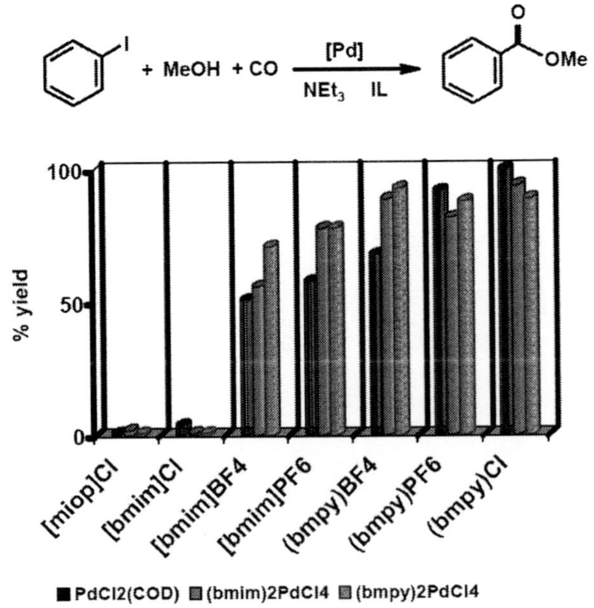

Scheme 9. Methoxycarbonylation of iodobenzene catalyzed by different Pd(II) precursors in ionic liquids.

This suggests that these anionic compounds may undergo similar transformations under catalytic reaction conditions to catalytically active forms, like monomolecular complexes of Pd(0) or nanosize Pd(0) colloids. For all palladium catalysts precursors used, a similar dependence of the yield of methoxycarbonylation products on the structure of the ionic liquid was observed. One of the most important observations was a significant inhibiting effect of imidazolium halides used as reaction media on all the various catalyst precursors used. For example, the yield of the reaction product obtained in [bmim]Cl never exceeded 5% (Scheme 9). However, when the chloride anion was replaced with BF_4^- or PF_6^- (in [bmim]BF_4 or [bmim]PF_6), the yield of methoxycarbonylation products increased to 70–80%. It is worth noting that in pyridinium salts, regardless of the kind of anion bonded to the $[C_5H_5NR]^+$ cation, methoxycarbonylation reaction always occurred with higher yield. One may conclude that pyridinium halides, but not imidazolium halides, are excellent reaction media for methoxycarbonylation [31]. It means that the observed inhibiting effect is not caused by the chloride anion itself but by its composition with the imidazolium cation. It was found in supplemental studies that the inhibiting effect of imidazolium chloride can be significantly reduced when alkyl substituents (like a methyl group) are attached to the C2 carbon in the imidazolium ring. Methoxycarbonylation reaction in [dmiop]Cl and [dbmim]Br yielded 47% and 58% of esters respectively, much more than in the similar ionic liquids [bmim]Cl and [miop]Cl, which have a mobile proton at the C2 carbon.

Methoxycarbonylation of iodobenzene in ionic liquids like [bmim]X and [bmpy]X (X = Cl, BF_4, PF_6) or in molten ammonium salts like $[R_4N]X$ (R = ^{n}Bu, Et, Me; X = Cl, Br, I) catalyzed by colloidal Pd(0) obtained by reduction of $[Pd(\eta^3\text{-allyl})_2Cl]_2$ with Vcp_2 as the reducing agent and stabilized with PVP (PVP = polyvinylopyrrolidone) were studied for comparison [32]. Like in the case of the palladium catalyst precursors discussed above, the yield of methoxycarbonylation reaction products depended on the structure of the ionic liquids used, but in all cases ionic liquids containing pyridinium cations were better than those with imidazolium cations.

An increase in the MeOH concentration in the reaction mixture caused a decrease in reaction yield from *ca.* 60% for [MeOH] : [bmpy] = 2.8 to 2% for [MeOH] : [bmpy] = 4.1. At the optimum concentration of methanol, the yield of methoxycarbonylation reaction catalyzed by nanosized Pd(0) colloid decreased in the order:

$[^nBu_4N]Br > [^nBu_4N]I > [^nBu_4N]Cl > [bmpy]PF_6 > [Et_4N]Br > [bmpy]Cl > [bmpy]BF_4 > [Et_4N]Cl > [bmim]PF_6 > [bmim]BF_4 > [bmim]Cl$.

The highest yield of methoxycarbonylation was obtained for reactions carried out in tetrabutylammonium salts; however, in recycling experiments the yields significantly dropped in successive entries. Things were different when [bmpy]BF_4 was used as the reaction medium, in which case the catalyst was reused many times in successive reactions with no significant decrease in activity [32].

4.2. The Sonogashira Reaction

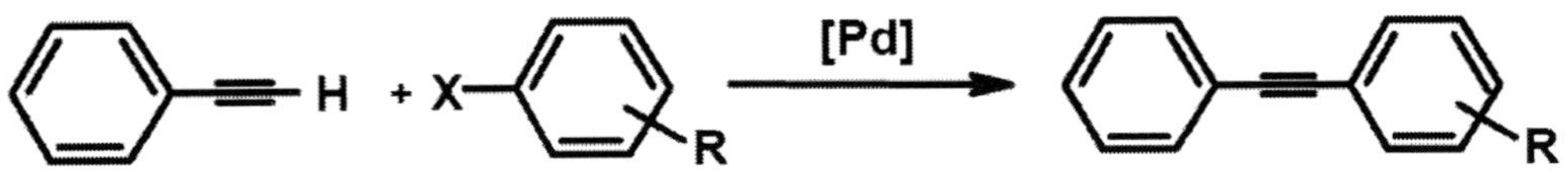

Scheme 10. The Sonogashira reaction.

An inhibiting effect of imidazolium halides was also observed in reaction of iodobenzene with phenylacetylene (Sonogashira reaction). When this reaction was carried out in imidazolium salts like [bmim]PF_6 and [mokt]PF_6 with $PdCl_2[P(OPh)_3]_2$ as the catalyst precursor, up to 83–100% of the products were obtained, whereas in [mokt]Cl the reaction was totally retarded (Scheme 11) [33].

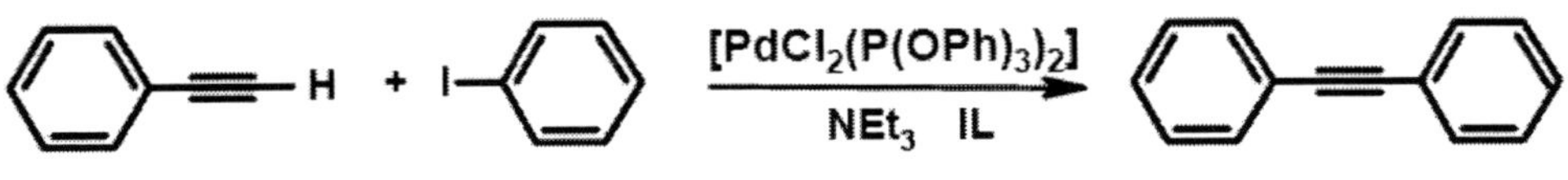

IL	% [Pd]	Yield %
[bmim]PF_6	1	83
[bmim]PF_6	2	100
[mokt]Cl	1	0
[mokt]PF_6	2	100

Scheme 11. Sonogashira reaction catalyzed by $PdCl_2[P(OPh)_3]_2$ in different ionic liquids [33].

A phosphorus-free palladium catalyst precursor, a Pd(II) complex with a bis-imidazole ligand, (bisimidazole)PdCl(Me) (Scheme 12), was successfully used in reaction of iodobenzene with phenylacetylene in [bmim]PF_6 at 120 °C without the addition of a copper cocatalyst or free phosphorus ligand [34] (Scheme 12).

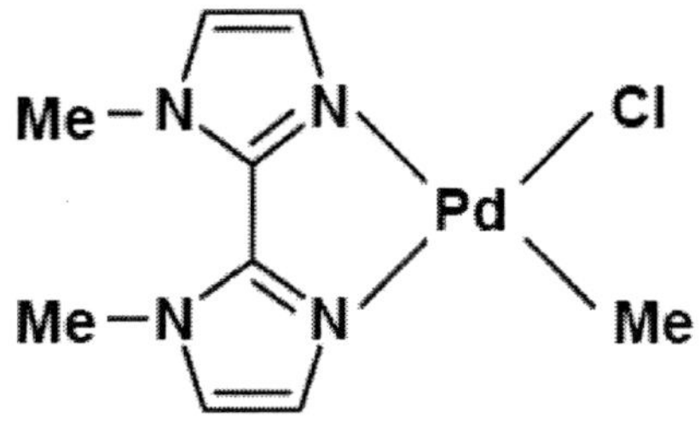

Scheme 12. Bis-imidazole Pd(II) complex [34].

The reaction course clearly depended on the kind of base used (i.e. Et_3N, $^iPr_2NH$, piperidine, K_2CO_3, Cs_2CO_3). The highest yields of diphenylacetylenes were obtained in reactions in which amines were used. The weaker effect of inorganic salts can be explained by their lower solubility in the reaction medium. This was shown by the fact that the application of piperidine as a base made it possible to conduct the recycling experiments with reused catalysts dissolved in ionic liquid with almost constant catalyst activity [34].

The coupling of iodoarenes with phenylacetylene with $PdCl_2$ as the catalyst precursor in different imidazolium ionic liquids containing anions such as Cl, Br, BF_4, PF_6, or ClO_4 was successfully performed under ultrasound irradiation (50 kHz, 30 °C) [35]. Relatively high yields of diphenylacetylenes were obtained (85–93%) regardless of the kind of ionic liquid used, and an inhibiting effect of imidazolium halides was observed. Five consecutive experiments were carried out with the same ionic liquid as the reaction medium and reaction yield decreased only slightly, from 93% to 85% in the fifth attempt. On the basis of TEM measurements it was shown that during the reaction the $PdCl_2$ catalyst precursor undergoes reduction to Pd(0) nanoparticles, which catalyze the reaction [35].

A conventional catalytic system for the Sonogashira reaction, composed with $Pd(OAc)_2$, PPh_3, and CuI, was successfully used for reaction of iodobenzene with phenylacetylene in different imidazolium ionic liquids with PF_6 and BF_4 anions at 80 °C [36]. High yields of the product were also obtained in recycling experiments [36].

Several palladium complexes, both with phosphorus ligands, like $PdCl_2(PPh_3)_2$, and without phosphorus ligands, like $Pd(OAc)_2$, $PdCl_2$, or $PdCl_2(PhCN)_2$, were used as catalyst precursors in reaction of iodobenzene with phenylacetylene in [bmim]PF_6 with $^iPr_2NH$ as a base [37]. In several cases the yields of the reaction in ionic liquids were higher than in organic solvents, toluene or THF. The same ionic liquid, [bmim]PF_6, was found to be a good solvent for reaction of iodobenzene derivatives with phenylacetylene. The fact that [bmim]PF_6 is not miscible with water was taken advantage of to separate the products from the catalyst. It was possible to wash out the side product (ammonium salt) with water from the ionic liquid phase containing the catalyst, which simplified its use in recycling experiments. Successive recycling experiments were carried out in a continuous process in a minireactor [37].

Imidazolium ionic liquids were found not to be good solvents for the Sonogashira reaction catalyzed by carbapalladacycle of 4-hydroxyacetophenone oxime as a ligand (Scheme 13) [38].

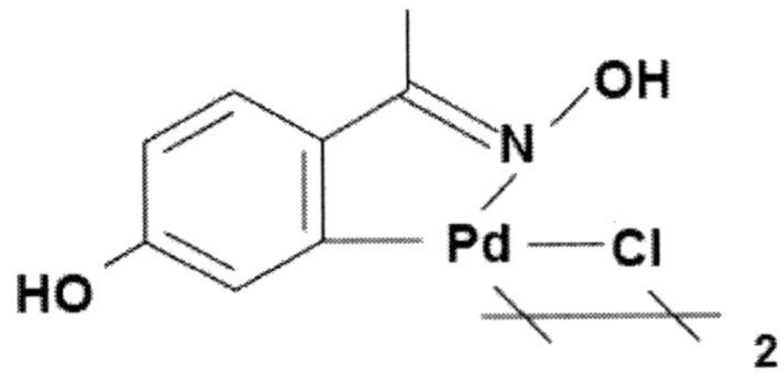

Scheme 13. Dimeric Pd(II) complex with oxime 4-hydroxyacetophenone [38].

In a model reaction of 4-iodo- or 4-bromoacetophenone with phenylacetylene in [bmim]PF_6 the palladium(II) precursor was reduced to a Pd(0) colloid or "palladium black" and diphenylacetylene was formed with a yield of 57%. However, in [bmim]Cl the palladium(II) precursor was not reduced to Pd(0) and the formation of an anionic $PdCl_4^{2-}$ complex was observed instead. As a consequence, only 6% of the reaction product was obtained. Better results (38–52 %) were obtained when the imidazolium salt [dbmim]PF_6 was used, in which a mobile proton at C2 was replaced by a methyl group. In this medium, the catalyst precursor was more stable and the authors concluded that reactions of the catalyst precursor with ionic liquids are not favorable for the catalytic process [38].

Highly effective coupling of activated and deactivated aryl bromides with phenylacetylene in the ionic liquid ([bmim]BF_4) has been reported for a system composed from $Pd(\eta^3\text{-}C_3H_5)Cl]_2$ /3 PPh_3 [39]. Good recycling possibility was demonstrated for 4-bromoacetophenone and 4-bromotoluene, whereas for 4-bromoanisole the yield of product decreased to 20% in third the run [39]. Novel array reactor was used for palladium catalyzed Sonogashira and Heck reactions in ionic liquids, such as [bmpy]BF_4, [bmim]NTf_2 (NTf_2 = bis(trifluoromethylsulfonyl)imide), [bmim]BF_4 and [bmim]PF_6 [40]. In these reactions Pd(II) carbene complex, Pd(bmim-y)$Cl_2(PPh_3)$, soluble in ionic liquids, was used as catalyst [40].

4.3. The Heck reaction

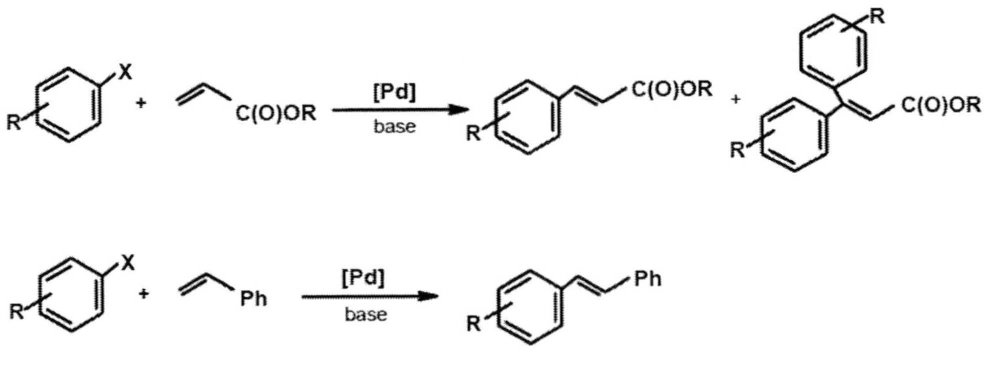

Scheme 14. The Heck reaction.

The first well-documented results of the application of ionic liquids (tetraalkylammonium or phosphonium, $[Bu_3PC_{16}H_{33}]Br$, salts) in the Heck reaction of bromobenzene with butyl acrylate were published in 1996 [41]. Using $PdCl_2(PPh_3)_2$ as a catalyst precursor and Et_3N as a base and the eventual reducing agent for Pd(II), a relatively high yield of trans-butylcinnamate was obtained. It was concluded that ammonium and phosphonium salts stabilized the reactive form of the palladium catalyst. Molten, anhydrous salts like $[Bu_4N]Br$ were found to be very good reaction media for the Heck reaction, particularly for the less reactive chloroarenes (like chlorobenzene) catalyzed by the palladacyclic complex (trans-di(μ-acetato)-bis[o-(di-o-toluilphosphino)benzyl]dipalladium(II)) (Scheme 15) and other Pd(II) complexes as well as by Pd(0) species [42].

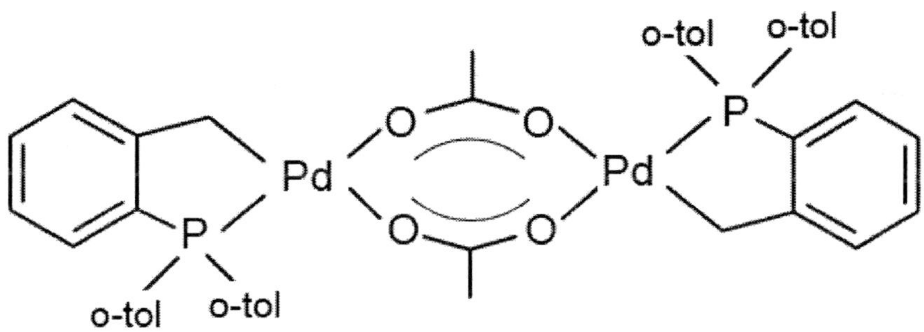

Scheme 15. Carbapalladacyclic complex [25].

In a similar reaction of chlorobenzene with styrene, imidazolium salts were found useless as the reaction medium regardless of the kind of anion. Palladium precursors containing phosphine ligands like $PdCl_2(PPh_3)_2$ were usually better than those without phosphorus ligands when used in $[Bu_4N]Br$, although 20% of stilbene was obtained with the application of phosphane-free $Pd(OAc)_2$ precursor [42].

It is worth noting that quite good results in the Heck reaction of bromobenzene with butylacrylate were obtained using simple Pd(II) precursors like $PdCl_2$, $PdCl_2(PhCN)_2$, or $Pd(OAc)_2$ in molten $[Bu_4N]Br$ [43]. In this reaction, carried out in the absence of a base, only trans-butylcinnamate was found as the product, and its yield nonlinearly depended on palladium concentration. In the presence of bases, in most of the catalytic systems studied, the main product was butylphenylcinnamate; however, after proper selection of the base, it is possible to obtain a mixture of both products (butylcinnamate and butylphenylcinnamate). Most probably the reactions are catalyzed by nanoparticles of Pd(0), formed *in situ* under the Heck reaction conditions and stabilized with tetrabutylammonium salt. It cannot be ruled out that the product of oxidative addition of bromobenzene to Pd(0) and its subsequent reaction with ammonium salt leads to the formation of soluble square planar Pd(II) complexes of the general formula $[Bu_4N]_2[Pd(Ph)Br_3]$. Finally, several different reactive forms of palladium participated in the catalytic process [44].

In Heck reactions in which electron-rich alkenes, such as vinyl ethers, enamides, or alkyltrimethylsilane, are used as substrates, it is difficult to obtain one product, α or β, with high selectivity. It has been proposed by several authors [45–47] that a neutral alkene complex of the formula [Pd(P-P)(Ar)(X)(alkene)] (where P-P – bidentate phosphorus ligand) may be considered an intermediate form responsible for the formation of a mainly linear product (β-substituted alkene), whereas the contribution of an ionic complex of the formula $[Pd(P\text{-}P)(Ar)(alkene)]^{+}X^{-}$ leads to a branched product α (Scheme 16).

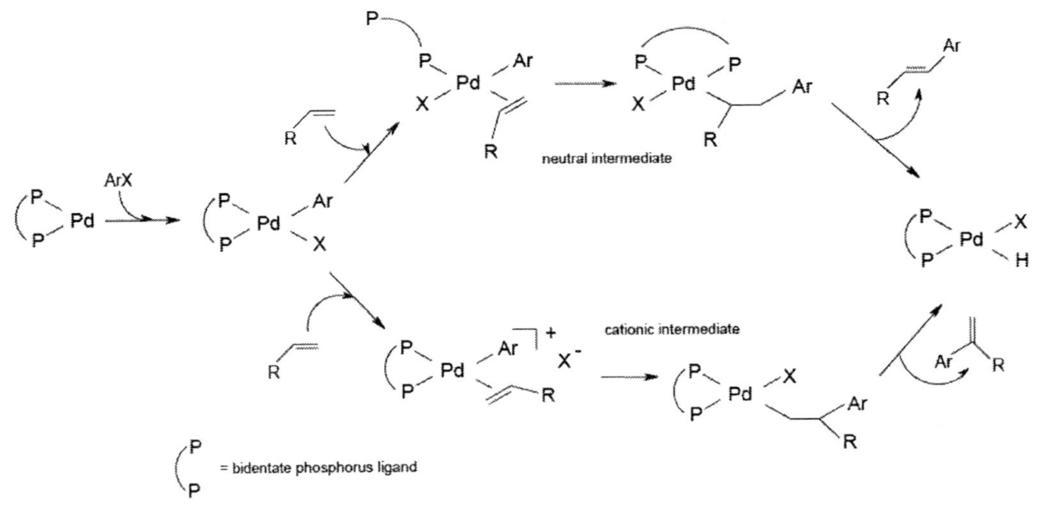

Scheme 16. The mechanizm of isomeric arylated alkenes formation in Heck reaction catalyzed by palladium complex [47].

It was expected that the application of ionic liquids would lead to the formation of cationic rather than neutral intermediate alkene complexes. In fact [bmim]BF_4 appeared to be a very good solvent, in which high regioselectivity (>99%) of the coupling reaction of bromobenzene and iodobenzene derivatives with vinyl ethers was achieved [47]. In the same catalytic system containing $Pd(OAc)_2$ and diphenylphosphinopropane (dppp), the α-arylated product of 1-bromonaphtalene coupling with butyl vinyl ether in [bmim]BF_4 was obtained with a selectivity of *ca.* 99%. It is worth noting, for the sake of comparison, that in typical organic solvents a mixture of α and β isomers was usually obtained [48].

Arylation of allyl alcohols in [bmim]BF_4 gave little conversion of substrates. The reaction proceeded much better in a mixture of [bmim]BF_4 and DMSO, the same in which enamides were also arylated with high regioselectivity [49]. In fact, this modification made it possible to obtain products of the Heck coupling of allyl alcohol and long chain (CH_2=CH-$(CH_2)_n$-OH) alcohols with different derivatives of bromobenzene and bromonaphtalene with high regioselectivity [49].

In some Heck reactions, [bmim]Br appeared to be a better reaction medium than [bmim]BF_4. Although there are not many examples of such cases, it is worth mentioning the reaction of bromo- or iodobenzene with acrylates or styrene catalyzed by $Pd(OAc)_2$ as one of them [50]. The higher reaction yields obtained in [bmim]Br were explained as an effect of transformation of the starting palladium complex into a carbene one, identified in the reaction mixture and then isolated. Interestingly, the carbene complex appeared to be a good catalyst in [bmim]Br, where 64–94% of the product was formed, whereas in [bmim]BF_4 the reaction yield was significantly lower [50].

A carbene complex of Pd(II) with a benzotiazole ligand (Scheme 7) was used as a catalyst precursor in the Heck reaction of aryl bromides and activated aryl chlorides with allyl alcohols in [bmim]Br; however, better results were obtained in reactions carried out in $[Bu_4N]Br$. In reaction of 4-bromotoluene with *trans*-ethylcinnamate, the reaction yield (22%) was also much lower than in $[Bu_4N]Br$ [51]. Using the TEM image it was shown that during the catalytic reaction the catalyst precursor underwent reduction to Pd(0) nanoparticles, which most probably play the role of catalyst in the Heck reaction [51].

Pd(0) nanoparticles, immobilized on chitosan (natural polymer), catalyzed the Heck reaction of bromobenzene with butylacrylate in an ionic liquid ($[bmpy]BF_4$) medium with a yield of *ca.* 76% after 2.5 h [52]. When $[bmim]BF_4$ or [bmim]Br was used instead of $[bmpy]BF_4$, the reaction yield was below 1% even after 24 h. However, in tetrabutylphosphonium bromide, $[Bu_4P]Br$, the reaction yield was 95%.

Colloidal Pd(0) stabilized on PVP (polyvinylpyrrolidone) with the average nanoparticle size of 19.8 nm catalyzed reaction of iodobenzene with butylacrylate in the molten salt $[Bu_4N]Br$ [53]. By appropriately selecting the parameters of the reaction conditions, such as temperature, reaction time, and kind of base (Et_3N, $NaHCO_3$, HCOONa), it was possible to obtain a very high yield of the product and selectivity of monoarylated (butylcinnamate) or diarylated (butyl phenylcinnamate) products [53].

Good results of the Heck reaction were achieved for a Pd(II) complex with bis-imidazole as the ligand in an ionic liquid medium [54].

A palladium complex with a hemilabile pyrazolyl-functionalized carbene ligand is an excellent catalyst for Heck and Suzuki reactions in the ionic liquids $[bmim]BF_4$ and $[bmim]PF_6$. At 120 °C with 2% mol of the catalyst, *ca.* 90% of butylcinnamate was obtained from iodobenzene and up to 50% from bromobenzene. An ionic liquid containing Pd(II) catalyst was recycled five times using different substituted iodobenzenes and styrene as substrates. The yield was *ca.* 90% in each of the five runs [55].

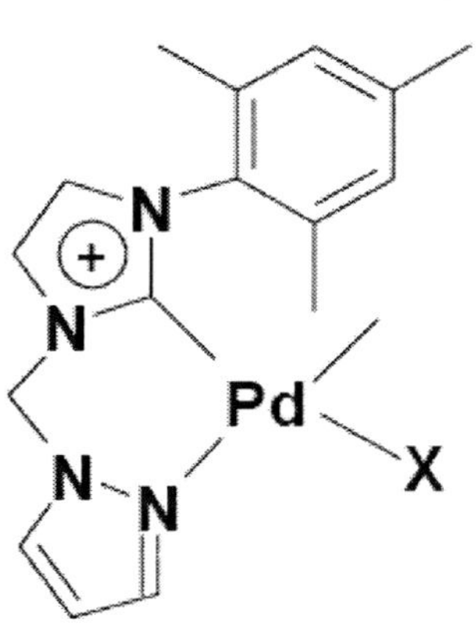

Scheme 17. Pd(II) complex with pyrazolyl-functionalized N-heterocyclic carbene [55].

$Pd(OAc)_2$ was immobilized on amorphous silica covered with $[bmim]PF_6$, and it was proposed that a solution of $Pd(OAc)_2$ in the ionic liquid exists in the pores of the silica. $Pd(PPh_3)_4$ and Pd black were immobilized in a similar manner. In a test reaction, iodo- and bromobenzenes reacted with cyclohexylacrylate giving substituted cyclohexylcinnamates [56].

An original ionic liquid with a glass transition temperature of –42.76 °C, the product of the reaction of 2,2'-bis-imidazole with iodobutane (Scheme 1), acted as solvent and ligand for a palladium catalyst precursor. The system was recyclable three times at 100 °C not only for coupling of iodobenzene with methyl acrylate but also for coupling chlorobenzene with styrene [57].

An interesting functionalized ionic liquid, [bmim][TPPMS] (TPPMS = diphenyl(3-sulfonatophenylphosphine), was used together with [bmim]OAc as a base and $PdCl_2(CH_3CN)_2$ in the Heck synthesis of ethyl cinnamate from bromobenzene. The main advantage of the system was a synergic effect of the two imidazolium salts and no accumulation of sodium or ammonium bromides. An excellent recycling experiment demonstrated almost constant productivity (*ca.* 60% yield in 10 runs with leaching of Pd to the extracted organic phase beyond the limit of ICP detection) [58].

4.4. The Suzuki Reaction

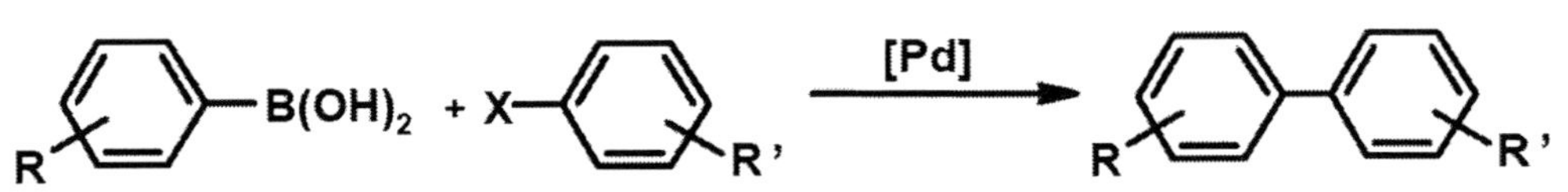

Scheme 18. The Suzuki reaction.

Palladium complexes with phosphorus ligands or complexes with a phosphorus-free coordination sphere or even Pd(0) colloids are frequently used for the synthesis of biphenyl compounds in the Suzuki reaction proceeding between aryl halides and phenylboronic acid. Depending on the catalyst precursor structure and the reaction conditions, the catalysts used present different effectiveness as activators of the aryl–X bond. The most active catalysts can be used in the reaction of chlorobenzene and its derivatives as substrates. In the Suzuki reaction of bromobenzene at 110 °C in different ionic liquids, palladium(II) complexes with PPh_3 coordinated or added to the catalytic system as a free ligand were used as catalyst precursors [59]. In this case the reaction required initiation, which could be achieved by heating the catalyst precursor for 1 h in the presence of bromobenzene prior to the introduction of the second substrate, phenylboronic acid. The results obtained made it possible to distinguish two types of ionic liquids. In $[bmim]BF_4$ and $[bbmim]BF_4$, the $Pd(PPh_3)_4$ catalyst precursor was dissolved and did not undergo decomposition, whereas reactions carried out in $[dbmim]BF_4$ and [bmim]Cl were not initiated, probably because a catalytically active form of the palladium complex was not formed. It was established that this active form of the palladium catalyst is a carbene complex of the formula $[(PPh_3)_2Pd(im\text{-}y)X]^+$, which catalyzes the Suzuki reaction in $[bmim]BF_4$ with yields of *ca.* 68.2% [59].

The carbene complex $[Pd(bbim\text{-}y)_2] \cdot 2\ BF_4$ has also been proposed as an intermediate in a catalytic system based on $Pd(OAc)_2$ and used for the Suzuki reaction of aryl halides with phenyl boronic acid carried out in a solvent mixture ($[bbim]BF_4$/MeOH) under ultrasonic conditions [60]. Under such conditions, at 30 °C, the Suzuki reaction yield was 95% for iodobenzene derivatives and *ca.* 50% for chlorobenzene derivatives [60]. 2,2'-bisimidazolium ionic liquids were obtained and applied as reaction media for the Suzuki reaction of substituted bromobenzene, at 100 °C with $PdCl_2$ as a catalyst precursor [61]. The catalyst dissolved in that ionic liquid was used 14 times in successive Suzuki reactions with different substrates giving relatively high yields of the products, ranging from 82 to 90%. After each reaction test, the organic products were separated from the ionic liquid phase containing dissolved catalyst, and a new portion of the substrates was introduced to the reactor for the next experiment. Very high yields in all the 14 experiments suggest stabilization of the catalyst under these particular reaction conditions [61]. These results also confirm that the catalyst is not extracted from the ionic liquid during the product separation procedure.

A carbapalladacyclic complex in which oxime of 4-hydroxyacetophenone was present as a chelating ligand (Scheme 13) was found to be a very active catalyst precursor in the Suzuki reaction of aryl halides in water, but the best results were obtained in polyethylene glycol (PEG) [38]. However, this catalyst precursor used in the ionic liquids [bmim]PF_6 and [dbmim]PF_6 was inactive. As an example of a new synthetic methodology in the Suzuki process, the reaction was carried out in water with the addition of [Bu_4N]Br under microwave conditions with a Pd/C catalyst [62]. A similar reaction medium (H_2O/[R_4N]BF_4, where R = alkyl group) appeared to be convenient for a reaction catalyzed by $PdCl_2$; however, the reaction proceeded in a two-phase system [62].

Pd(0) colloid synthesized by heating of $Pd(OAc)_2$ dissolved in [Bu_4N]Br mixed with an excess of [Bu_4N]OAc was used as a catalyst precursor in the Suzuki reaction of bromo- and chlorobenzenes [63]. The reactions were carried out in different tetraalkylammonium salts with good results, including a recycling experiment [63].

Interesting results were obtained when different imidazoles were added to the catalytic system with $PdCl_2(CH_3CN)_2$ as the precursor and ionic liquids as solvents [64]. The highest yields (85.3 and 83.4%) were obtained in those ionic liquids that cannot form carbene complexes, i.e. in [dbmim]Cl and [bmpy]BF_4, whereas in [bmim]Cl and [bmim]OSO_2CH_3 reaction did not occur. It was also noted that reactions carried out in [bmim]BF_4 were inhibited in the presence of imidazoles containing N–H bonding or substituents in the C2 position. The best result (88.2%) was obtained when 1-methylimidazole was added [64].

5. Mechanism of the Inhibiting Effect of Imidazolium Halides

The literature data presented above warrant the conclusion that good results in catalytic reactions are very seldom obtained in imidazolium halides used as the reaction medium. Many more applications have been found for imidazolium salts with BF_4 and PF_6 anions. A particularly strong inhibiting effect of imidazolium chlorides has been demonstrated in methoxycarbonylation of iodobenzene for all tested catalyst precursors, both monomolecular Pd(II) and nanosized Pd(0) colloid. Investigations of palladium catalyst precursor reactions with imidazolium halides made it possible to identify anionic square planar Pd(II) complexes as their products (Scheme 8). However, their formation does not explain the inhibition caused by imidazolium halides, because in other solvents the catalytic activity of anionic complexes of the type $[IL]_2[PdX_4]$ (IL = ionic liquid cation) used as precursors is high and similar to that of the $PdCl_2$(cod) complex. It was observed during those studies that a very important role can be ascribed to the proton attached to the C2 carbon of the imidazolium ring (Scheme 19).

Imidazolium chlorides in which this proton was replaced by the methyl group (Scheme 1) demonstrated much weaker inhibition, and the yield of the methoxycarbonylation product was *ca.* 50%. This suggests involvement of carbene complexes in the inhibition mechanism of catalytic reaction. The reaction mechanism presented in Scheme 19 explains the inhibiting action of imidazolium halides and describes the processes occurring in the catalytic system.

The most important stage of this mechanism (Scheme 19) is the interaction of imidazolium halide with the Pd(II) complex formed as a product of oxidative addition of aryl halide to Pd(0) as an active form of the catalyst. This particular stage of the reaction is common also to other processes in which a new C–C bond is formed (Suzuki, Heck, Sonogashira reactions). The intermediate Pd(II) complex containing an [X-Pd(II)-Ph] 12-electron core interacts with imidazolium halide producing benzene (chromatographically proved). This unexpected reaction eliminates the intermediate aryl-palladium complex from the main stream of the catalytic reaction leading to C–C bond formation. According to Scheme 19, a carbene palladium complex is formed instead of a C–C coupling product. In order to discover the role of the palladium carbene complex as a catalyst precursor, such complexes have been prepared separately in reaction of $Pd(OAc)_2$ with imidazolium halides. In all experiments the catalytic activity of carbene complexes was lower than that of $PdCl_2(cod)$ used as reference.

It may be assumed that the inhibiting action of imidazolium halides can be explained by two effects:

i. Decomposition of the [X-Pd(II)-Ph] intermediate complex and formation of benzene
ii. Formation of a stable Pd(II) carbene complex that appears a much weaker catalyst precursor in ionic liquids other than imidazolium halide

A slightly different mechanism can be proposed for the Sonogashira reaction, in which the $PdCl_2[P(OPh)_3]_2$ complex is used as a catalyst precursor in reaction of iodobenzene with phenylacetylene. In this reaction, the product of the oxidative addition of PhI to Pd(0), formed *in situ,* reacts with [bmim]Cl and forms the phosphonium salt $[P(Ph)(OPh)_3]Cl$ and a palladium(II) anionic complex of formula $[bmim][Pd(I)_x(Cl)_{3-x}(P(OPh)_3)]$ [31]. Decomposition of the intermediate complex and elimination of the aryl ligand from the coordination sphere of palladium(II) explain well the deactivation of the catalytic system by imidazolium halide (Scheme 19) [31].

Scheme 19 presents the overall mechanism of the inhibiting effect of imidazolium chlorides in the above-discussed catalytic systems (for methoxycarbonylation and Sonogashira reactions). In both cases imidazolium halide causes decomposition of the intermediate [X-Pd(II)-Ph] complex and subsequent transformation of a coordinated phenyl ligand into benzene or to phosphonium salt with a phenyl group respectively.

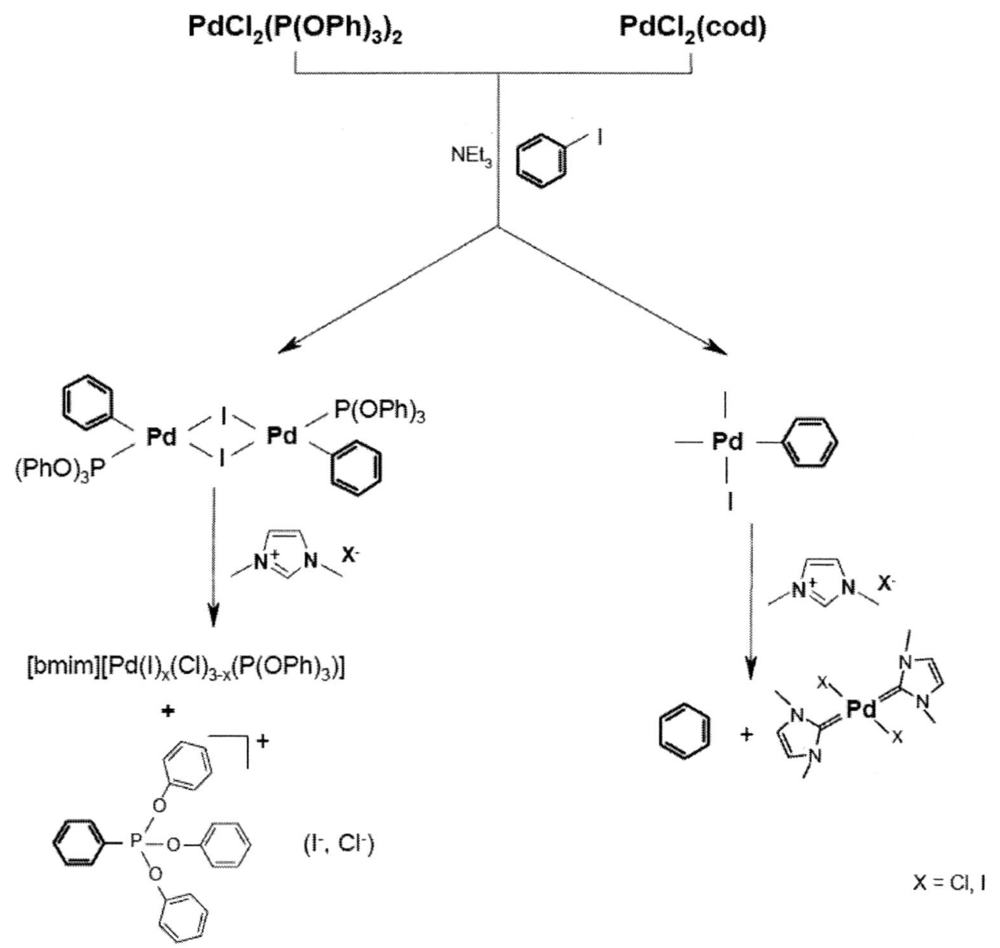

Scheme 19. The mechanizm of inhibition of C-C bond coupling reaction by imidazolium halides [31, 73].

6. Impact of Ionic Liquids on the Catalytic Reaction Mechanism - Conclusion

The examples presented above illustrate the advantages and disadvantages of the use of ionic liquids as solvents for organic reactions catalyzed by palladium. In many cases in both methoxycarbonylation and the Sonogashira reaction carried out in ionic liquids, quite good results have been obtained. The application of ionic liquids, which are environmentally friendly solvents, additionally simplifies the product–catalyst separation procedure, and the palladium catalyst immobilized in ionic liquids remains highly active in successive reactions.

Analysis of the examples discussed above also leads to the conclusion that there is no single ionic liquid suitable for a particular reaction and the overall effect of a catalytic reaction depends both on the kind of ionic liquid as well as the type of palladium precursor used. However, it is worth noting that one of the most effective ways to increase the yield of the Heck reaction was to apply so-called Jeffery conditions, i.e. introduce tetraalkylammonium salts to the reaction medium [65, 66]. The use of tetraalkylammonium salts has made it possible to obtain very good results in a number of Heck reactions [67 - 69] as well as in carbonylation and Suzuki reactions [70]. The role of tetraalkylammonium salts and their interactions with palladium catalysts has been discussed by many authors, who regarded ammonium salts as stabilizers of metal nanoparticles formed *in situ* as colloids, preventing their aggregation into bigger size particles, usually inactive or less catalytically active [71, 72]. The effect of tetraalkylammonium salts as phase-transfer agents has also been considered [65, 66]. One of the ideas for explaining the positive effect of tetraalkylammonium salts on the course of catalytic reactions is based on steric considerations. According to this model, bulky tetraalkylammonium cations have a better stabilizing effect than planar imidazolium or pyridinium cations, particularly with respect to Pd(0) nanoparticles and to 12-electron Pd(II) transient species of the [Ar-Pd-X] type formed *in situ* during the reaction.

This explanation reflects well the properties of a catalytic system based on a carbene complex of Pd(II) with a benzotiazole ligand, which is very active in tetraalkylammonium salts but shows very low activity in other ionic liquids. However, the explanation, which is applicable to the Heck reaction, cannot be used for methoxycarbonylation, in which good results have been obtained for reactions catalyzed by a Pd(0) colloid in pyridinium salts with different anions and in imidazolium salts with nonhalogenic anions. On the other hand, in the Heck reaction catalyzed by carbene Pd(II) complexes, results obtained in imidazolium halides were better than in ammonium salts.

When considering the effect of ionic liquids on the course of the catalytic reaction, it is important to keep in mind that ionic liquids can interact with the catalyst precursor. In the case of Pd(II) catalyst precursors, this interaction may lead to the formation of anionic Pd(II) complexes of the type $[IL]_2[PdX_4]$ or $[IL][PdX_3P]$ (P = phosphorus ligand). The formation of such anionic complexes may complicate Pd(II) to Pd(0) reduction, which occurs in all of the reactions discussed above prior to the activation of aryl halides (ArX). The formation of Pd(0) by *in situ* reduction of any Pd(II) catalyst precursor is an extremely important reaction stage, determining the oxidative addition of aryl halide and the formation of an [Ar-Pd-X] 12-electron intermediate. XPS investigations have shown that at elevated temperatures Pd(II) in $PdCl_2(PhCN)_2$ undergoes reduction with $NaHCO_2$, but reduction was not observed in the presence of $[Bu_4N]Br$ [44]. The opposite effect has been observed when another base, $NaHCO_3$, was used. In that case, reduction of Pd(II) in $PdCl_2(PhCN)_2$ was observed only in the presence of $[Bu_4N]Br$ [44]. The above examples show that ionic liquids (in this case tetraalkylammonium salts) are not only the reaction medium, but also interact with the catalyst precursor, changing its catalytic function.

To better understand the overall function of ionic liquids in all of the C–C cross-coupling reactions discussed above, one should first of all consider the interaction of a Pd(0) colloid with aryl halides in an ionic liquid reaction medium. The *in situ* formation of Pd(0) nanocolloids in methoxycarbonylation, Heck, Suzuki, and Sonogashira reactions is at present well proved experimentally. The contribution of Pd(0) colloidal nanoparticles to the catalytic process is being investigated intensively. The high catalytic activity of Pd(0) colloidal nanoparticles, which has been announced frequently in recent reports, suggests that their activity is at least comparable to that of monomolecular Pd(0) complexes. Results obtained by several authors have proved that Pd(0) on the surface of nanoparticles interacts with aryl halides (ArX) according to the oxidative addition mechanism, which can lead to the formation of reactive and unstable 12-electron species [Ar-Pd-X]. An ionic liquid can stabilize them as 16-electron anionic Pd(II) complexes of the type $[PdArX_3]^{2-}$, whose negative charge is compensated for by two cations originating from the ionic liquid. The presence of such intermediates has been proved using the XPS method for a Pd(0) colloid stabilized by PVP or supported on Al_2O_3.

When similar reactions are carried out in typical organic solvents, the formation of soluble Pd(II) complexes is less intensive. The initial size of Pd(0) nanoparticles (19.8 nm) heated with aryl halide in an organic solvent (dioxane) decrease only slightly, which suggests that Pd(II) complexes are not formed. In contrast, a significant decrease in Pd(0) nanoparticle size (from 19.8 to 7.6 nm) after reaction with iodobenzene in $[Bu_4N]Br$ points to the formation of soluble monomolecular complexes from Pd(0) immobilized on organic or inorganic support (Scheme 20) [53].

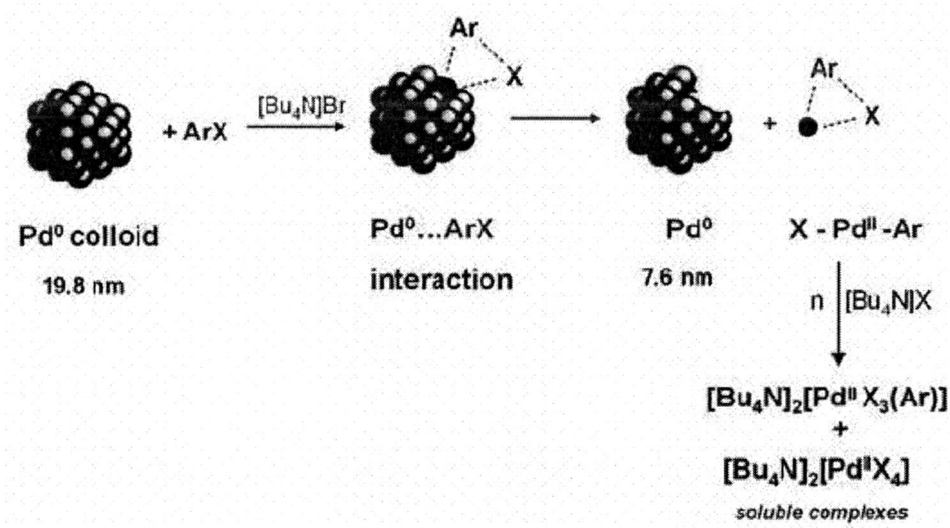

Scheme 20. Interaction of Pd(0) nanoparticles with aryl halides in ionic liquid ($[Bu_4N]Br$) medium [53].

The following important conclusion emerges from these observations: when heterogeneous or heterogenized Pd(0) catalysts are used for C–C cross-coupling reactions in ionic liquids, a large part of Pd(0) from the support is transferred to the solution as soluble Pd(II) complexes. Therefore, initially heterogeneous systems become mixed, hetero-homogeneous ones. This helps to increase the reaction rate and the yield of product but is disadvantageous for product–catalyst separation and for the reuse of the heterogeneous catalyst. Of course, some amounts of soluble Pd(II) complexes may be bounded again to the surface and then reduced to Pd(0) nanoparticles attached to the support.

Imidazolium halides appeared less useful for methoxycarbonylation and Sonogashira reactions, in contrast to all pyridinium ionic liquids and imidazolium salts with non-coordinating anions (like BF_4, PF_6). The inhibiting effect of imidazolium halides is due to decomposition of the [Ar-Pd-X] intermediate complex and the formation of benzene (when a Pd–phosphane-free catalyst precursor is used) or the phosphonium salt $[P(OPh)_3Ph]Cl$ (when

$PdCl_2(P(OPh)_3)_2$ is used as the catalyst precursor) (Scheme 19). During the decomposition of the [Ar-Pd-X] intermediate, Pd(II) carbene complexes are formed in addition to benzene [31]. Those complexes are less active in methoxycarbonylation but have found very wide use in other C–C bond forming reactions, mainly the Suzuki coupling. Palladium complexes with N-heterocyclic carbene ligands are considered an excellent alternative to palladium precursors with phosphorus ligands. However, negative consequences of their formation in some catalytic reactions cannot be ruled out.

Acknowledgement

The authors thank PhD students Iweta Pryjomska-Ray, Andrzej Gniewek, Wojciech Wojtków, Wojciech Zawartka and Michał Szulmanowicz who contributed to the results.

The financial support with the Grant PBZ-KBN-118/T09/2004 is gratefully acknowledged.

References

[1] Anastas P. T. *Green Chem.* 2003, *5*, G29

[2] Anastas P.T., Warner J. *Green Chemistry. Theory and Practice,* Oxford Univ. Press, Oxford, 1998

[3] Winterton N. *Green Chem.* 2001, *3*, G73

[4] Li C.-J., Chan T.-H. *Organic reactions in aqueous media,* J. Wiley and Sons Inc., New York, 1997

[5] Joo F., Katho A. *J. Mol. Catal. A: Chem.* 1997, *116*, 3

[6] Cornils B.,. Herrmann W. A. *Aqueous Phase Organometallic Catalysis*, Wiley, Weinheim, 1998

[7] Trzeciak A. M., Ziółkowski J. J. *Coord. Chem. Rev.* 1999, *190-192*, 883

[8] Holbrey J. D., Seddon K. R. *Clean Products and Processes. I* 1999, 223

[9] Olivier-Bourbigou H.,. Magna L. *J. Mol. Catal. A: Chem.* 2002, *182*, 419

[10] Wasserscheid P., Keim W. *Angew. Chem. Int. Ed.* 2000, *39*, 3772

[11] Sheldon R., *Chem. Comm.* 2001, 2399

[12] Dupont J., De Souza R. F., Suarez P. A. Z., *Chem. Rev.* 2002, *102,* 3667

[13] Gordon C. *M. Appl. Catal. A: General.* 2001, 222, 101

[14] Welton T. *Chem. Rev.* 1999, *99*, 2071

[15] Pernak J. *Przem. Chem.* 2000, *79(5),* 150

[16] Pernak J., Sobaszkiewicz K., Mirska I. *Green Chem.* 2003, *5*, 52

[17] Pernak J., Czepukowicz A., Poźniak R. *Ind. Ing. Chem. Res.* 2001, *40*, 2379

[18] Walden P. *Bull. Acad. Imper. Sci.* (St. Petersburg) 1914, 1800

[19] Sellin M. F., Webb P. B., Cole-Hamilton D. J. *Chem. Comm.* 2001, 781

[20] Herrmann W. A. *Angew. Chem. Int. Ed.* 2002, *40*, 1290

[21] Weskamp T., Bőhm V. P.W., Herrmann W. A. *J. Organomet. Chem.* 2000, *600*, 12

[22] Herrmann W. A., Fisher J., Artus G. R. J. *Angew. Chem. Int. Ed. Engl.* 1995, *34,* 2371

[23] Altenhoff G., Goddard R.,. Lehmann C.W, Glorius F., *J. Am. Chem. Soc.* 2004, *126,* 15195
[24] Viciu M. S., Kissling R. M., Stevens F. D., Nolan S. P. *Org. Lett.* 2002, *4*, 2229
[25] Herrmann W. A., Kocher C. *Angew. Chem. Int. Ed. Engl.* 1997, *36*, 2162
[26] Cavell K. J., McGuiness D. S. *Coord. Chem. Rev.* 2004, *248,* 671
[27] Trzeciak A. M., Ziółkowski J. J. *Coord. Chem. Rev.* 2005, *249*, 2308
[28] Mizushima E., Hayashi T., Tanaka M. *Green Chemistry.* 2001, *3,* 76
[29] Calo V., Giannoccaro P., Nacci A., Monopoli A. *J. Organomet. Chem.* 2002, *645,* 152
[30] Mizushima E., Hayashi T., Tanaka M. *Topics in Catalysis.* 2004, *29*, 166
[31] Zawartka W.,. Trzeciak A. M,. Ziółkowski J. J, Lis T., Ciunik Z., Pernak J. *Adv. Synth. Catal.* 2006, *348*, 1689
[32] Wojtków W., Trzeciak A. M., Choukroun R., Pellegatta J. L. *J. Mol. Catal. A: Chem.* 2004, *221,* 81
[33] Sans V., Trzeciak A. M., Luis S., Ziółkowski J. J. *Catal. Lett.* 2006, *109*, 37
[34] Park S. B., Alper H. *Chem. Commun.* 2004, 1
[35] Gholap A. R., Venkatesen K., Pasricha R., Daniel T., Lahoti R. J., Srinivasen K. V. *J. Org. Chem.* 2005, *70*, 4869
[36] Klementova I., Gotov B., Gajda V., Toma S. *Monatshefte f. Chem.* 2003, *134*, 545
[37] Fukuyama T., Shinmen M., Nihitani S., Sato Ryu M. *Org. Lett.* 2002, *4(10),* 1691
[38] Corma A., Garcia H., Leyva A. *Tetrahedron.* 2005, *61*, 9848
[39] Hierso J.-M., Boudon J., Piquet M., Meunier P., *Eur. J. Org. Chem.* 2007, 583
[40] Rahman Md. T., Fukuyama T., Ryu I., Suzuki K., Yonemura K., Hughes P. F., Nokihara K. *Tetrahedron Lett.* 2006, *47,* 2703
[41] Kaufmann D. E., Nouroozian M., Henze H. *Synlett.* 1996, 1091
[42] Bőhm V. P.W., Herrmann W. A. *Chem. Eur. J.* 2000, *6(6),* 1017
[43] Pryjomska-Ray I., Trzeciak A. M., Ziółkowski J. J. *J. Mol. Catal. A: Chem.* 2006, *257,* 3
[44] Pryjomska-Ray I., Gniewek A., Trzeciak A. M., Ziółkowski J. J., Tylus W. *Topics in Catal.* 2006, *40*, 173
[45] Cabri W., Candiani I., Bedeschi A. *J. Org. Chem.* 1993, *58*, 7421
[46] Ozawa F., Kubo A., Hayashi T. *J. Am. Chem. Soc.* 1991, *113,* 1417
[47] Mo J., Xu L., Xiao J. *J. Am. Chem. Soc.* 2005, *127,* 751
[48] Xu L., Chen W., Ross J. Xiao J. *Org. Lett.* 2001, *3*, 295
[49] Mo J., Xu L., Ruan J., Liu S. *Chem. Comm.* 2006, 3591
[50] Xu L., Chen W., Xiao J. *Organomet.* 2000, *19*, 1123
[51] Calo V., Nacci A., Monopoli A., Spinelli M. *Eur. J. Org. Chem.* 2003, 1382
[52] Calo V., Nacci A., Monopoli A. *Eur. J. Org. Chem.* 2006, 3791
[53] Gniewek A., Trzeciak A. M., Ziółkowski J. J., Kępiński L., Wrzyszcz J., Tylus W. *J. Catal.* 2005, *229*, 332
[54] Park S. B., Alper H. *Org. Lett.* 2003, *5(18),* 3209
[55] Wang R., Twamley B., Shreeve J.M. *J. Org. Chem.* 2006, *71*, 426
[56] Hagiwara H., Sugawara Y., Isobe K., Hoshi T., Suzuki T. *Org. Lett.* 2004, *6(14),* 2325
[57] Xiao J.-C., Twamley B., Shreeve J.M. *Org. Lett.* 2004, *6(21),* 3845
[58] Liu Y., Li M., Lu Y., Gao G.-H., Yang Q., He M.-Y. *Catal. Comm.* 2006, *7*, 985
[59] McLachlan F., Mathews C. J., Smith P. J., Welton T. *Organomet.* 2003, *22*, 5350
[60] Rajagopal R., Jarikote D. V., Srinivasan K. V. *Chem. Comm.* 2002, 616

[61] Xiao J.-C., Shreeve J. M. *J. Org. Chem.* 2005, *70*, 3072
[62] Arvela R. K., Leadbeater N. E. *Org. Lett.* 2005, *7(11)*, 2101
[63] Calo V., Nacci A., Monopoli A., Montingelli F. *J. Org. Chem.* 2005, *70*, 6040
[64] Mathews C. J., Smith P. J., Welton T. *J. Mol. Catal. A: Chem.* 2004, *214*, 27
[65] Jeffery T., Galland J.-C. *Tetrahedron Lett.* 1994, *35(24)*, 4103
[66] Jeffery T., David M. *Tetrahedron Lett.* 1998, *39*, 5751
[67] Calo V., Nacci A., Lopez L., Napola A. *Tetrahedron Lett.* 2001, *42*, 4701
[68] Calo V., Nacci A., Monopoli A., Detomaso A., Iliade P. *Organomet.* 2003, *22*, 4193
[69] Bauquillon S., Ganchegui B., Estrine B., Henin F., Muzart J. *J. Organomet. Chem.*, 2001, *634*, 153
[70] Zou G., Wang Z., Zhu J., Tang J., He M. Y. *J. Mol. Catal. A: Chem.* 2003, *206*, 193
[71] Bonnemann H., Brinkmann R., Koppler R., Neiteler P., Richter J. *Adv. Mater.* 1992, *4*, 804
[72] Klingelhofer S., Heitz W., Greiner A., Oestereich S., Forster S., Antonietti M. *J. Am. Chem. Soc.* 1997, *119*, 10116
[73] Trzeciak A.M. *Wiadomości Chem.* 2006, *60*, 819

In: Advances in Organometallic Chemistry Research ISBN: 978-1-60021-779-1
Editor: Kenji Yamamoto, pp. 201-225

Chapter 8

MAKING CRYSTALS FROM CRYSTALS: METHYL-MIGRATIONAL DISMUTATION OF TRIMETHYLTIN COMPLEXES

Andrea Deák* and Gábor Tárkányi
Institute of Structural Chemistry, Chemical Research Center of the Hungarian Academy of Sciences, P. O. Box 17, 1525 Budapest, Hungary

ABSTRACT

The dismutation (degradation) via alkyl group removal of triorganotins has been detected in the environment and has been attributed to the action of UV light, chemical cleavage and biological degradation by bacteria, although its chemistry has never been elucidated. Interestingly, in contrast to thermally induced solid-state methyl migrations in organic molecules, which have been the subject of extensive investigations, the solid-state methyl transfer phenomenon in triorganotin(IV) compounds have been hardly exploited. Recently, we have shown that the single crystals of Me_3Sn(O,O-donor) complexes with bidentate-chelating O,O-donor ligands such as *N*-nitroso-*N*-phenylhydroxylaminato and tropolonato, demethylated into single crystals of Me_2Sn(O,O-donor)$_2$ and volatile Me_4Sn, as a result of a crystal-to-crystal supramolecular structural transformation. In solution, the solvent-induced changes in the tin coordination number followed by specific conformational changes were held responsible for the second order dismutation kinetics, in which the role of the monomeric triorganotin(IV)-solvates could be verified by various temperature- and concentration dependent NMR experiments. We have concluded that in most of the cases, this methyl-migrational dismutation process can be a potential source of the instability of trimethyltin(IV)-complexes, which prevents the preparation of devised structures via crystallization from coordinating solvents.

During the last few decades a large number of triorganotin(IV)-carboxylates have been prepared and a multitude of structural types have been discovered. These complexes have a tendency to self-assemble by dative Sn–O bonds (2.2–2.3 Å) and/or Sn···O secondary interactions (> 2.5 Å) into supramolecular structures. Moreover, the triorganotin(IV) carboxylates have been one of the most extensively studied class of

* Andrea Deák: Fax: 00-36-1-438-4141/201; Tel: 00-36-1-438-4141/327; E-mail: deak@chemres.hu

anticancer compounds since it was observed that they significantly reduced the growth rates of tumors. Although, a series of triorganotin(IV)-benzoates has been reported, the trimethyltin(IV)-benzoate, $Me_3Sn(bz)$, remained uncharacterized. This discrepancy between the extremely studied nature of triorganotin(IV)-benzoates and the absence of trimethyltin(IV)-derivative suggests that this might be due to its dismutational reactivity which prevents its isolation. Herein, we report the successful synthesis of $Me_3Sn(bz)$ complex, as well as, its X-ray structural determination and NMR analyses both in solution- and solid-state. The data obtained for $Me_3Sn(bz)$ revealed that the essentially planar triorganotin moieties are bridged by anisobidentate carboxylato groups forming one short Sn–O bond and one relatively short Sn···O secondary interaction. In the resulting polymeric chain the difference between these Sn–O and Sn···O bond lengths is quite large (0.429 Å), therefore the helix is stretched in *S*-configuration and the Sn atoms are no longer collinear along the chain. We also report the first thermally accelerated crystal-to-crystal supramolecular structural transformation of a $Me_3Sn(bz)$ polymer into self-assembled dimeric dibenzoato-tetraorganodistannoxane, $\{[Me_2Sn(bz)]_2O\}_2$. The parallels and deviations in the mechanism of the various dismutation processes of the Me_3Sn(O,O-donor) complexes will be highlighted.

INTRODUCTION

The application of a solventless synthetic methodology for organic compounds, which were conventionally synthesized in a solvent medium, becomes more common in recent years. The thermally induced solid-state methyl migrations in organic molecules have also been the subject of extensive investigations. Thus, methyl rearrangement of methyl esters of cyanurates and thiocyanurates [1], methyl *p*-dimethylaminobenzenesulfonate [2], methyl-2-(methylthio)-benzenesulfonate [3], methoxy triazines [4], and Chapman-like rearrangement of 5-methoxy-2-aryl-1,3,4-oxadiazoles [5] have been found to occur in the solid-state even in the absence of a solvent. Whereas *solid-state organic synthesis* represents a well-established area of research [1–5], thermally induced solid-state reactions of organometallic molecules based on methyl-migration have been hardly exploited. We have recently demonstrated that the methyl migration of trimethyltins can be thermally induced in the solid state and as a result of this dismutation the trimethyltin is converted into dimethyltin and tetramethyltin. We have shown [6, 7] that the single crystals of the Me_3Sn(O,O-donor) complexes of bidentate O,O-donor ligands (*N*-nitroso-*N*-phenyl-hydroxylaminato (cupferronato) and tropolonato) underwent the methyl transfer process to give $Me_2Sn(O,O\text{-donor})_2$ and volatile Me_4Sn (Scheme 1).

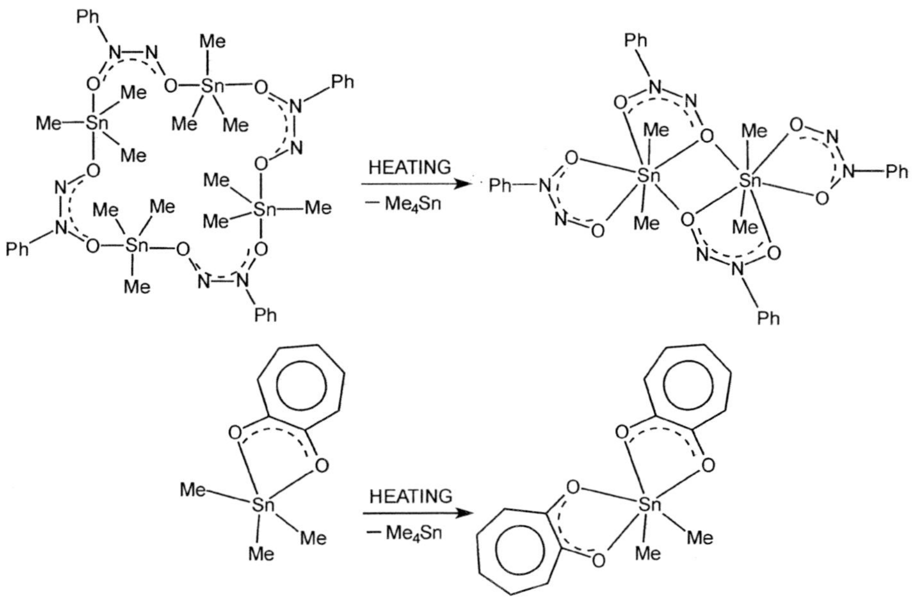

Scheme 1.

Interestingly, the dismutation (degradation) via alkyl group removal of triorganotins has also been detected in the environment and has been attributed to the action of UV light, chemical cleavage and biological degradation by bacteria, although its chemistry has never been elucidated [8]. Recently, it was shown that a linear peptide derived from the membrane protein stannine is able to dealkylate the trimethyltins and the toxic organotins are progressively dealkylated to inorganic Sn(IV) in mammalian organs [9]. An attractive feature of this solid- and liquid-state dismutation is that the resulting dimethyltin and tetramethyltin complexes can be prepared without solvent. The solid-state reactions are generally induced photochemically [10] and/or thermally [11], and they proceed either by topochemical or non-topochemical mechanism [12]. The topochemical principle states the reactions occur with minimum atomic and molecular motion in the crystals [13], whereas recently it was stated that the molecules can migrate to great distances within the crystal as a precondition of reactivity, thus solid-state reaction reactions without molecular migrations are very rare and not all typical [14]. Chemical reactions in the solid state provide and extreme case for evaluating the effect of intermolecular forces on a reaction and their influence on a reaction mechanism [10g]. In the crystal, most bimolecular reactions occur between identical partners, therefore they require precise distances and orientations between reaction partners [2b]. Unfortunately, the topochemical "allowed" distances, i.e. appropriate intermolecular distances, do not guarantee that a reaction will take place [10a]. At the first instance, the knowledge of the crystal structure of trimethyltins provides a direct rationalization of the solid-state dismutation. However, it is difficult to predict the packing motifs for bimolecular reactions in a reliable manner, since the thermal activation of the dismutation reaction is likely a prerequisite for a reorientation of the molecules relative to each other in order to bring the reacting functions as close as possible to react. It was recently showed that several so-called solid-state reactions proceed via phase change from solid to liquid and the existence of a liquid phase is a prerequisite for reaction [15]. Moreover, when heating is required, the *solid-state reactions* took place in a liquid melt, and in the melt the molecular mobility is not restricted by the crystal lattice, thus a large number of productive molecular collisions take place. Based on the NMR kinetic data we concluded that the $Me_3Sn(trop)$ complex rearranges faster in the solid and liquid than in dilute $CDCl_3$ solution [7]. On the basis of the stacking pattern of the $Me_3Sn(trop)$ complex it was apparent that this system requires only a small degree of reorientation to transfer in concerted fashion the methyl- and tropolonato-groups between two molecules to form $Me_2Sn(trop)_2$ and Me_4Sn. Moreover, both $Me_3Sn(trop)$ and $Me_2Sn(trop)_2$ crystallized in similar space group $P2_1/c$, which also suggested the topochemically-controlled nature of the solid-state dismutation. In contrast to this, the $[Me_3Sn(cupf)]_4$ demethylation in solid-state is infinitely slow, and the dismutation requires the formation of the reactive monomeric $Me_3Sn(cupf)$ species [6]. The formation of these species can be triggered by melting and/or dissolution: the $[Me_3Sn(cupf)]_4$ dismutation into $[Me_2Sn(cupf)_2]_2$ dimer and Me_4Sn proceeds with significant change of the molecular and crystal structure as it requires the breaking up of the tetramer. On the basis of these examples, we discuss the possibility to synthesize (discover) another trimethyltin complex showing methyl-migrational dismutation reactivity and the implication of these findings on the organotin chemistry.

Despite being intensively studied the chemistry of organotin carboxylates still represent considerable interest not only for their biological activity [16], but also for their pivotal position in the development or understanding of structural organotin chemistry [17–31]. In recent years it was observed that the majority of organotin carboxylates possessed higher in vitro activity than that of cisplatin, nevertheless, their activity often was comparable with or better than that of doxorubicin [16a]. Moreover the tri-*n*-butyl-, triphenyl- and tricyclohexyltin-carboxylates show specific activity towards aquatic organism, although, they were incorporated in antifouling paints for marine transport vessels and were also utilized as PVC stabilizers [32]. In addition to their biological activity and industrial application, the organotin carboxylates are also of interest in view of the considerable structural diversity that they possess. In triorganotin carboxylates, tin(IV) can have a coordination number of 4 or 5, and the carboxylato ligands may bind in several different ways (monodentate, bidentate chelate versus bridge) so there is potential for many structural types in triorganotin(IV) carboxylates. The organotin carboxylates have a tendency to self-assemble by dative Sn–O bonds (2.2–2.4 Å) and/or Sn···O secondary interactions (> 2.5 Å) into supramolecular structures. Chart I shows some different structures, such as tetra- (A) and penta-coordinated (B) monomer [18, 19, 23i], solvated monomer (C) [20], solvated end-stopped dimer (D) [21,18b], cyclic tetramer (E) and hexamer (F) [22], as well as coordination polymer (G) [23, 18b, 18j].

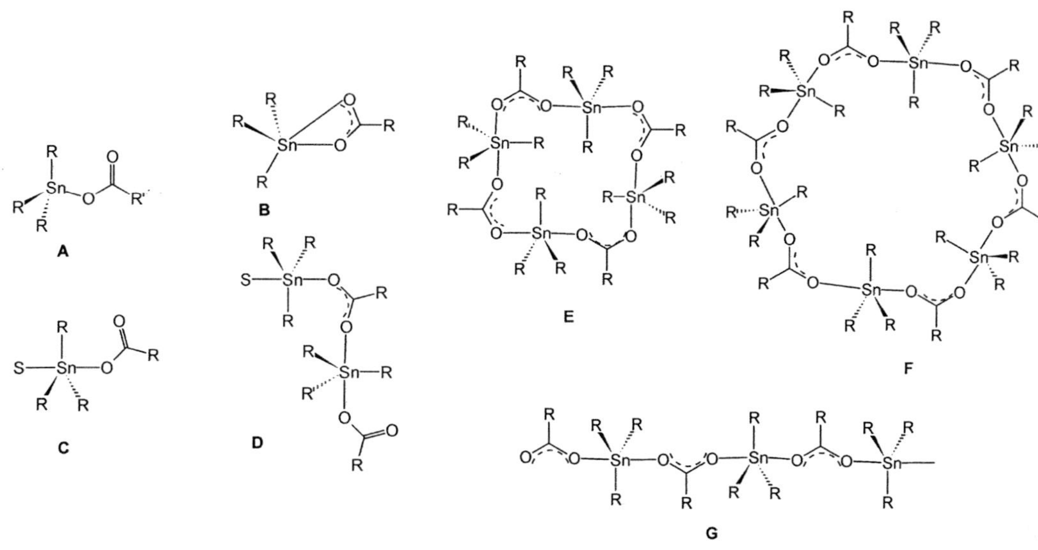

Chart 1. The major structural types of triorganotin carboxylates in the solid state: tetra- (A) and penta-coordinated (B) monomer, solvated monomer (C), solvated end-stopped dimer (D), tetramer (E) and hexamer (F) and coordination polymer (G).

The structural diversity of diorganotin dicarboxylates is also attractive. In these compounds, the tin(IV) coordination number ranges from 5 to 7, and the carboxylato anion may bind in monodentate, bidentate chelate versus bridge as well as bridging-chelating modes. As shown in Chart 2, the main structural types are the tetra- (A) and hexa-coordinated (B) monomeric [24, 25, 29b], solvated monomeric (C) [26], carboxylato- (D) and stannoxane-bridged (E) dimeric [27, 28] and infinite sheet structures (F) [29].

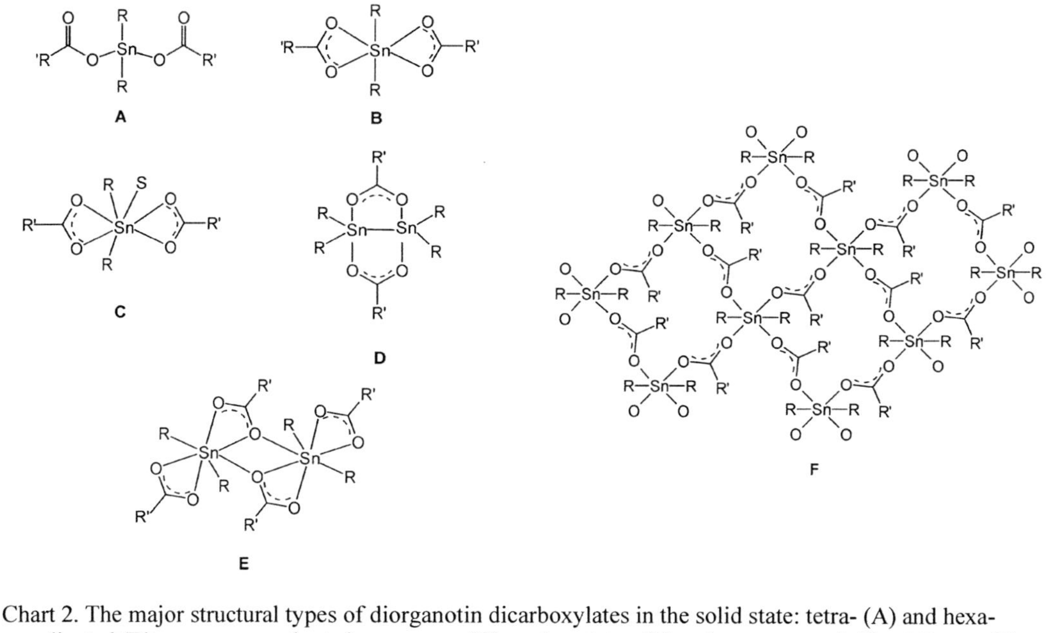

Chart 2. The major structural types of diorganotin dicarboxylates in the solid state: tetra- (A) and hexa-coordinated (B) monomers, solvated monomers (C), carboxylato- (D) and stannoxane-bridged dimers (E) and infinite sheets (F).

Usually, these complexes are unstable with respect to hydrolysis and form dimeric dicarboxylato tetraorganodistannoxanes, which depending on the mode of binding of the carboxylato ligand has been classified into four major structural types (Chart 3) [30].

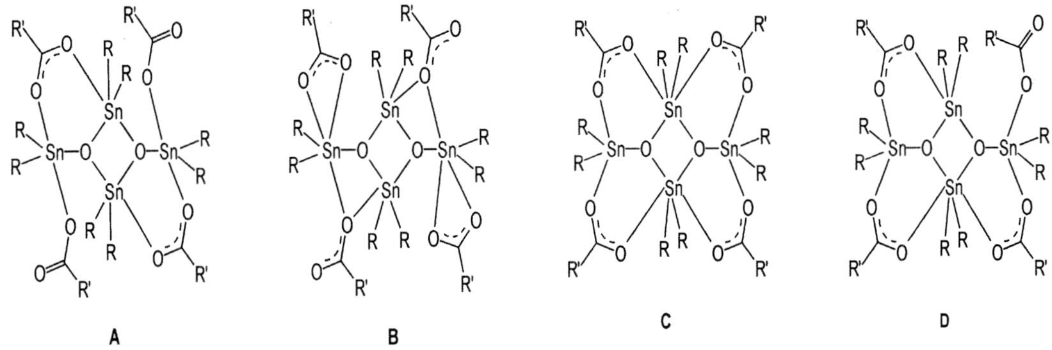

Chart 3. The four major structural types of dimeric dicarboxylato tetraorganodistannoxanes found in the solid state.

Interestingly, however a series of organotin benzoates has been synthesized [16c, 18j, 23b], the trimethyltin benzoate was only obtained in the thermolysis of benzoyl peroxide in the presence of trimethylphenyltin [31], and it remained structurally uncharacterized so far. This discrepancy between the extremely studied nature of organotin benzoates, and the absence of the conventional synthesis as well as structural characterization of trimethyltin benzoate, hereinafter $Me_3Sn(bz)$, suggested that this might be due to its high dismutational reactivity. Here we report the synthesis as well as X-ray and multinuclear NMR characterization of $Me_3Sn(bz)$. The methyl-migrational reactivity of $Me_3Sn(bz)$ was investigated in the solution-, solid and liquid-states. Typically after 9 hours of heating of the $Me_3Sn(bz)$ compound the single crystals of the $\{[Me_2Sn(bz)]_2O\}_2$ dismutation product crystallized from the melt. The X-ray structure of $\{[Me_2Sn(bz)]_2O\}_2$ has already been published without detailed spectroscopic characterization, as the compound was fortuitously obtained in the dimethylphenyltin benzoate synthesis [30c]. The incomplete synthetic, as well as, spectroscopic features of the methyltinbenzoates and the lack of structural changes accompanying the dissolution of these species have led us to examine both the solid and solution structures.

Results and Discussion

Synthesis and Structural Characterization of $Me_3Sn(bz)$

The $Me_3Sn(bz)$ complex was prepared by metathetical reaction between trimethyltin(IV) chloride and ammonium benzoate. Single crystals of $Me_3Sn(bz)$ were successfully obtained from *n*-heptane. The X-ray structural data obtained for $Me_3Sn(bz)$ revealed that the tin adopts a slightly distorted trigonal bipyramidal geometry with the methyl-groups in equatorial position and the O atoms of benzoato ligands are in axial sites (figure 1). The distortion from the idealized TBP geometry can be quantified by the τ descriptor [33], where values of 0 and 1 are indicative of perfect square pyramidal (SP) and trigonal bipyramidal (TBP) geometries, respectively. The calculated value of τ = 0.83 also revealed a strong TBP character of the tin coordination environment. The essentially planar triorganotin moieties are bridged by anisobidentate carboxylato functions forming one short Sn–O bond (2.137(2) Å) and one relatively short Sn···O secondary interaction (2.564(2) Å). The difference in the Sn–O and Sn···O bond lengths is 0.429 Å, and as a result, the Sn centers are no longer collinear along the chain of molecules and the *coordination polymer* (Chart 1, structural type G) is stretched in *S-configuration.* The repeat distance (the distance between two neighboring Sn atoms) is 5.24 Å, and this value falls in the range of 5.19 ± 0.21 Å observed in polymeric triorganotin carboxylates [34]. The bridging coordination mode reported here for $Me_3Sn(bz)$ is one of the major coordination motifs found in triorganotin carboxylates [23], as shown in Chart 1, structural types D–G. In contrast, triphenyltin benzoate, $Ph_3Sn(bz)$, exist in a discrete monomeric form, where the tin is tetra-coordinated and the carboxylato anion is bound to the tin center in monodentate manner (Chart 1, structural type A) forming a 2.073(3) Å Sn–O bond [18j]. The greater electronegativity of the phenyl group over the methyl group has been cited as a factor influencing the formation of a discrete form, since the more electronegative phenyl group occupies an axial position of a TBP more readily than a methyl group [18b].

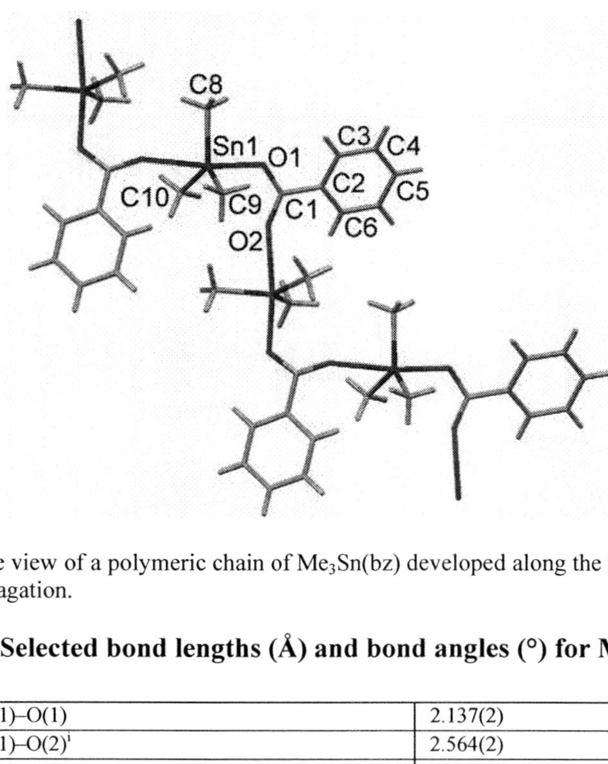

Figure 1. A perspective view of a polymeric chain of $Me_3Sn(bz)$ developed along the 2_1 screw axis coincident with the *b* axis of propagation.

Table 1. Selected bond lengths (Å) and bond angles (°) for $Me_3Sn(bz)$

Sn(1)–O(1)	2.137(2)
Sn(1)–O(2)^i	2.564(2)
Sn(1)–C(8)	2.111(3)
Sn(1)–C(9)	2.107(4)
Sn(1)–C(10)	2.101(4)
O(1)–Sn(1)–O(2)^i	173.3(1)
C(8)–Sn(1)–C(9)	117.0(2)
C(8)–Sn(1)–C(10)	117.4(2)
C(9)–Sn(1)–C(10)	123.6(2)

Equivalent atom (i) generated by (1– x, ½ + y, ½ – z).

The $Me_3Sn(bz)$ complex was also characterized by NMR methods in the solid-state and in solution as well. Its ^{119}Sn NMR spectrum in $CDCl_3$ reveals a single tin resonance at 136.5 ppm, which is the characteristics of trialkyltin(IV) derivatives with tin coordination number of 4 or 5 [35]. Additional structural characterization of the solvated complex is possible from the determination of the relevant one-bond $^{1}J_{(13C-119Sn)}$ scalar coupling constant. Lockhart et al. established that the absolute values of the coupling constants ($^{1}J_{(13C-119Sn)}$ or $^{2}J_{(1H-119Sn)}$) can be used to estimate the Me–Sn–Me bond angles [36]. Accordingly, the relatively small $^{1}J_{(13C-119Sn)}$ value of 398.7 Hz obtained from the ^{1}H decoupled ^{13}C NMR spectrum for the methyl satellites, yields a 111.7° Me–Sn–Me bond angle. Thus, the Me–Sn–Me angle assumes a value of 111.7° in solution as opposed to 119.3(2)° (average value) in the solid-state, suggesting that the molecular geometry is affected by dissolution. Although bidentate carboxylate ligands via bridging coordination to the tin centers could form coordination oligomers (Chart 1, structures E and F) or polymers (Chart 1, structure G) in the solid-state, however, the coordination mode of the carboxylate can transform into monodentate one (Chart 1, structure A) upon dissolution. This implies the predominance of monomeric solvates in solution [18j]. Thus, the oligomeric or polymeric organotin carboxylates, present in the solid state and/or concentrated solutions, can decompose into their monomers upon dilution [22]. In case of $Me_3Sn(bz)$ this is manifested in the reduction of the tin coordination number from 5 to 4, and the shift towards a distorted tetrahedral geometry (T_h) which is in better agreement with the measured ^{119}Sn chemical shift (136.5 ppm). To verify this, it is more appropriate to compare the ^{119}Sn chemical shift of the dissolved $Me_3Sn(bz)$ with the isotropic δ_{iso} shift available from solid-state NMR experiments. Above many other parameters, the isotropic δ_{iso} shift, as determined from magic angle spinning (MAS) ^{119}Sn NMR measurements, is of primary importance in organotin coordination chemistry [37]. Its value is closely related to the tin coordination number and it is directly comparable to the ^{119}Sn chemical shifts recorded in solution. This is a common approach to probe structural changes and to set up solvent coordination models. Figure 3 shows the ^{119}Sn MAS experiments for $Me_3Sn(bz)$ complex at two spinning rates. The shielding parameters have been determined and summarized in table 2. The near zero asymmetry parameter ($\eta \approx 0$) which indicates axial symmetry of the chemical shielding around the tin nucleus is the consequence of the TBP geometry in the solid state. The solid state spectrum of the $Me_3Sn(bz)$ complex has its isotropic chemical shift at δ_{iso} = 32.1 ppm and indicates the TBP geometry at the tin. In contrast, the ^{119}Sn resonance is shifted to higher frequency (δ = 136.5 ppm) in $CDCl_3$ solution. This can be best understood by assuming that the polymeric complex dissociates into monomeric species featuring T_h coordination geometry in a non-coordinating solvent. To study the effect of the solvent, the $Me_3Sn(bz)$ complex was dissolved in an oxygen donor coordinating solvent, methanol. The observed ^{119}Sn chemical shift became 26.1 ppm, which is close to the solid-state value of 32.1 ppm and seems to confirm the existence of solvated monomeric species (Chart 1, structure C). Signal broadening of the ^{119}Sn resonance in methanol is associated with the equilibrium between structures A and C seen in Chart 1. The diffusion NMR studies presented below have also confirmed the monomeric nature of $Me_3Sn(bz)$ yielding an appropriately small r hydrodynamic radius (table 4). Finally, this simple comparison of $CDCl_3$ and CD_3OD recorded ^{119}Sn NMR spectra suggests that the trimethyltin carboxylates may exhibit chemical shift alterations as a result of self-association also in non-coordinating solvents. Thus, depending on the degree of the self-association, the

^{119}Sn chemical shifts measured in $CDCl_3$ may vary as a function of the concentration and/or temperature.

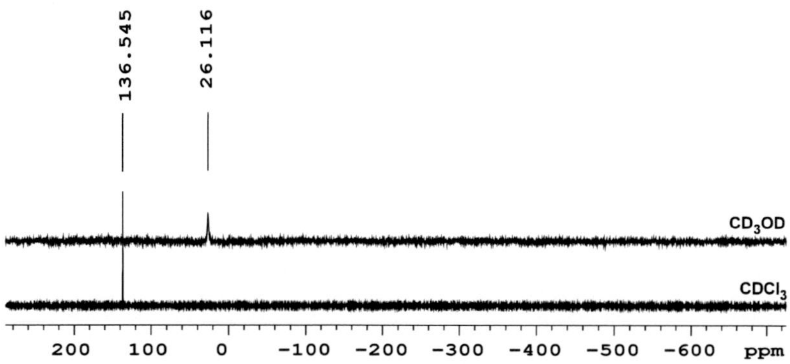

Figure 2. The ^{119}Sn NMR of dissolved $Me_3Sn(bz)$ complex in coordinating (2 mM, CD_3OD) and non-coordinating (10 mM, $CDCl_3$) solvents at 25 °C.

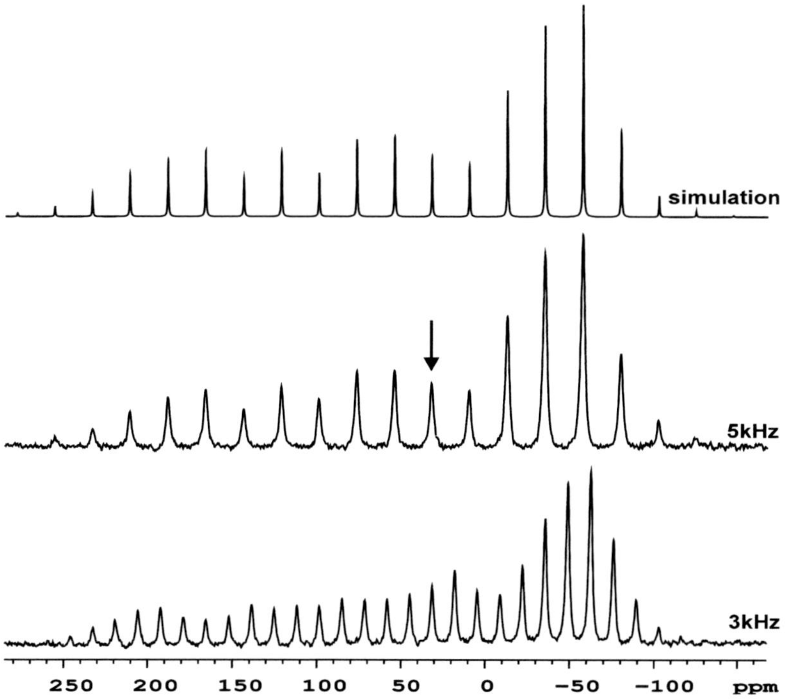

Figure 3. The solid-state ^{119}Sn NMR spectra of $Me_3Sn(bz)$ complex at two different MAS spinning rates. The isotropic chemical shift is denoted with an arrow. The spectrum on the top is a simulated one (5 kHz) used to determine the principle components of the chemical shielding tensor.

Table 2. ^{119}Sn chemical shielding data for $Me_3Sn(bz)$ and $\{[Me_2Sn(bz)]_2O\}_2$ complexes obtained from spin simulation under MAS conditions

	^{119}Sn chemical shielding δ (ppm)				
	δ_{iso}	δ_{11}	δ_{22}	δ_{33}	η
$Me_3Sn(bz)$					
solid MAS NMR	32.1	139.3	139.2	−182.3	0.0005
liquid NMR (MeOD)	26.1				
liquid NMR ($CDCl_3$)	136.5				
$\{[Me_2Sn(bz)]_2O\}_2$					
solid MAS NMR	– 192.2 – 194.1	254	– 104	– 730	0.66
liquid NMR ($CDCl_3$)	– 179.5 – 189.7				

^{119}Sn chemical shielding tensor principal values are given according to the convention $|\delta_{33} - \delta_{iso}| \geq |\delta_{11} - \delta_{iso}| \geq |\delta_{22} - \delta_{iso}|$ with estimated errors of less than 1 ppm; the asymmetry parameter η is defined as $\eta = (\delta_{22} - \delta_{11})/ (\delta_{33} - \delta_{iso})$.

Methyl-Migrational Reactivity of $Me_3Sn(bz)$ in the Solution-, Solid and Liquid-States

When heating the single crystals of $Me_3Sn(bz)$ at 150°C, a small quantity of crystals sublimed to the top of open-ended vial and the colorless product crystallized from the melt after 9 hours (figure 4).

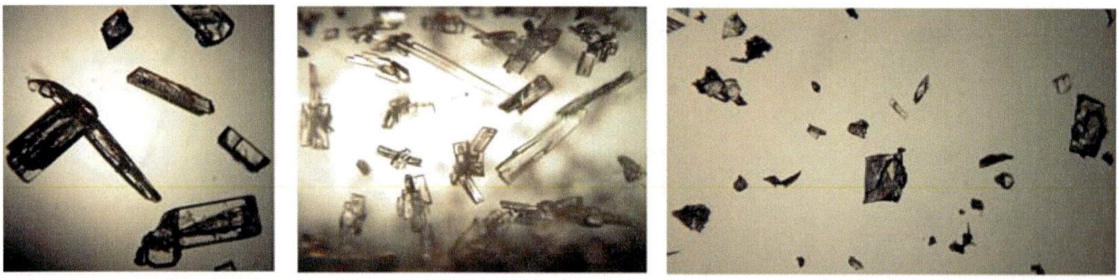

A B C

Figure 4. The images of the (single) crystals of $Me_3Sn(bz)$ before heating (A) and after heating sublimed on the top of the end-opened vial (B), and the melt-crystallized $\{[Me_2Sn(bz)]_2O\}_2$ compound after 9 hours of heating at 150 °C (C) as observed through a microscope.

The X-ray structure determination confirmed the formation of $\{[Me_2Sn(bz)]_2O\}_2$ (figure 5), and thus, the demethylation of $Me_3Sn(bz)$. As it was introduced earlier, trimethyltin complexes could undergo dismutation into dimethyltin and tetramethyltin under various experimental conditions: in solution, in solid and in the melt [6, 7, 38]. In the case of trimethyltin carboxylates, it is expected that the decomposition process does not stop at the primary $Me_2Sn(bz)_2$ product, since further chemical reactions are involved. As it will be discussed in the following pages, we have started to investigate the mechanism of $Me_3Sn(bz)$ dismutation, as well as the structural features of the secondary product $\{[Me_2Sn(bz)]_2O\}_2$.

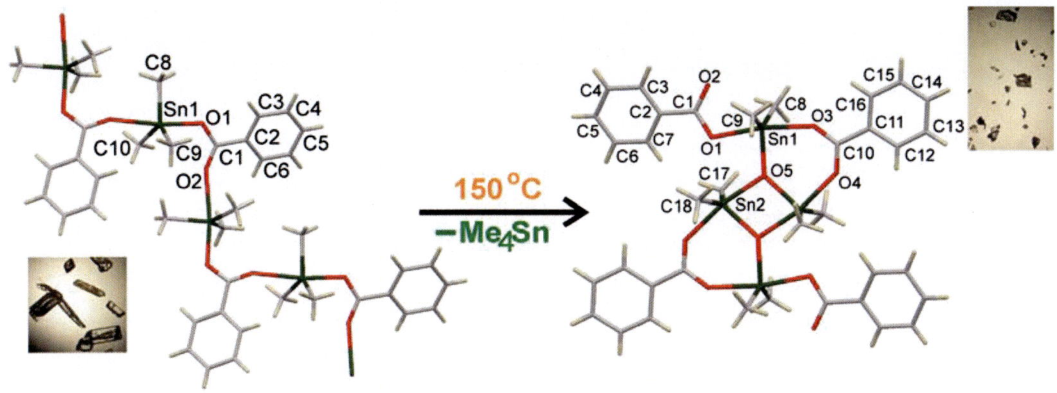

Figure 5. Thermally induced demethylation of $Me_3Sn(bz)$ crystals into crystals of $\{[Me_2Sn(bz)]_2O\}_2$.

The relevant geometric parameters of $\{[Me_2Sn(bz)]_2O\}_2$ established by X-ray diffraction were collected in table 3. The crystal and structural parameters obtained for this melt crystallized dismutation product $\{[Me_2Sn(bz)]_2O\}_2$ correspond closely to the results reported previously for $\{[Me_2Sn(bz)]_2O\}_2$ fortuitously obtained by metathetic reaction between dimethylphenyltin chloride and silver benzoate in an attempt of dimethylphenyltin benzoate synthesis [31]. The molecular structure of this $\{[Me_2Sn(bz)]_2O\}_2$ complex conforms to the common structural type A (Chart 3) of dimeric dicarboxylato tetraorganodistannoxanes. In the dismutation product $\{[Me_2Sn(bz)]_2O\}_2$, the central four-membered stannoxanic Sn_2O_2 ring is planar (torsion angle is 0°), where the corresponding O–Sn–O and Sn–O–Sn angles are 76.8(1) and 103.1(1)°, respectively and the difference in the stannoxanic Sn–O bond lengths is only 0.114(3) Å. The fourteen tin and oxygen atoms are nearly coplanar and the r. m. s. deviation of these atoms from the best plane is 0.0237 Å. For the penta-coordinated endocyclic Sn(1) and exocyclic Sn(2) centers the calculated τ values of 0.52 and 0.40, indicate that both tin centers are in a geometry intermediate between SP and TBP. Each type of tin is, however, subjected to a sixth Sn···O interaction in the form of a carboxyl oxygen atom (2.845(3) Å and 2.803(3) Å, respectively), thus, the molecular geometries about the tin centers are best described as skew-trapezoidal. In this coordination geometry, the dimethyltin skeleton shows a C–Sn–C angle of 139.8(3) and 145.3(2)°, respectively. These endo- and exocyclic tin centers are linked by two isobidentate benzoato anions, and the difference in the Sn–O bond length is almost 0. The other benzoate anions form one short Sn–O bond (2.159(3) Å) and one longer Sn···O secondary interaction (2.845(3) Å) with the exocyclic tin centers. Nevertheless, these oxygen atoms participating in Sn···O contacts are also engaged in weak C–H···O hydrogen bonds [C(15)···O(2)^{i} 3.473(8) Å; H(15)···O(2)^{i} 2.57 Å, C(15)–H(15)···O(2)^{i} 164.0°; symmetry code (i): 1 + x, − ½ − y, ½ + z], which assemble the molecules into layers.

Table 3. Selected bond lengths (Å) and bond angles (°) for $\{[Me_2Sn(bz)]_2O\}_2$

Sn(1)–O(1)	2.159(3)
Sn(1)–O(3)	2.231(4)
Sn(1)–O(5)	2.010(3)
Sn(1)–C(8)	2.097(6)
Sn(1)–C(9)	2.089(5)
Sn(2)–O(5)	2.152(3)
Sn(2)–O(5)^i	2.038(3)
Sn(2)–O(4)^i	2.225(4)
Sn(2)–C(17)	2.093(5)
Sn(2)–C(18)	2.090(6)
Sn(2)···Sn(2)^i	3.284(2)
Sn(2)···Sn(1)^i	3.743(1)
C(9)–Sn(1)–C(8)	139.8(3)
O(1)–Sn(1)–O(3)	170.9(1)
O(5)–Sn(1)–O(1)	80.6(1)
O(5)–Sn(1)–O(3)	90.3(1)
C(17)–Sn(2)–C(18)	145.3(2)
O(5)–Sn(2)–O(5)^i	76.8(1)
O(4)^i–Sn(2)–O(5)^i	92.8(1)
O(5)–Sn(2)–O(4)^i	169.6(1)
Sn(1)–O(5)–Sn(2)	121.6(1)
Sn(2)–O(5)–Sn(2)^i	103.1(1)
Sn(1)–O(5)–Sn(2)^i	135.2(1)

Equivalent atom (i) generated by (– x + 1, – y, – z + 2).

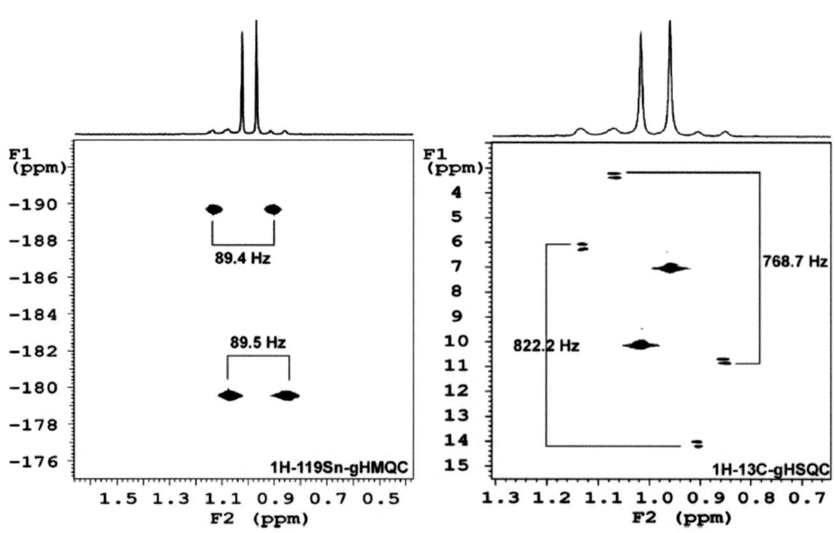

Figure 6. The proton aliphatic part of the ^1H-^{119}Sn-gHMQC (left) and ^1H-^{13}C-gHSQC (right) spectra of $\{[Me_2Sn(bz)]_2O\}_2$ recorded in $CDCl_3$ (10 mM, 25 °C). The horizontal scale represents the ^1H spectrum whereas the vertical scales show the ^{119}Sn and ^{13}C chemical shift dimensions, respectively.

The dimeric $\{[Me_2Sn(bz)]_2O\}_2$ distannoxane could also be prepared from the reaction of dimethyltin oxide and benzoic acid in good yields. Accordingly, the complex was prepared for purposes of NMR investigations in a quantitative manner with modification of the standard procedures described in the literature [30]. Figure 6 shows the ^1H-^{119}Sn-gHMQC spectrum [39] of $\{[Me_2Sn(bz)]_2O\}_2$ complex recorded in $CDCl_3$. The appearance of the pair of equal intensity doublet crosspeaks ($\delta_{Sn(2)}$ = – 179.5 and $\delta_{Sn(1)}$ = –189.7 ppm) is the characteristics of a bis-dicarboxylatotetraorganodistannoxane, in which both the exo- and the endocyclic tin atoms are pentacoordinated. The same information is reflected by the almost identical two-bond scalar coupling $^2J_{(1H-119Sn)}$ values (89.4 and 89.5 Hz) available from the proton dimension. The $^1J_{(13C-119Sn)}$ coupling constants were determined from the ^{119}Sn satellites seen in F1 traces of the gradient enhanced ^1H-^{13}C-gHSQC experiments [40] (figure 6). The $^1J_{(13C-119Sn)}$ values correspond to approx. 148° and 144° Me–Sn–Me bond angle for the endo- and exocyclic tins, respectively. As expected, the $\{[Me_2Sn(bz)]_2O\}_2$ complex exhibits large ^{119}Sn chemical shift anisotropy [39d] in the solid-state NMR spectra (figure 7). The two chemically non-equivalent tin sites at – 192.2 ($\delta_{Sn(2)}$) and – 194.1 ppm ($\delta_{Sn(1)}$), were identified by measuring the solid-state ^{119}Sn MAS NMR spectra at two different MAS rates. The chemical shifts of the endo- and exo-cyclic tins show marginal difference only owing to the highly similar chemical environment experienced by the two ^{119}Sn nuclei, as compared to literature published on the assignment of ^{119}Sn resonances in solid state NMR spectra [39d]. We must refer back to the X-ray molecular structure of $\{[Me_2Sn(bz)]_2O\}_2$, where the nearly planar disposition of the $Sn_4O_{10}C_4$ skeleton has been assessed. Although the chemical shifts difference for the tins grows form 2 to 10 ppm after dissolution, the overall geometry of the complex is apparently maintained in solution.

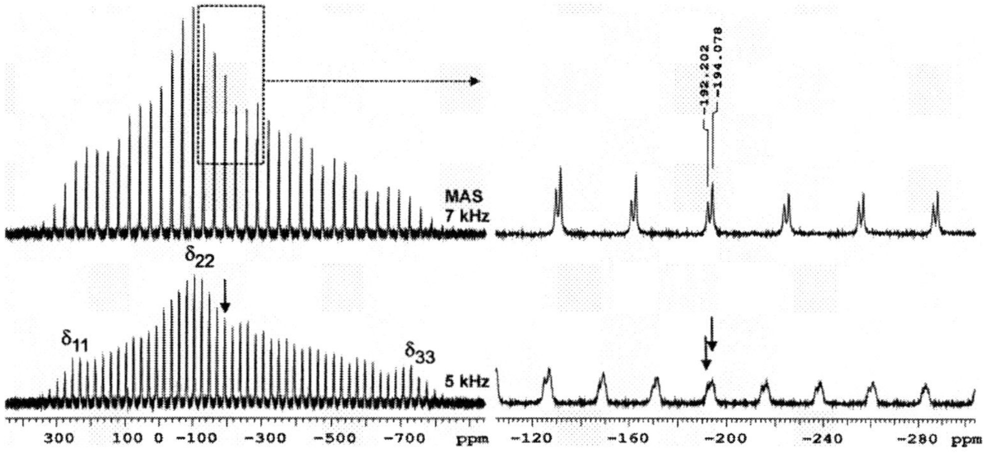

Figure 7. The solid-state ^{119}Sn MAS NMR spectra of $\{[Me_2Sn(bz)]_2O\}_2$ at two different spinning rates. The isotropic chemical shifts are denoted with arrows. The spectra on the right show magnified regions around the isotropic shifts.

To investigate the thermal induction of the methyl transfer, crystals of $Me_3Sn(bz)$ complex were powdered and placed in 2 mm diameter capillaries and the samples were heated for different durations in an oven at 150°C. The molar ratios of the methyltin moieties were monitored by ^{1}H-NMR spectroscopy after cooling and dissolving the samples in $CDCl_3$ (25°C). The $Me_2Sn(bz)_2$ / $Me_3Sn(bz)$ and $\{[Me_2Sn(bz)]_2O\}_2$ / $Me_3Sn(bz)$ molar ratios were monitored by ^{1}H-NMR spectroscopy and the progress with the time is shown in figure 8; the NMR studies confirmed growing amounts of $Me_2Sn(bz)_2$ and $\{[Me_2Sn(bz)]_2O\}_2$ components as a function of the incubation time of the $Me_3Sn(bz)$ complex.

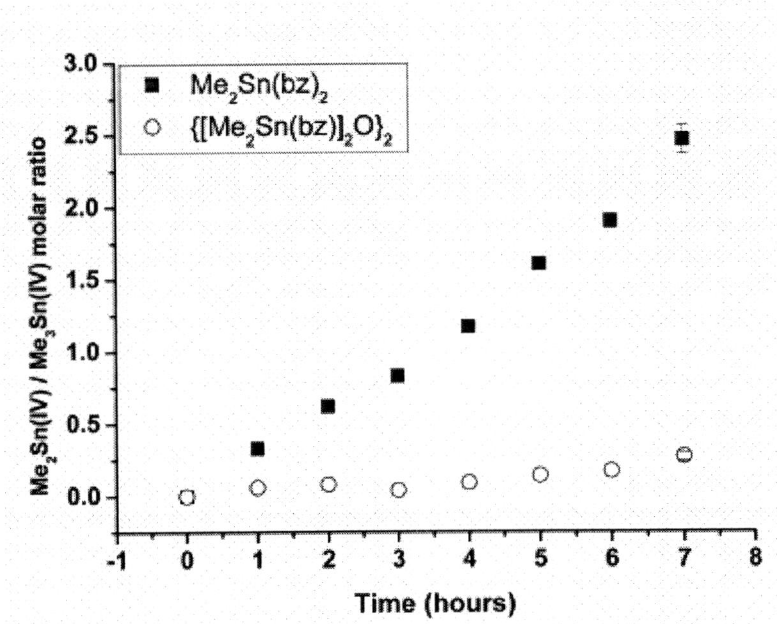

Figure 8. Time course of the $Me_3Sn(bz)$ methyl-transfer induced by heating. The $Me_2Sn(bz)_2$ / $Me_3Sn(bz)$ and $\{[Me_2Sn(bz)]_2O\}_2$ / $Me_3Sn(bz)$ molar ratios were monitored by ^{1}H-NMR (25°C, $CDCl_3$).

When heating was performed in sealed capillaries, the ^{1}H-NMR analysis also confirmed the presence of the volatile Me_4Sn byproduct. ^{1}H-DOSY analysis [41] has been performed on such a mixture containing four components (figure 9). Similar diffusion NMR experiments have become popular recently in supramolecular chemistry carrying the information of the molecular dimensions via the measurement of the D diffusion coefficient [42]. In case of the four component mixture in figure 9, the diffusion properties of the various methyltin(IV)-species are essentially rather different. Molecules having the smallest size ($CHCl_3$ and Me_4Sn) tend to diffuse faster. In case of the methyltinbenzoates, the growing number of bulky benzoate ligands on the tin yield smaller diffusion coefficients in the row: $Me_3Sn(bz)$ > $Me_2Sn(bz)_2$ > $\{[Me_2Sn(bz)]_2O\}_2$. The approximate r hydrodynamic radii [43], as calculated from the diffusion coefficients using the Stokes-Einstein relation, are listed in table 4. The small radius for $Me_3Sn(bz)$ complex confirms its monomeric nature in solution. In the slow exchange regime this method can be successfully applied as an independent analysis of organotin multicomponent mixtures containing symmetrical supramolecular assemblies.

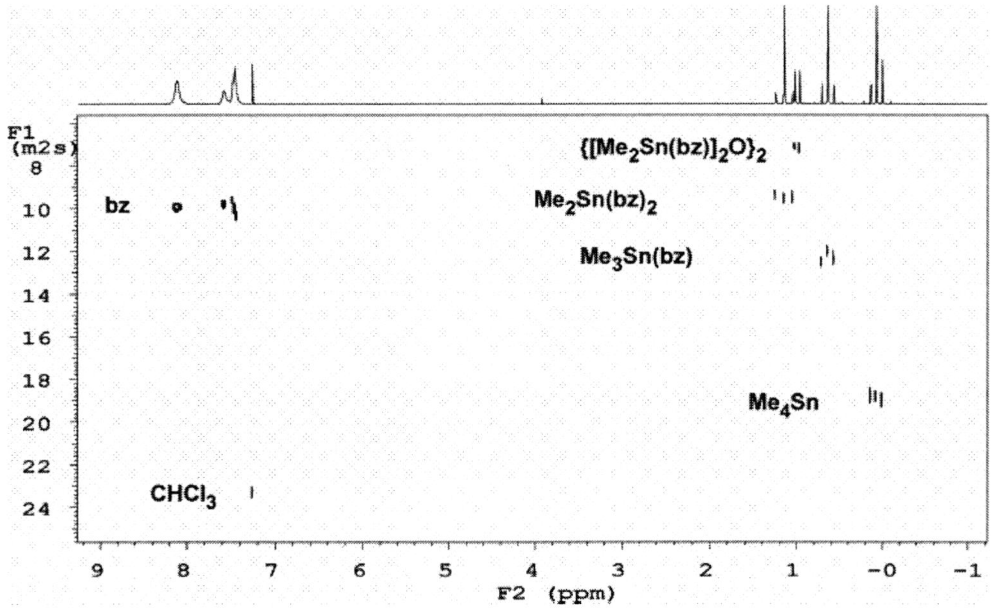

Figure 9. ^{1}H-DOSY NMR of $Me_3Sn(bz)$ after 48 hours of incubation at 150°C in a sealed capillary ($CDCl_3$, 25°C). The horizontal scale represents the ^{1}H spectrum whereas the vertical scale shows the diffusion coefficient ($\times 10^{-10}$ m^2/s).

Table 4. Measured diffusion coefficients and calculated hydrodynamic radii [43] for the reaction mixture components after heating $Me_3Sn(bz)$ in a sealed capillary

Compound	*D*, diffusion coefficient ($\times 10^{-10}$ m^2/s, 25 °C)	*r*, hydrodynamic radius (Å) in $CDCl_3$
$\{[Me_2Sn(bz)]_2O\}_2$	7.0 ± 0.1	5.7 ± 0.1
$Me_2Sn(bz)_2$	9.5 ± 0.3	4.2 ± 0.2
$Me_3Sn(bz)$	12.2 ± 0.2	3.3 ± 0.1
Me_4Sn	19.0 ± 0.2	2.1 ± 0.1
$CHCl_3$	23.2 ± 0.0	1.7 ± 0.0

Having unambiguously identified all components in the dismutation reaction of the $Me_3Sn(bz)$ complex, we turned our attention to the mechanism of the distannoxane formation. In our view distinction has to be made regarding the mechanism in the absence (e.g. solid powder heated in capillary) and in the presence of water (e.g. in methanol and acetonitrile). Although slowly, $Me_3Sn(bz)$ also transforms into $\{[Me_2Sn(bz)]_2O\}_2$ in methanol at room temperature (figure 10). The $Me_2Sn(bz)_2$ complex, as its $Me_2Sn(O_2CMe)_2$ acetate analogue [30a], is unstable to hydrolysis: the dissolution of $Me_2Sn(bz)_2$ in acetonitrile (containing traces of water) resulted in the quantitative formation of the bis[dibenzoatotetraorganodistannoxane], $\{[Me_2Sn(bz)]_2O\}_2$. Most likely, having formed via dismutation the $Me_2Sn(bz)_2$ complex rapidly undergoes partial hydrolysis [44] in the presence of water (Scheme 2).

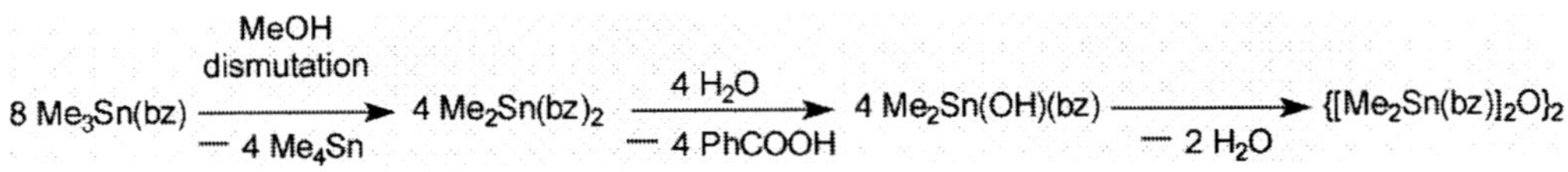

Scheme 2.

The $Me_2Sn(OH)(bz)$ intermediate however, is not stable under the given circumstances and the process finally yields $\{[Me_2Sn(bz)]_2O\}_2$. Water and Lewis bases play a catalytic role in the reaction, they promote the transformation via rapid Me_2Sn^{2+} formation. Figure 10 shows how $Me_3Sn(bz)$ transforms into $\{[Me_2Sn(bz)]_2O\}_2$ in CD_3OD as it was left for one month in an NMR tube. This reaction however is infinitely slow in high purity solvents (in the absence of base catalysis): traces of distannoxane can be identified only after one month. We have also seen that the $Me_2Sn(bz)_2$ complex has the worst solubility in methanol among the methyltinbenzoates, and this behavior might be a further reason for its high degradability. This is in contrast with the tetrameric cupferronato $[Me_3Sn(cupf)]_4$ complex, which dismutation was relatively fast in CD_3OD and the reaction stopped at a solvated dimethyltin species, $Me_2Sn(cupf)_2(CD_3OD)$ [38]. Thus here the dismutation product has a very high solubility in CD_3OD, and it is able to form a stable monomeric $Me_2Sn(cupf)_2(CD_3OD)$ solvate [38].

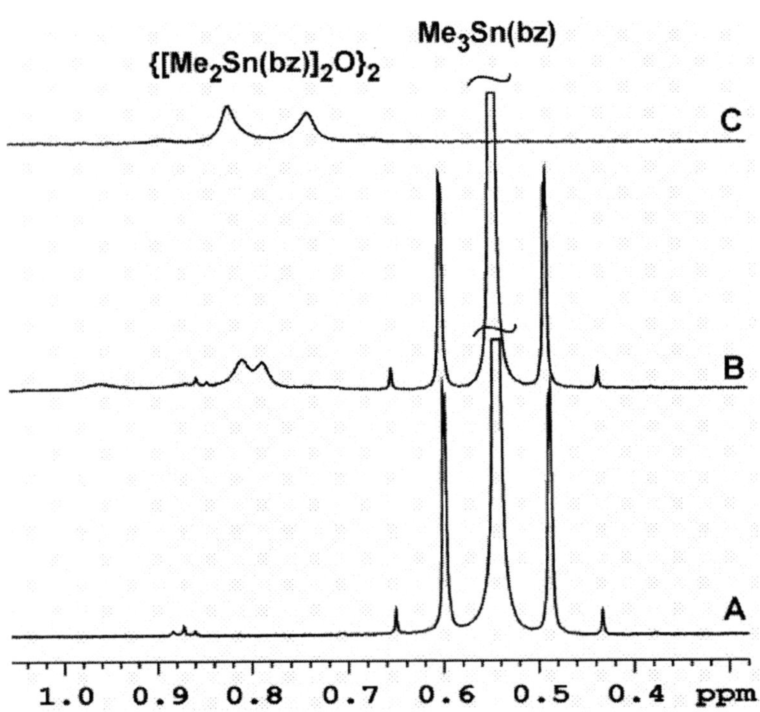

Figure 10. $Me_3Sn(bz)$ complex dissolved in CD_3OD measured by ^1H-NMR (aliphatic region) on the first day (A) and one month after dissolution (B). Above is the pure reference compound $\{[Me_2Sn(bz)]_2O\}_2$ in the same solvent (C). Spectra are recorded at 600 MHz and 35°C. Traces of ethanol at 0.87 ppm can be observed.

In contrast to hydrolytic conditions, complete exclusion of water realized by high temperature (150°C) incubation the $Me_3Sn(bz)$ complex might follow a somewhat different chemical route, however yields the same byproduct, $\{[Me_2Sn(bz)]_2O\}_2$. According to the post-reaction ^{1}H-NMR measurements ($CDCl_3$, 25°C), the $Me_2Sn(bz)_2$ complex was found to be the primary reaction product in each instance. Analysis has also shown that as the reaction proceeds, increasing amounts of $\{[Me_2Sn(bz)]_2O\}_2$ compound were detected in the mixtures (figure 11).

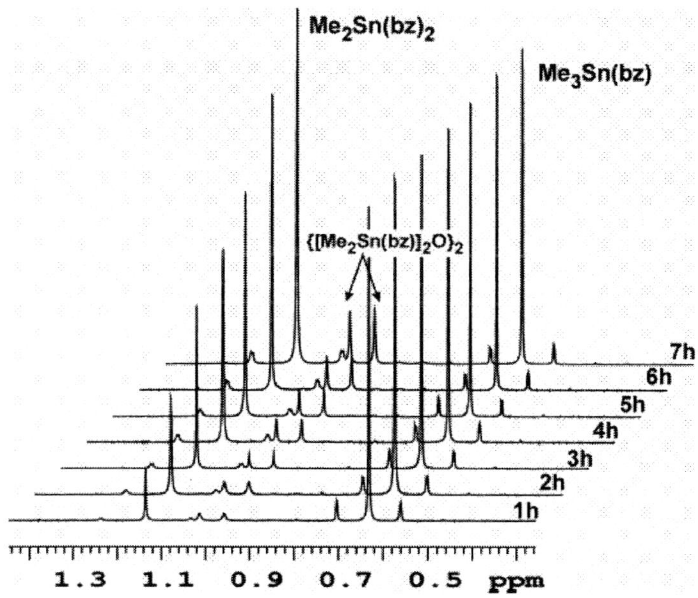

Figure 11. ^{1}H-NMR spectra (400 MHz) recorded after dissolution of the contents of the $Me_3Sn(bz)$ filled 2 mm capillaries which were heated at 150°C for various durations.

Unless high temperatures are set, the dismutation of $Me_3Sn(bz)$ complex into $\{[Me_2Sn(bz)]_2O\}_2$ complex requires several days. Having the lowest boiling point (74-75°C), the Me_4Sn is removed immediately from the reaction mixture. The rest of the reaction may favor the formation of dimethyltin oxide and benzoic anhydride (Scheme 3). The relatively low melting (38-42°C) and high boiling point (360°C) of the benzoic anhydride accounts for a slow ligand removal. When two moles of benzoic anhydride leave the reaction mixture one mole $\{[Me_2Sn(bz)]_2O\}_2$ byproduct formation takes place through a coupling reaction between $Me_2Sn{=}O$ and $Me_2Sn(RCOO)_2$. Scheme 3 shows the $Me_3Sn(bz)$ degradation into the dimeric $\{[Me_2Sn(bz)]_2O\}_2$ distannoxane when the heating is realized under "dry" conditions.

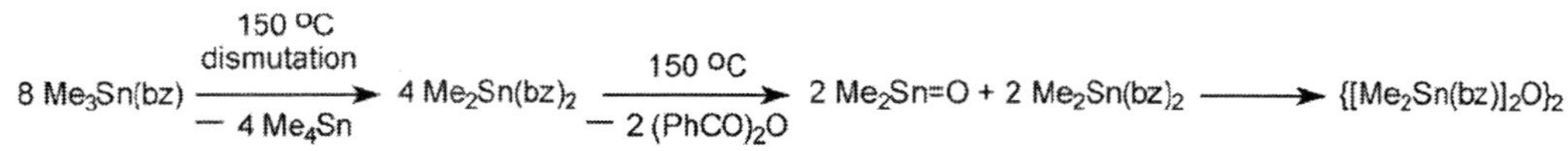

Scheme 3.

Compared to other recently studied ligands [6, 7, 38], the benzoate possess this unique property to facilitate the formation of dimeric $\{[Me_2Sn(bz)]_2O\}_2$ distannoxane structure in thermally induced, as well as, hydrolytic demethylation processes. In the case of tetrameric $[Me_3Sn(cupf)]_4$ cupferronate, the radical decomposition of the ligand started at temperatures much less than 150°C, whereas in the process of the $Me_3Sn(trop)$ tropolonate dismutation into $Me_2Sn(trop)_2$, a further transformation into a dimeric distannoxane seems to be unfavorable because of both steric and electronic reasons.

CONCLUSION

In this chapter we have presented the complete characterization of Sn–O and Sn···O bonded assemblies of methyltinbenzoates by single crystal X-ray diffraction, solid state and multinuclear NMR spectroscopy. The results were used to compare the solid-state and solution structures of methyltinbenzoates to provide new insight into structural changes that occur upon crystallization. The synthesis of the trimethyltin benzoate demonstrated that the selection of an appropriate solvent is critical to a successful synthesis and crystallization. We have shown that the thermally induced liquid- and solid-state methyl-transfer reaction is general for trimethyltin complexes containing an O,O-donor ligand. We have demonstrated how the dismutational reactivity of the trimethyltin complexes can be employed in the production of a wide range of structural types including monomeric, dimeric and distannoxane architectures. In particular, it is noteworthy that the single crystals of dimeric dibenzoato tetramethyldistannoxane were obtained as a result of a further transformation (benzoic acid or benzoic anhydride release) of the parent dimethyltin dibenzoate.

EXPERIMENTAL SECTION

General Remarks

Trimethyltin(IV) chloride, dimethyltin(IV) dichloride and ammonium benzoate were purchased from Aldrich as pure compounds and used as received. Dimethyltin(IV) oxide was prepared from dimethyltin(IV) dichloride by following a literature method [45]. The preparation and purification of the complexes were carried out in open air or in dry atmosphere.

Synthesis of $Me_3Sn(bz)$

To a solution containing ammonium benzoate (0.56 g, 4 mmol) in 5 mL distilled water was added a solution of Me_3SnCl (0.8 g, 4 mmol) in 5 mL distilled water while stirring. Precipitation of the complex occurred immediately. The precipitate was filtered off and washed with cold distilled water and dried in air. Yields, 0.98 g (3.438 mmol, 86%). The product was recrystallized from hot *n*-heptane under ambient conditions as large colorless block crystals. m.p. 120–126 °C, and the complex sublimes at 98 °C; ^{1}H-NMR (599.72 MHz,

15 mM, $CDCl_3$): δ 0.64 (s, $^{2}J(^{117}Sn/^{119}Sn-^{1}H)$ = 56.1/58.4 Hz, 9H, CH_3), 7.38–7.43 (m, 2H, CH), 7.48-7.52 (m, 1H, CH), 8.04–8.08 (m, 2H, CH); ^{13}C-NMR (150.82 MHz, $CDCl_3$): δ – 2.3 (s, $^{1}J(^{117}Sn/^{119}Sn-^{13}C)$ = 380/398 Hz, CH_3), 128.1 (s, CH), 130.1 (s, CH), 131.8 (s, C), 132.1 (s, CH), 171.6 (s, C); ^{119}Sn-NMR (223.58 MHz, $CDCl_3$): δ 136.5 ppm.

Synthesis of $Me_2Sn(bz)_2$

To a solution containing ammonium benzoate (0.56 g, 4 mmol) in 5 mL distilled water was added a solution of Me_2SnCl_2 (0.44 g, 2 mmol) in 5 mL distilled water while stirring. Precipitation of the complex occurred immediately. The precipitate was filtered off and washed with cold distilled water and dried in air. Yields, 0.64 g (1.637 mmol, 82%). The product was recrystallized from hot *n*-heptane under ambient conditions as large colorless block crystals. m.p. 162 °C; ^{1}H-NMR (599.72 MHz, 10 mM, $CDCl_3$): δ 1.15 (s, $^{2}J(^{117}Sn/^{119}Sn-^{1}H)$ = 81.1/81.1* Hz, 9H, CH_3), 7.44–7.50 (m, 2H, CH), 7.57-7.62 (m, 1H, CH), 8.13–8.17 (m, 2H, CH); ^{13}C-NMR (150.82 MHz, $CDCl_3$): δ 4.7 (s, $^{1}J(^{117}Sn/^{119}Sn-^{13}C)$ = 628/657 Hz, CH_3), 128.3 (s, CH), 129.8 (s, C), 130.5 (s, CH), 133.3 (s, CH), 176.1 (s, C); ^{119}Sn-NMR (223.58 MHz, $CDCl_3$): δ – 121.3 ppm. *unresolved.

Preparation of $\{[Me_2Sn(bz)]_2O\}_2$: Route A

To an ethanolic solution of Me_2SnO (0.66 g, 4 mmol) was added a solution of benzoic acid (0.488 g, 4 mmol) in ethanol. The reaction mixture was heated at 100 °C for 9 hours. The reaction mixture was concentrated and cooled to yield the white solid of $\{[Me_2Sn(bz)]_2O\}_2$. Yields, 0.68 g (0.612 mmol, 61.3%). Recrystallization from *n*-heptane gave the product as colorless blocky crystals. m. p. 242 °C; ^{1}H-NMR (599.72 MHz, 4 mM, CD_2Cl_2): δ 0.95 (s, $^{2}J(^{117}Sn/^{119}Sn-^{1}H)$ = 87.2/87.2* Hz, 12H, CH_3), 1.03 (s, $^{2}J(^{117}Sn/^{119}Sn-^{1}H)$ = 91.0/91.0* Hz, 12H, CH_3), 7.46–7.52 (m, 8H, CH), 7.55-7.60 (m, 4H, CH), 8.20–8.07 (m, 8H, CH); ^{13}C-NMR (150.82 MHz, CD_2Cl_2): δ 7.2 (s, $^{1}J(^{117}Sn/^{119}Sn-^{13}C)$ = 739/778 Hz, CH_3), 10.4 (s, $^{1}J(^{117}Sn/^{119}Sn-^{13}C)$ = 786/825 Hz, CH_3), 128.8 (s, CH), 130.4 (s, CH), 132.8 (s, CH), 133.8 (s, C), 173.5 (s, C); ^{119}Sn-NMR (223.58 MHz, CD_2Cl_2): δ – 180.7, –191.9 ppm. *unresolved.

Route B

To 15 mL chloroformic solution of $Me_2Sn(bz)_2$ (0.39 g, 1 mmol) 25 mL of acetonitrile was added. After few days, colorless blocky crystals of $\{[Me_2Sn(bz)]_2O\}_2$ were obtained.

X-Ray Structure Determination

Crystal data and refinement parameters are summarized in table 5. Intensity data of $Me_3Sn(bz)$ were collected on an Enraf-Nonius CAD-4 diffractometer with graphite monochromated Mo-K_α radiation (λ = 0.71073 Å) using the ω-2θ scan technique. Three standard reflections were monitored every hour; these remained constant within experimental error. Crystal of $\{[Me_2Sn(bz)]_2O\}_2$ was mounted in Paratone-N oil within a conventional cryo-loop, and intensity data were collected on a Rigaku R-AXIS RAPID image plate diffractometer (λ(Mo-K_α radiation)= 0.71070 Å). Several scans in the φ and ω direction were made to increase the number of redundant reflections, which were averaged over the refinement cycles. The structures were solved by direct method (*SIR92*) [46] and refined by full-matrix least-squares (*SHELXL-97*) [47]. All calculations were carried out using the *WinGX* package of crystallographic programs [48]. All non-hydrogen atoms were refined anisotropically in F^2 mode. Hydrogen atomic positions were generated from assumed geometries. The riding model was applied for the hydrogen atoms. In addition, the riding methyl hydrogen atoms were allowed to rotate freely during refinement.

Table 5. Summary of X-ray diffraction data for Me_3Snbz and $\{[Me_2Sn(bz)]_2O\}_2$ complexes

	Me_3Snbz	$\{[Me_2Sn(bz)]_2O\}_2$
Empirical formula	$C_{10}H_{14}O_2Sn$	$C_{36}H_{44}O_{10}Sn_4$
Formula mass	284.90	1111.47
Crystal size [mm]	0.20 × 0.20 × 0.40	0. 14 × 0.31 × 0.31
Color	colorless	colorless
Crystal system	monoclinic	monoclinic
Space group	$P2_1/c$	$P2_1/c$
θ range for data collection (°)	$2.33 \leq \theta \leq 34.22$	$3.02 \leq \theta \leq 25.03$
a [Å]	6.541(1)	9.387(3)
b [Å]	10.083(1)	20.982(6)
c [Å]	17.649(4)	11.862(4)
β [°]	98.32(1)	117.15(2)
V [Å^3]	1151.8(3)	2078.9(12)
Z	4	2
d_{calc} [Mg/m^3]	1.643	1.776
μ [mm^{-1}]	2.188	2.425
F(000)	560	1080
Index ranges (°)	$-10 \leq h \leq 10$	$-11 \leq h \leq 11$
	$-15 \leq k \leq 9$	$-24 \leq k \leq 24$
	$-27 \leq l \leq 14$	$-14 \leq l \leq 14$
No. of reflections collected	5033	15288
No. of indep. reflns./R_{int}	4766 / 0.0128	3669 / 0.067
No. of obsd. reflns. $I > 2\sigma(I)$	2557	2992
No. of parameters	121	230
GOOF	0.946	1.067
$R1$ (obsd. data)	0.0348	0.0363
$wR2$ (all data)	0.0873	0.0713
Largest diff. peak/ hole (e Å^{-3})	– 1.109/0.492	– 0.495/0.486

NMR Methods

NMR experiments were carried out on *Varian Inova* (400 MHz for ^{1}H) and *Varian NMR SYSTEM* (600 MHz for ^{1}H) spectrometers using a switchable broadband X{^{1}H}, and a direct detection double tuneable pfg 5mm X{^{1}H-^{19}F} probe (X = ^{13}C, ^{119}Sn), respectively. Solid-state MAS spectra were recorded by using a 3.2 mm HXY Varian Chemagnetics and a 6.0 mm HXY Varian Chemagnetics tunable MAS probes in HX double resonance (DR) mode. For ^{119}Sn MAS experiments recycle delay of 300 seconds and TPPM ^{1}H decoupling field strength of 57 kHz were used. Chemical shifts in the solid state were referenced to tetracyclohexyltin ($\delta_{119Sn} = -97.4$ ppm). Liquid ^{1}H and ^{13}C chemical shifts were referenced to the residual solvent signals. ^{119}Sn shifts are given relative to Me_4Sn. Solution ^{119}Sn-NMR spectra were recorded by using 5 seconds recycle delay and WALTZ proton decoupling during acquisition. Room temperature ^{119}Sn chemical shift data were partially obtained from ^{1}H-^{119}Sn-gHMQC experiments. The DOSY experiment (400 MHz) was carried out in a 2 mm capillary placed in a 3 mm tube at 25°C to minimize convection effects. Performa I gradient amplifier was used. The gradient strength was calibrated by using 5w/w % sucrose in D_2O at 25°C ($D = 5.22 \times 10^{-10}$ m^2/s). The bipolar pulse-pair stimulated echo pulse-sequence was used for acquiring diffusion data with 50 ms diffusion delay, 16 squared increments for gradient levels and 16 transients. The Varian DOSY package was used for the processing. Chloroform viscosity $\eta = 5.42$ mPa·s was used for the calculation of the hydrodynamic radii at 25°C. Deuterated (99.98 atom%) solvents were purchased from Merck® GmbH, Germany. Spectra were processed in VnmrJ® 2.1B. Solid-state simulations were performed by a Varian software: STARS®.

ACKNOWLEDGEMENTS

This work was supported by the Hungarian Scientific Research Funds (OTKA) F 60496 and Hungarian GVOP-3.2.1.-2004-04-0210/3.0 projects. A diffractometer purchase grant from the National Office for Research and Technology (MU-00338/2003) is gratefully acknowledged. The authors are indebted to Tünde Tunyogi and Péter Király for their help.

REFERENCES

[1] (a) (a) Paolini, L.; Tosato, M. L.; Cignitti, M. *Heterocycl. Chem.* 1968, *5*, 533–544. (b) Tosato, M. L. *J. Chem. Soc. Perkin Trans. 2* 1979, 1371–1375. (c) Tosato, M. L. *J. Chem. Soc. Perkin Trans. 2* 1982, 1321–1326. (d) Tosato, M. L. *J. Chem. Soc. Perkin Trans. 2* 1984, 1593–1599.

[2] (a) Sukenik, C. N.; Bonapace, J. A. P.; Mandel, N. S.; Bergman, R. G.; Lau, P.; Wood, G. *J. Am. Chem. Soc.* 1975, *97*, 5290–5291. (b) Sukenik, C. N.; Bonapace, J. A. P.; Mandel, N. S.; Lau, P.; Wood, G.; Bergman, R. G. *J. Am. Chem. Soc.* 1977, *99*, 851–853.

[3] Venugopalan, P.; Venkatesan, K.; Klausen, J.; Novotny-Bregger, E.; Leumann, C.; Eschenmoser, A.; Dunitz, J. D. *Helv. Chim. Acta* 1991, *74*, 662–669.

[4] Greenberg, M.; Shteiman, V.; Kaftor, M. *Acta Cryst.* 2001, *B57*, 428–434.

[5] (a) Dessolin, M.; Eisenstein, O.; Golfier, M.; Prange, T.; Sautet, P. *J. Chem. Soc. Chem. Commun.* 1992, 132–134. (b) Dessolin, M.; Golfier, M. *J. Chem. Soc. Chem. Commun.* 1986, 38–39.

[6] Deák, A.; Tárkányi, G. *Chem. Commun.* 2005, 4074–4076.

[7] Deák, A.; Király, P.; Tárkányi, G. *Dalton Trans.* 2007, 234–239.

[8] (a) Guard, H. E.; Cobet, A. B.; A. B. Coleman III, A. B. *Science* 1981, *213*, 770–771. (b) Hallas, L. E.; Means, J. C.; Cooney, J. J. *Science* 1982, *215*, 1505–1507.

[9] (a) Buck, B. A.; Mascioni, A.; Cramer, C. J.; Veglia, G. *J. Am. Chem. Soc.* 2004, *126*, 14400–14410. (b) Buck, B.; Mascioni, A.; Que Jr., L.; Veglia, G. *J. Am. Chem. Soc.* 2003, *125*, 13316–13317.

[10] (a) Garcia-Garibay, M. A. *Acc. Chem. Res.* 2003, *36*, 491–498. (b) Papaefstathiou, G. S.; Zhong, Z.; Geng, L.; MacGillivray, L. R. *J. Am. Chem. Soc.* 2004, *126*, 9158–9159. (c) Sampedro, D.; Soldevilla, A.; Rodriguez, M. A.; Campos, P. J.; Olivucci, M. *J. Am. Chem. Soc.* 2005, *127*, 441–448. (d) Devic, T.; Batail, P.; Avarvari, N. *Chem.Commun.* 2004, 1538–1539. (e) Enkelmann, V.; Wegner, G.; Novak, K.; Wagener, K. B. *J. Am. Chem. Soc.* 1993, *115*, 10390–10391. (f) Shin, S. H.; Keating, A. E.; Garcia-Garibay, M. A. *J. Am. Chem. Soc.* 1996, *118*, 7626–7627. (g) Shin, S. H.; Cizmeciyan, D.; Keating, A. E.; Khan, S. I.; Garcia-Garibay, M. A. *J. Am. Chem. Soc.* 1997, *119*, 1859–1868. (h) Nomura, S.; Itoh, T.; Nakasho, H.; Uno, T.; Kubo, M.; Sada, K.; Inoue, K.; Miyata, M. *J. Am. Chem. Soc.* 2004, *126*, 2035–2041. (i) Cave, G. W. V.; Raston, C. L.; Scott, J. L. *Chem. Commun.* 2001, 2159–2160. (j) Odani, T.; Matsumoto, A.; Sada, K.; Miyata, M. *Chem. Commun.* 2001, 2004–2005.

[11] (a) Byrn, S. R.; Curtin, D. Y.; Paul, I. C. *J. Am. Chem. Soc.* 1972, *94*, 890–898. (b) Paul, I. C.; Curtin, D. Y. *Acc. Chem. Res.* 1973, *6*, 217–224. (c) Sekiya, R.; Kiyo-oka, K.; Imakubo, T.; Kobayashi, K. *J. Am. Chem. Soc.* 2000, *122*, 10282–10288. (d) Kim, J. H.; Lindeman, S. V.; Kochi, J. K. *J. Am. Chem. Soc.* 2001, *123*, 4951–4959. (e) Toda, F.; Tanaka, K.; Tamashima, T.; Kato, M. *Angew. Chem. Int. Ed.* 1998, *37*, 2724–2727. (f) Bogadi, R. S.; Levendis, D. C.; Coville, N. J. *J. Am. Chem. Soc.* 2002, *124*, 1104–1110. (g) Barea, E.; Navaro, J. A. R.; Salas, J. M.; Masciocchi, N.; Galli, S.; Sironi, A. *J. Am. Chem. Soc.* 2004, *126*, 3014–3015. (h) Lotsch, B. V.; Senker, J.; Schnick, W. *Inorg. Chem.* 2004, *43*, 895–904. (i) Dunitz, J. D.; Harris, K. D. M.; Johnston, R. L.; Kariuki, B. M.; MacLean, E. J.; Psallidas, K.; Schweizer, W. B.; Tykwinski, R. R. *J. Am. Chem. Soc.* 1998, *120*, 13274–13275. (j) MacLean, E. J.; Harris, K. D. M.; Kariuki, B. M.; Kitchin, S. J.; Tykwinski, R. R.; Swainson, I. P.; Dunitz, J. D. *J. Am. Chem. Soc.* 2003, *125*, 14449–14451. (k) Sureshan, K. M.; Murakami, T.; Miyasou, T.; Watanabe, Y. *J. Am. Chem. Soc.* 2004, *126*, 9174–9175. (l) Elizabé, L.; Kariuki, B. M.; Harris, K. D. M.; Tremayne, M.; Epple, M.; Thomas, J. M. *J. Phys. Chem. B* 1997, *101*, 8827–8831. (m) Braga, D.; Grepioni, F. *Chem. Commun.* 2005, 3635–3645. (m) Chandrasekhar, V.; Baskar, V.; Boomishankar, R.; Gopal, K.; Zacchini, S.; Bickley, J. F.; Steiner, A. *Organometallics* 2003, *22*, 3710–3716.

[12] Tanaka, K.; Toda, F. *Chem. Rev.* 2000, 1025–1074.

[13] (a) Cohen, M. D.; Schmidt, G. M. J. *J. Chem. Soc.* 1964, 1996–2000. (b) Schmidt, G. M. J. *J. Chem. Soc.* 1964, 2014–2021.

[14] Kaupp, G. *CrystEngComm* 2003, *5*, 117–133.

[15] Rothenberg, G.; Downie, A. P.; Raston, C. L.; Scott, J. L. *J. Am. Chem. Soc.* 2001, *125*, 8701–8708.

[16] (a) Gielen, M.; Biesemans, M.; de Vos, D.; Willem, R. *J. Inorg. Biochem.* 2000, *79*, 139–145. (b) Di Stefano, R.; Scopelliti, M.; Pellerito, C.; Casella, G.; Fiore, T.; Stocco, G. C.; Vitturi, R.; Colomba, M.; Ronconi, L.; Sciacca, I. D.; Pellerito, L. *J. Inorg. Biochem.* 2004, *98*, 534–546. (c) Tranter, C. J.; Berners Price, S. J.; Cutts, J.; Parsons, P. G.; Rintoul, G.; Young, D. J. *Main Group Chem.* 1995, *1*, 165–167.

[17] Chandrasekhar, V.; Nagendran, S.; Baskar, V. *Coord. Chem. Rev.* 2002, *235*, 1–52.

[18] (a) Willem, R.; Verbruggen, I.; Gielen, M.; Biesemans, M.; Mahieu, B.; Basu Baul, T. S.; Tiekink, E. R. T. *Organometallics* 1998, *17*, 5758–5766. (b) Holmes, R. R.; Day, R. O.; Chandrasekhar, V.; Vollano, J. F.; Holmes, J. M. *Inorg. Chem.* 1986, *25*, 2490–2494. (c) Smith, P. J.; Day, R. O.; Chandrasekhar, V.; Holmes, J. M.; Holmes, R. R. *Inorg. Chem.* 1986, *25*, 2495–2499. (d) Swisher, R. G.; Vollano, J. F.; Chandrasekhar, V.; Day, R. O.; Holmes, R. R. *Inorg. Chem.* 1984, *23*, 3147–3152. (e) Harrison, P. G.; Lambert, K.; King, T. J.; Majee, B. *J. Chem. Soc., Dalton Trans.* 1983, 363. (f) Calogero, S. P. Ganis, Peruzzo, V.; Tagliavini, G. *J. Organomet. Chem.* 1980, *191*, 381–390. (g) Baul, T. S. B.; Dhar, S.; Pyke, S. M.; Tiekink, E. R. T.; Rivarola, E.; Butcher, R.; Smith, F. E. *J. Organomet.Chem.* 2001, *633*, 7–17. (h) Molloy, K. C.; Purcell, T. G.; Mahon, M. F.; Minshall, E. *Appl. Organomet. Chem.* 1987, *1*, 507–519. (i) James, B. D.; Kivlighon, L. M.; Skelton, B. W.; White, A. H. *Appl. Organomet. Chem.* 1998, *12*, 13–23. (j) Willem, R.; Bouhdid, A.; Mahieu, B.; Ghys, L.; Biesemans, M.; Tiekink, E. R. T.; de Vos, D.; Gielen, M. *J. Organomet. Chem.* 1997, *531*, 151–158. (k) Alcock, N. W.; Timms, R. E. *J. Chem. Soc. A* 1968, 1876–1878. (l) Chandrasekhar, V.; Baskar, V.; Boomishankar, R.; Gopal, K.; Zacchini, S.; Bickley, J. F.; Steiner, A. *Organometallics* 2003, *22*, 3710–3716. (m) Ud-Din, I.; Molloy, K. C.; Mazhar, M.; Dastgir, S.; Ali, S.; Mahon, M. F. *Appl. Organomet. Chem.* 2003, *17*, 781–787.

[19] (a) Das, B. C.; Biswas, G.; Maji, B. B.; Ghatak, K. L.; Ganguly, S. N.; Iitaka, Y.; Banerjee, A. *Acta Crystallogr. C* 1993, *49*, 216–219. (b) Hirano, M.; Akita, M.; Tani, K.; Kumagai, K.; Kasuga, N. C.; Fukuoka, A.; Komiya, S. *Organometallics* 1997, *16*, 4206–4213. (c) Kovala-Demertzi, D.; Dokorou, V.; Ciunik, Z.; Kourkoumelis, N.; Demertzis, M. A. *Appl. Organomet. Chem.* 2002, *16*, 360–368. (d) Fang, X.; Song, X.; Xie, Q. *J. Organomet. Chem.* 2001, *619*, 43–48. (e) Gibson, D. H.; Ye, M.; Sleadd, B. A.; Mehta, J. M.; Mbadike, O. P.; Richardson, J. F.; Mashuta, M. S. *Organometallics* 1995, *14*, 1242–1255.

[20] (a) Kemmer, M.; Ghys, L.; Gielen, M.; Biesemans, M.; Tiekink, E. R. T.; Willem, R. *J. Organomet. Chem.* 1999, *582*, 195–203. (b) Das, V. G. K.; Wei, C.; Ng, S. W.; Mak, T. C. W. *J. Organomet. Chem.* 1987, *322*, 33–47. (c) Ng, S. W. *Acta Crystallogr. C* 1999, *55*, 523–531. (d) Parvez, M.; Ali, S.; Mazhar, M.; Bhatti, M. H.; Khokhar, M. N. *Acta Crystallogr. C* 1999, *55*, 1280–1282. (e) Alcock, N. W.; Roe, S. M. *J. Chem. Soc., Dalton Trans.* 1989, 1589–1598. (f) Ng, S. W.; Kuthubutheen, A. J.; Das, V. G. K.; Linden, A.; Tiekink, E. R. T. *Appl. Organomet. Chem.* 1994, *8*, 37–42. (g) Lo, K. M.; Ng, S. W.; Das, V. G. K. *Acta Crystallogr. C* 1997, *53*, 545–546. (i) Yeap, L.-L.; Teoh, S.-G. *J. Coord. Chem.* 2003, *56*, 701–708.

[21] (a) Ng, S. W.; Hook, J. M. *Acta Crystallogr. C* 1999, *55*, 312–316. (b) Ng, S. W. *Main Group Metal Chemistry* 1996, *19*, 113–120.

[22] Gielen, M.; El Khloufi, A.; Biesemans, M.; Kayser, F.; Willem, R.; Mahieu, B.; Maes, D.; Lisgarten, J. N.; Wyns, L.; Moreira, A.; Chattopadhay, T. K.; Palmer, R. A. *Organometallics* 1994, *13*, 2849–2854.

[23] (a) Molloy, K. C.; Quill, K.; Nowell, I. W. *J. Chem. Soc. Dalton Trans.* 1987, 101–106. (b) Molloy, K. C.; Blunden, S. J.; Hill, R. *J. Chem. Soc. Dalton Trans.* 1988, 1259–1266. (c) Molloy, K. C.; Purcell, T. G.; Quill, K.; Nowell, I. W. *J. Organomet. Chem.* 1984, *267*, 237–247. (d) Willem, R. R.; Bouhdid, A.; Mahieu, B.; Ghys, L.; Biesemans, M.; Tiekink, E. R. T.; de Vos, D.; Gielen, M. *J. Organomet. Chem.* 1997, *531*, 151–158. (e) Amini, M. M.; Ng, S. W.; Fidelis, K. A.; Heeg, M. J.; Muchmore, C. R.; van der Helm, D.; Zuckerman, J. J. *J. Organomet. Chem.* 1989, *365*, 103–110. (f) Hneihen, A. S.; Bruno, J. W.; Huffman, J. C. *J. Organomet.Chem.* 1990, *382*, 361–373. (g) Ng, S. W.; Das, V. G. K.; Syed, A. *J. Organomet. Chem.* 1989, *364*, 353–362. (h) Begley, M. J.; Sowerby, D. B.; Kapoor, P.; Kapoor, R. *Polyhedron* 1995, *14*, 1937–1941. (i) Song, X.; Cahill, C.; Eng, G. *Appl. Organomet. Chem.* 2003, *17*, 743–744.

[24] (a) Alcock, N. W.; Culver, J.; Roe, S. M. *J. Chem. Soc., Dalton Trans.* 1992, 1477–1484. (b) Harston, P.; Howie, R. A.; Wardell, J. L.; Doidge-Harrison, S. M. S. V.; Cox, P. J. *Acta Cryst. C* 1992, *48*, 279–281. (c) Baul, T. S. B.; Dhar, S.; Rivarola, E.; Smith, F. E.; Butcher, R.; Song, X.; McCain, M.; Eng, G. *Appl. Organomet. Chem.* 2003, *17*, 261–267.

[25] (a) Lockhardt, T. P.; Calabrese, J. C.; Davidson, F. *Organometallics* 1987, *6*, 2479–2483. (b) Chandrasekhar, V.; Day, R. O.; Holmes, J. M.; Holmes, R. R. *Inorg. Chem.* 1988, *27*, 958–964. (c) Tiekink, E. R. T. *J. Organomet. Chem.* 1991, *408*, 323–327. (d) Kapoor, R.; Gupta, A.; Kapoor, P.; Venugopalan, P. *Appl. Organomet. Chem.* 2003, *17*, 600–606. (e) Dokorou, V.; Demertzis, M. A.; Jasinski, J. P.; Kovala-Demertzi, D. *J. Organomet. Chem.* 2004, *689*, 317–325.

[26] Dakternieks, D.; Kuan, F. S.; Tiekink, E. R. T. *Main Group Met. Chem.* 2001, *24*, 291–292.

[27] (a) Faggiani, R.; Johnson, J. P.; Brown, I. D.; Birchall, T. *Acta Cryst. B* 1979, *35*, 1227–1229. (b) Adams, S.; Drager, M.; Mathiasch, B. *J. Organomet. Chem.* 1987, *326*, 173–186.

[28] (a) Tiekink, E. R. T. *J. Organomet. Chem.* 1991, *408*, 323–327. (b) Narula, S. P.; Bharadwau, Y.; Sharda, S. K.; Day, R. O.; Howe, L.; Holmes, R. R. *Organometallics* 1992, *116*, 2206–2211. (c) Ng, S. W.; Das, V. G. K.; Yip, W.-H.; Wang, R.-J.; Mak, T. C. M. *J. Organomet. Chem.* 1990, *393*, 201–204.

[29] (a) Misty, F.; Rettig, S. J.; Trotter, J.; Aubke, F. *Acta Cryst. C* 1990, *46*, 2091–2093; (b) Misty, F.; Rettig, S. J.; Trotter, J.; Aubke, F. *Z. Anorg. Allg. Chem.* 1995, *621*, 1875–1882.

[30] (a) Lockhart, T. P.; Manders, W. F.; Holt, E. M. *J. Am. Chem. Soc.* 1986, *108*, 6611–6616. (b) Beckmann, J.; Dakternieks, D.; Kuan, F. S.; Tiekink, E. R. T. *J. Organomet. Chem.* 2002, *659*, 73–83. (c) Amini, M. M.; Abadi, S. H.; Mirzaee, M.; Lügger, T.; Hahn, F. E.; Ng, S. W. *Acta Crystallogr. E* 2002, *E58*, m697–m699.

[31] Kozuka, S.; Nakagami, J.-I.. *J. Organomet. Chem.* 1977, *135*, C1–C5.

[32] Davies, A. G. *Organotin Chemistry;* VCH: Weinheim, 1997.

[33] Addison, A. W.; Rao, T. N.; Reedijk, J.; van Rijn, J.; Verschoor, G. C. *J. Chem. Soc. Dalton Trans.* 1984, 1349–1356.

[34] Ng, S. W.; Wei, C.; Kumar Das, V.G. *J. Organomet. Chem.* 1988, *345*, 59–64.

[35] (a) Nádvorník M.; Holeček J.; Handlíř, K. *J. Organomet. Chem.* 1984, *275*, 43–51. (b) Danish, M.; Alt, H. G.; Badshah, A.; Ali, S.; Mazhar, M. *J. Organomet. Chem.* 1995, *486*, 51–56.

[36] (a) Lockhart, T. P.; Manders, W. F.; Schlemper, E. O. *J. Am. Chem. Soc.* 1985, *107*, 7451–7453. (b) Lockhart, T. P.; Manders, W. F. *Inorg. Chem.* 1986, *25*, 892–895. (c) Lockhart, T. P.; Davidson, F. *Organometallics* 1987, *6*, 2471–2478.

[37] (a) Deák, A.; Radics, L.; Kálmán, A.; Párkányi, L.; Haiduc, I. *Eur. J. Inorg. Chem.* 2001, 2849–2856. (b) Wrackmeyer, B.; Sebald, A.; Merwin, L. H. *Magn. Reson. Chem.* 1991, *29*, 260–263.

[38] Tárkányi, G.; Deák, A. *Organometallics* 2005, *24*, 3784-3791.

[39] (a) Kayser, F.; Biesemans, M.; Boualam, M.; Tiekink, E. R. T.; Khloufi, A. E.; Meunier-Piret, J.; Bouhdid, A.; Jurkschat, K.; Gielen, M.; Willem, R. *Organometallics* 1994, *13*, 1098–1113. (b) Willem, R.; Bouhdid, A.; Kayser, F.; Delmotte, A.; Gielen, M.; Martins, J. C.; Biesemans, M.; Mahieu, B.; Tiekink, E. R. T. *Organometallics* 1996, *15*, 1920–1929. (c) Willem, R.; Bouhdid, A.; Meddour, A.; Camacho-Camacho, C.; Mercier, F.; Gielen, M.; Biesemans, M.; Ribot, F.; Sanchez, C.; Tiekink, E. R. T. *Organometallics* 1997, *16*, 4377–4385. (d) Ribot, F.; Sanchez, C.; Meddour, A.; Gielen, M.; Tiekink, E. R. T.; Biesemans, M.; Willem, R. *J. Organomet. Chem.* 1998, *552*, 177–186; (e) Martins, J. C.; Biesemans, M.; Willem, R. *Progr. Nucl. Magn. Res.* 2000, *36*, 271–322.

[40] (a) Kupče, Ē.; Wrackmeyer, B.; Granger, P. *J. Magn. Res.* 1969, *100*, 401–405. (b) Wrackmeyer, B.; Kupče, Ē.; Kümmerlen, J. *Magn. Res. Chem.* 1992, *30*, 403–407. (c) Wrackmeyer, B.; Kupče, Ē.; Kehr, G.; Sebald, A. *Magn. Res. Chem.* 1992, *30*, 964–968. (d) Wrackmeyer, B.; Horchler von Locquenghien, K.; Kupče, Ē.; Sebald, A. *Magn. Res. Chem.* 1993, *31*, 45–50.

[41] (a) Morris, K. F.; Johnson Jr., C. S. *J. Am. Chem. Soc.* 1992, *114*, 3139–3141. (b) Johnson Jr., C. S. *Progr. Nucl. Magn. Res.* 1999, *34*, 203–256. (c) Edward, J. T. *J. Chem. Educ.* 1970, *47*, 261–270; (d) Wu, D.; Chen, A.; Johnson Jr., C. S. *J. Magn. Reson. A* 1995, *115*, 260–264. (e) Deák, A.; Tárkányi, G. *J. Organomet. Chem.* 2006, *691*, 1693–1702.

[42] Pregosin, P. S. *Progr. Nucl. Magn. Res.* 2006, *49*, 261–288.

[43] Delpuech, J.-J. *Dynamics of Solutions and Fluid Mixtures by NMR,* John Wiley and Sons Ltd.: New York, 1995.

[44] Houghton, R. P.; Mulvaney, A. W. *J. Organomet. Chem.* 1996, *517*, 107–113.

[45] Ingham, R. K.; Rosenberg, S. D.; Gilman, H. *Chem. Rev.* 1960, *60*, 459–539.

[46] Altomare, A.; Cascarano, G.; Giacovazzo, C.; Guagliardi, A.; Burla, M. C.; Polidori, G.; Camalli, M. *J. Appl. Cryst.* 1994, *27*, 435–436.

[47] Sheldrick, G. M. *SHELXL97*-Program for the Refinement of Crystal Structures, University of Göttingen, Germany, 1997.

[48] Farrugia, L. J. *J. Appl. Cryst.* 1999, *32*, 837–838.

In: Advances in Organometallic Chemistry Research
Editor: Kenji Yamamoto, pp. 227-256
ISBN: 978-1-60021-779-1

Chapter 9

Siloxane Based Polymeric Structures Containing Complexed Metals

Maria Cazacu[*]
"Petru Poni" Institute of Macromolecular Chemistry,
Aleea Grigore Ghica Voda 41 A, RO-700487 Iasi, Romania

Abstract

The incorporation of the metals into the structure of synthetic polymers offers considerable potential for the preparation of processable materials with properties that significantly differ from those of conventional organic polymers. The presence of metal centers (especially those of transition elements) that, as a consequence of their electronic structure, generally exist in a variety of oxidation states, has been shown to facilitate access to a diverse range of interesting and useful properties. Therefore, the polymer-metal complexes are now a growing research field.

Concurrent with the development in this area, there has also been increased interest in the chemistry of metal complexes containing Si-O bonds. Polysiloxanes are involved either directly with siloxane bond (i.e. metal siloxides containing Si-O-M linkages) or through the attached organic chelating functions in 2D or 3D metal-containing structures. Depending on the used ligands or synthesis pathway, polymers with metals either in the main chain or in the side position can be obtained. Precomplexed metals (e.g., metallocenes) as functionalized structures can be used as monomers in polycondensation reaction with siloxane macromers. Also, polysiloxanes in crosslinking state can serve as matrix for encapsulating of metals nanoparticles or metal ions complexed by a proper chelating groups immobilized on the porous solid siloxane polymers. As a result of the incorporation of the metal atoms, such materials could exhibit attractive electrical, electrochemical, magnetic, optical and catalytic properties. In addition, these materials would also be of interest as pyrolytic precursors to transition metal containing ceramic products.

This chapter is an overview of the recent reports on the silicones involving in some complex structures containing metals (siloxides, coordination polymers, crosslinked structures). Own contributions are also enclosed.

[*] Maria Cazacu: e-mail: mcazacu@icmpp.ro

Keywords: polysiloxanes, complexes, ligands, polymer-metal, siloxane-metal, functionalized polysiloxanes.

INTRODUCTION

Coordination compounds constitute a very important field in chemistry because of their thermal stability, electrical conductivity and catalytic action, mainly in some of the chemical reactions occuring in biological materials [1]. In the framework of coordinative chemistry, the branch of polymer-metal complexes has been developed relatively recently as an interdisciplinary area involving chemistry, electrochemistry, metallurgy, environmental protection, and materials science [1,2], even though the first studies in this field were made in the middle of the 20th century [3-5]. Polymer chemistry provides many opportunities for designing materials with desired properties by the use of selected monomers, different functionalities and combinations. Therefore, studies on the polymer-metal complexes are of great practical importance. The potential applications of such materials are, for example, as surface coatings on metals and glasses, adhesives, high-temperature lubricants, electrical insulators, semiconductors. Ligands that selectively form complexes with some metal ions are also of particular importance in various application fields, such as recycling and refining of metals, purification of solutions, improving the environment and treatment of poisoned metals, as well as, qualitative and quantitative analysis of metals [2,6].

Most of the coordination polymers have only short chains, possess very poorly defined structures, are often insoluble in common organic solvents, are infusible or melted at high temperatures as a result of intermolecular forces between chains or crosslinking by bridging coordination sites [7,8]. These led to a poor characterization of these materials and hinder their incorporation into useful devices [7]. Some of these problems could be avoided if the chelate moieties were covalently incorporated into flexible polymer systems. The siloxane insertion in such structures could provide improved solubility for coordination polymers because they are well known as having low intermolecular forces, which are responsible for a low solubility parameter [9,10]. Siloxanes also offer possibilities for variations in viscosities, softening temperatures and rheological characteristics, their backbone being one of the most flexible of all known polymers. Even at very low temperatures, these materials maintain their elasticity. Another important property of polysiloxanes is their exceptional stability against heat, oxidation, and UV radiation. Polysiloxanes have very low surface free energies and therefore the materials on their basis are widely used as coatings, mould-release agents, surface modifiers, or separation membranes. Another important feature of siloxane polymers is their unusually high gas permeability. The high free volume in the siloxanes compared to hydrocarbons explains the high diffusion coefficient of gases [11]. The chemical inertness and high gas permeability makes them attractive in biomedicine [12]. The high flexibility of the polymer backbone can cause changes in geometry of the polymer bound metal complex that is an important parameter in the case of their use as catalyst [13]. Therefore, it is expected that by the siloxanes insertion, the metal-containing polymeric structures having improved solubility and ability to form films with potential applications as separation membranes, catalyst, semiconductors or artificial oxygen carrier by reversible oxygen binding to be obtained. Even though this area is not so developed, the siloxane presence in metal-containing complex more or less macromolecular structures is reported in a series of articles in the last time.

Several groups of polymer-metal complexes containing siloxanes, differing by the metal position on the chain, which is determined by the preparation technique, were reviewed in this chapter.

THE MAIN TYPES OF SILOXANE-BASED POLYMERIC STRUCTURES CONTAINING METALS

The combination of metals and polymers in hybrid materials is a research area of great current interest. Several different possible types of metal-containing polymer structures exist, depending on the metal incorporating site and the nature of the linkages between them. In the case of linear polymers, metallo-centers can be either directly in the main chain or in the side-group structure. These situations represent extremes. It is possible to prepare materials with metals both in the side-position and the main chain. Dendrimers and hyperbranched polymers represent another structural class of growing interest. In the case of another structural class (i.e., dendrimers and hyperbranched polymers), the metalo-centers can be located throughout the structure or, alternatively, in the core or at the periphery. The linkages or "spacers" between the metallo-centers can either possess conjugated structures (involving delocalization of σ- or π-electrons) or localized electrons. These situations also represent extremes, and partial conjugation is often possible. Studies on the possibility to control interactions between the metals centers by the electronic structure of the linker constitute an important area of research and has important implications for the physical properties (e.g. conductivity, magnetic properties) and applications of the materials. It is useful to consider the types of characteristics expected for metal-containing polymers that provide a key motivation for making the materials [14].

A number of methods for controlling the positioning of metallic species within polymer matrices on the nanometer scale have been developed [15].

According to the literature reports, it can be emphasized the following types of polymeric structures containing metals and synthesis pathways for them [8,15]:

Coordination polymers that in macromolecular chemistry are defined as entities constructed by a supramolecular approach through metal coordination and consisting of a backbone, which is held together by metal-ligand interactions. These interactions must be strong enough to retain the polymer chain also in solution. The coordination polymers should exhibit properties that are characteristic for polymers, such as enhanced viscosity in comparison with their monomeric building blocks and/or a glassy solid state. The coordination bond is highly directional, the ligand structures can be varied in a desired manner by established organic chemistry, and the thermodynamic and kinetic stability can be controlled by the appropriate ligand types and metal ions. Supramolecular systems constructed from metal-ligand bonds can have different architectures (linear, cyclic, lattice, etc.) [16].

A classification of the different structures of coordination polymers is well presented in ref. [12].

Polymerization of metal-coordinating monomers constitutes a useful way to prepare metal-containing polymers that has already attracted much interest in the literature. In principal two different polymer systems can be distinguished: those in which the metal is part of the polymer chain and others in which the metal is coordinated to pendent groups. There are a huge number of possibilities using combinations of different ligands and metals. The possibilities to obtaining polymers with metals in the main chain, as well as, a selection of potential metal-coordinated monomers used in polymerization reactions were well presented in ref. [8].

Complexation of the metals by macromolecular ligands. Ligand-functional polymers can be prepared either by the direct polymerization of ligand-functional monomers or by the chemical modification of appropriate precursor polymers [15]. The major interactions that can be used for the incorporation of metals in polymers are the use of functional groups to form coordination compounds. All types of interactions suitable to form a metal complex can be applied [8].

Complex architectures having metal core. In recent years, a new type of metal-centered polymeric systems having complex architectures was developed, the so-called metal core macromolecules. In this approach, a metal is coordinated with ligands that include functional groups that are able to initiate a polymerization reaction, or that can be transformed to have such a functionality [8]. Silsesquioxane cubes represent an example for such compounds. These systems are obtained as mono- or octafunctionalized systems. A variety of polyhedral oligomeric silsesquioxanes (POSS), $R_8Si_8O_{12}$ nanoclusters can be obtained. Octafunctionalized cages as the anionic oxyfunctionalized system $(SiO_{2.5})_8^{8-}$ or the octahydrido species $(HSiO_{1.5})_8$ are obtained in reasonable yields and have been used as precursors for multifunctionalized crosslinking agents [8].

Attaching of the ligand groups on a solid polymer particle surface (i.e., silica) constitutes a research of high interest due to the possibility to obtain materials with multicoordinating capabilities used, either as such (in retention of the metallic ions) or with complexed metals, in catalysis, in principal [17,18].

Direct assembly of polymers with nanoscale metallic species. Preformed polymers or copolymers are directly blended with metallic nanoparticles, nanoclusters, or quantum dots. The induced specific interactions between polymers and nanoparticles, such as mutual hydrogen-bonding moieties, can permit the structural control in composites of this type [15]; incorporation of the metal into plasma polymers, and mechanochemical synthesis [8]. The other types of above presented polymers containing metals can be post-treated also to afford polymer-nanoparticle composites.

The widespread utility of the organosiloxane polymers has led to increasing interest in modification of their properties by incorporation of various moieties in the polymer chain. The chemical insertions of the metals in siloxane-based polymeric structures can permite to obtain materials with new properties combinations, taking into account the polysiloxanes oneness.

Therefore, the synthesis and study of silicon-containing polymers having transition metals in the structure is an area of growing interest since, as a result of the metal atoms incorporation, such materials exhibit attractive electrical, electrochemical, magnetic, and optical properties. In addition, these materials would also be of interest as pyrolytic precursors to transition metal containing ceramic products [19]. Applications of such compounds include nonlinear optical materials, specialty magnetic materials, liquid crystals, anticancer and antiviral drugs, treatment of arthritis, antibacterial drugs, antifouling materials, treatment of certain vitamin deficiencies, electrical conductors and semiconductors, piezoelectric materials, electrodes, UV absorption applications, super-strength materials, special lubricants and gaskets, and selective catalytic and multi-site catalytic agents [20].

Polysiloxanes can be involved either directly with siloxane bond (i.e. metal siloxides containing Si-O-M linkages) or through the attached organic chelating functions in 2D or 3D metal-containing structures [21]. The using of the siloxane derivative as co-monomere sequence alternating with metal-containing units is also possible.

The incorporation of some transition metals in siloxane polymers [22], as well as the catalytic activity of these metal-coordinated polymers already has been reported [23].

1. METALLASILOXANES

The synthesis of organosilicon compounds containing a Si-O-M group (M=hydrogen or a heteroelement) dates back to the end of the XIX century [24]. Molecular and macromolecular compounds, containing Si-O-M groups, where M=metal, received considerable attention in recent years because of their wide application in materials sciences and catalysis. These are used as mechanistic models for heterogeneous SiO_2-supported catalytic systems, as well as precursors for metal oxides and silicates [25]. The metallasiloxanes, formally derived from $[R_3SiO]^-$, $[R_2SiO_2]^{2-}$, $[(R_2SiO)_2O]^{2-}$, $[RSiO_3]^{3-}$, $[R_7Si_7O_{12}]^{3-}$, or $[(RSiO_2)_n]^{n-}$ silanolate functionalities, have attracted considerable interest among chemists, material scientist, and physicists. The presence of a metal in the siloxane framework often results in high thermal stability and improved catalytic and conducting properties. Moreover, the metallasiloxanes can be considered as precursors for silicone polymers containing metal centers in the polymer backbone [24].

Alkoxide and siloxide based ligands are assimilated to cyclopentadienyl complexes of transition metals [26]. Due to a more positive character of the silicon atom when compared to carbon, siloxide ligands are generally less basic than alkoxides, and therefore can bind to a transition metal with slightly more ionic character. Siloxy derivatives of the transition metals, e.g. ruthenium, rhodium, nickel, platinum, and to a minor extent iron, osmium, iridium, palladium and cobalt have been synthesized and characterized, and the structures of a few of them have been elucidated by X-ray techniques. A dimeric complex of cobalt(II), $[Co(OSiPh_3)_2\text{-}THF]_2$, was synthesized via a cobalt silylamide complex according to the following equation [27]:

$$[\mu\text{-}N(SiMe_3)_2CoN(SiMe_3)_2]_2 + 4Ph_3SiOH \rightarrow [\{Co(\mu\text{-}OSiPh_3)(OSiPh_3).THF\}_2] + 4NH(SiMe_3)_2$$

X-ray studies revealed the structure of this dimeric complex with two siloxy groups as bridging ligands. Each cobalt atom has a distorted tetrahedral geometry due to the coordination of one THF to each cobalt giving a central Co_2O_6 core. Synthesis, spectroscopic characterization and X-ray structure of a cobalt(I)-siloxide complex, $[Co(PPh_3)_3(OSiMe_3)]$ have also been reported [28]. The complex was synthesized by the reaction of $[CoCl(PPh_3)_3]$ with trimethylsilanolate [28].

There are several examples of species with M-Si-O linkages described in the literature, with different properties [29-34]. Such as metallasiloxanes derived from several types of silanols (silanediols and silanetriols) and alkali metals as precursors for new materials have been described [24].

Polymetallosiloxanes have been prepared by base-catalyzed (NH_4OH) hydrolytic polycondensation of ethyl silicate with zinc(II) or lead(II) salts. The effect of the anion of Pb and Zn salts on the thermal evolution of the structure was studied [35].

$O(SiMe_2O)_2^{2-}$ bidentate chelating siloxane anions were coordinated to Ba_6^{12-} layer framed by eight σ/π coordinated pyrazolate anions, acting as a molecular stample to two alternate Ba_3 triangles each (above and below) and favoring the layer arrangement rather than the three-dimensional cluster [36].

The synthesis of lithium-bridged cyclosiloxyniobium, $[Nb\{O\text{-}(SiPh_2O)_2\}_3]$-m-$[Li(Py)_2]$ (Py = pyridine) and cyclosiloxytantalum, $[Ta\{O(SiPh_2O)_2\}_3]$-m-$[Li(Py)_2]$ complexes have been reported [25]. Their structures in solid state or in solution were investigated. These complexes were tested as source precursors for $LiNbO_3$ and $LiTaO_3$. Tetraorganodisiloxanediolate ligands, $[(R_2SiO)_2O]_2$, in metallasiloxane complexes co-ordinate in a variety of modes, chelate-bridging being preferred [25].

Disiloxane-1,3-diols was reacted with diisopropoxidobis(2,4-pentanedionato)titanium(IV) or diisopropoxidobis (2,4-pentanedionato)zirconium(IV) to result cyclic titanasiloxanes and zirconasiloxanes having twelve-membered rings with the $(Si–O–Si–O–Ti–O)_2$ linkages and $(Si–O–Si–O–Zr–O)_2$ linkages, respectively. The pyrolysis of cyclic titanasiloxanes and zirconasiloxanes provided $2SiO_2–TiO_2$ and $2SiO_2–ZrO_2$ ceramics, respectively, consisting in amorphous silica and crystallized titania or zirconia [37].

The cage phenylcopper and phenylmanganese siloxanes were prepared and reacted with [(2-acetoxyetoxy) methyl]dimethylchlorosilane and 1-acetoxy-2-(dimethylchlorosilylmethoxy) benzene leading to the cleavage of the M—O—Si bond to give metal chlorides and six-unit cyclosiloxanes with an acetoxy group in the organic substituent at the silicon atom. Methanolysis of these acetoxy derivatives does not affect the ring structure and affords the corresponding polyhydric alcohols [38]. There also is a particularly interest for metal siloxides, i.e. compounds containing Si-O-M linkages, based on cage silsesquioxanes, those having very bulky substituents on silicon and also those with relatively simple arylsiloxide ligands such as $[(OPh_2Si_2)_2O]^{2-}$. The formation of both discrete molecular complexes and three-dimensional networks with metals (Ag, Ti) is possible depending on the silicon substitution [21]. A range of copper coordination polymers formed with ligands of the type $[iPr_2SiO(CH_2)_nPy]_2O$ with n=0, 1 or 3 has been reported [21]. It has been demonstrated that this type of siloxypyridine ligand is able to form large-ring polymeric arrays with a diversity of structures depending on the co-anion employed [21].

2. Polymeric Structures Containing Siloxanes and Metal Coordinated by Organic Chelating Functions

Diverse organic hydrophilic groups able to act as a chelating agent to bind metal can be attached to the hydrophobic polysiloxane backbone. This moiety can be hydrophilic prior or after binding with metal. The hydrophilic chelating groups can be on side or end position towards a short or long siloxane moiety [39]. Depending on the length of the siloxane sequence and chelating group nature, polymeric structures containing siloxanes and complexed metal can be obtained by:

- -direct coordination reaction of the bidentate ligand with metal;
- -complexing of a previously synthesized macromolecular ligand having side-chelating groups;
- -synthesis of a metal-functional monomers having complexed metal and functional groups able to polymerize itself or with other monomers.

2.1. Polymeric Structures Having Metal in the Main Chain

2.1.1. Polymers Obtained by Polycoordination Reactions

Coordination oligomers and polymers can be assembled by the interaction of metal ions with polydentate ligands containing two or more separated metal-binding domains. The simplest structural requirement for such ligand is to have two polydentate functionalities linked by a spacer group [40]. Polymers obtained by reaction of dicarboxylic acids with metal ions have been investigated either from the point of view of coordination chemistry or macromolecular chemistry. The polymeric nature of the metal dicarboxylate salts depends mainly on: the coordination number of the metal ion, the number of ligands from system, steric factors and preparation method [41]. Such type of polymer-metal complex has been synthesized since 1950 for the purpose of obtaining semiconducting organic materials, heat-resistant organic polymers or polymer catalysts [42].

The use of polysiloxane as a spacer in such ligands could provide improved solubility for coordination compounds but another important feature of the siloxane is their unusually high gas permeability. The high free volume in the siloxanes compared to hydrocarbons explains the high solubility and high diffusion coefficient of gases [11].

Such coordination polymers were synthesised by equilibrium polycoordination reactions between siloxane diacids as bridging ligands and Cu^{2+} ion. Three siloxane diacid types (i.e., 1,3-bis(3-carboxypropyl)tetramethyldisiloxane, α,ω-bis(3-carboxypropyl)oligodimethyl-siloxane, and 1,3-bis(sebacomethyl)tetramethyldisiloxane) and copper (II) acetate were used as reactants resulting in coordinated polymers of the type shown in Scheme 1 [43], confirmed by elemental analysis, IR and UV-Vis absorption spectra. The obtained polymers had reasonable thermal stabilities and were soluble in a wide range of organic solvents. Due to the highly flexible and nonpolar units, the packing of macromolecular chains through hydrogen bonding or by ionic-coordinative networks is probably reduced and, consequently, the solvent molecules can penetrate easily to solubilize the polymer chains [43]. The inherent viscosity values in the range 0.076-0.328 depending on the molecular weight but also on the siloxane sequences lengths were obtained, revealing a good degree of polycoordination. A very viscous blue-greenish product or transparent blue-greenish films depending on the siloxane sequence length were obtained [43].

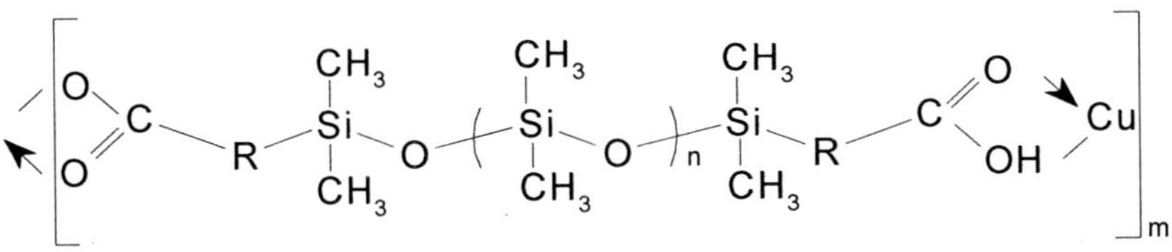

with n = 0 or 15 and R = $(CH_2)_3$ or $(CH_2)_8COOCH_2$.

Scheme 1 [43].

Di- or oligo-dimethylsiloxanes were also functionalized on the ends with acetyl-acetone in order to their using as complexing agents for various metals [44]. Thus, 1,3-bis-(chloromethyl)-1,1,3,3-tetramethyldisiloxane or 1,3-bis-(chloromethyl)-oligodimethyldisiloxane were chemically modified by treated, by their treatment with freshly prepared natrium acetyl acetonate in 1:2 molar ratio (Scheme 2).

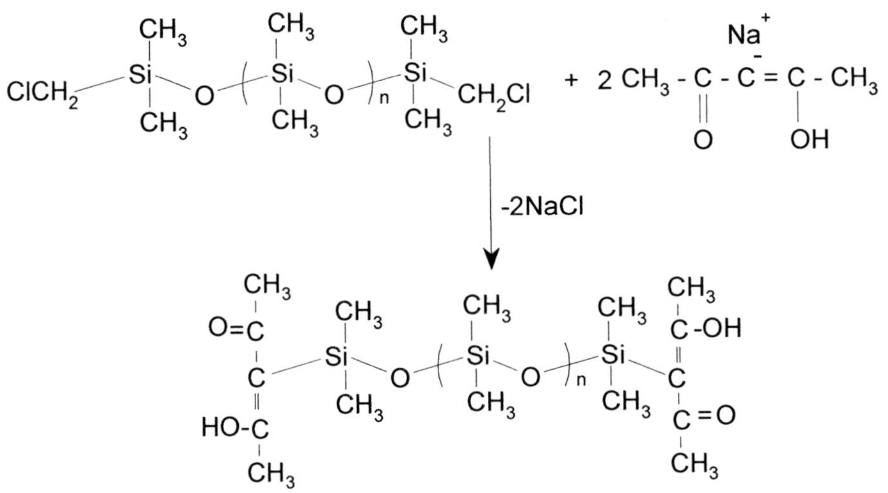

Scheme 2 [44].

The ending acetyl-acetone groups are able to chelate with metals resulting in coordination polymeric structures of the type presented in Scheme 3:

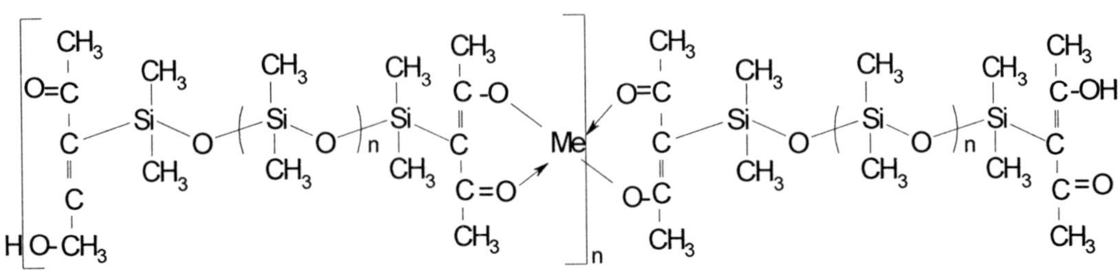

with: m having values in the range 0 and 10.
Me: transitional metal (Cu(II), Ni(II), Co(II), Cd(II), Zn(II))

Scheme 3 [44].

2.1.2. Polymers Obtained by Polymerization Procedure of the Siloxane Precursors with Pre-Complexed Metals (Metallocenes)

The polymers containing metallocene in the main chain are of high interest as catalysts, especially stereoregular catalysts. There are also a number of other potential uses including corrosion resistant coatings, thermally stable materials, liquid crystals (some metallocene derivatives possessing mesogenic properties), electrodes, laser interactions, high strength nanofibers, semiconductors, sensors, solar energy conversion and control-release applications [45]. Metallocenium radicals are fundamental building blocks in ferromagnetic materials and precursors for charge transfer complexes [46]. Metallocene-based compounds often display unexpected biological activity that resulted in their wide application in pharmacy (manufacture of anticancer and antimalarial drugs, and radiopharmaceutical agents), biochemistry (biosensors for DNA detection), etc. [47].

In the last time there is a considerable interest in polymers, which either incorporate ferrocene units heteroannulary, or as pendant groups [48]. There are many reports on macromolecular systems containing, in special, ferrocenyl moiety besides highly flexible dimethylsiloxane and more rigid organic [49,50] or silane sequences synthesized in general by transition metal-catalyzed ring-opening polymerization of the ferrocenophane [50-54] or by hydrosilylation of vinyl-functionalized ferrocene with methyl-H-siloxanes [55-58]. Such materials exhibit novel electrical, electrochemical, and optical properties [48]. Self-assembling ability to micellar structures with a variety of morphologies has also been emphasized [51,59-60]. Cyclosiloxane and silsesquioxane-based ferrocenyl and permethylferrocenyl polymers have been used as mediators in amperometric enzyme electrodes for the detection of glucose (biosensors) [61].

It has been studied redox-active organometallic dendritic and polymeric materials containing ferrocenyl moieties besides linear and cyclic siloxanes, cubic silsesquioxane and silicon- and nitrogen-based dendrimers as framework. In these multimetallic systems, the ferrocenyl moieties behave as independent, electrically isolated units, and the macromolecules undergo a simultaneous multielectron transfer at the same potential [62]. The combination of the architectural features of dendritic molecules together with the electronic communication between metal centers and the well-known, remarkable features and properties of polysiloxanes (such as chemical stability and low toxicity) is a promising strategy for controlling the physical and redox properties of the resulting hybrid macromolecules [62].

A three-membered Si-O-Si bridge, usually 1,1,3,3-tetramethyldisiloxane-1,3-diyl has been used for construction of ansa-metallocene compounds [63]. The discovery of the living anionic ROP of [1]ferrocenophanes has, for the first time, permitted access to a range of multiblock copolymers containing skeletal transition metal atoms [59].

Poly(ferrocenyldimethylsilane-b-dimethylsiloxane)s (Scheme 4) were prepared either by sequential living anionic ring-opening polymerization of the silicon-bridged [1]ferrocenophane and hexamethylcyclotrisiloxane or by addition of [1]ferrocenophane to the Si-H end groups of a polydimethylsiloxane in presence af a hydrosilylation catalyst (Karstedt`s catalyst) [51].

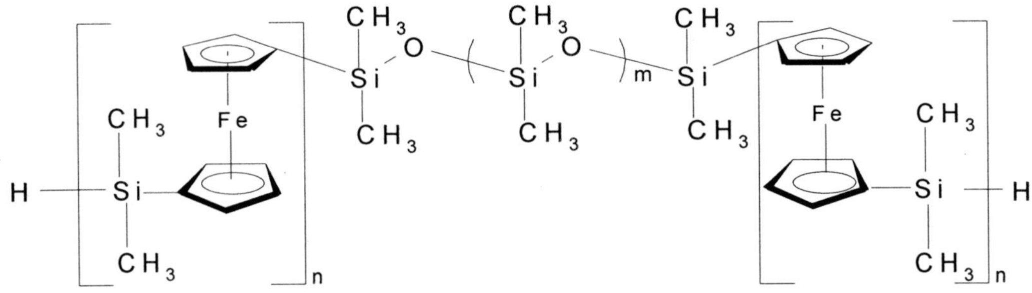

Scheme 4 [51].

Such block copolymers self-assemble to micellar structures with a variety of morphologies, offering potential access to supramolecular structures with redox activity, high refractive index contrast, and preceramic properties. It has been shown that crystalline-coil poly(ferrocenyldimethylsilane-b-dimethylsiloxane) (PFS-b-PDMS), diblock copolymers (block ratio 1:6) form cylindrical micelles with an organometallic core and a polysiloxane corona in selective solvents for the PDMS block such as hexanes. As PFS homopolymers become semiconducting on oxidative doping and also function as high yield precursors to ceramics, these cylinder are of interest as semiconducting nanowires and as nanoscopic etching resists [60,64].

Ferrocene grafted siloxanes were prepared in high yields via Rh-catalyzed dehydrogenative coupling of a series of monomeric, polymeric, and cyclic hydrosiloxanes with ferrocenemethanol. Wilkinson's catalyst was the most efficient between the other catalyst (Karstedt`s catalyst, Speier, etc.) taking into account the yield and selectivity [65].

There are only few literature references on the synthesis of siloxane-ferrocene macromolecular systems by polycondensation procedure [66,67]. Condensation polymerization by formation of amide internal function bonds has been used to prepare polymers, which contain ferrocene and disiloxane units [19]. The synthesis, structural characterization and redox properties of a new family of siloxane-based poly(ferrocenyl) dendronized molecules are also reported [55,62].

The obtaining of some condensation polymers (polyesters, polyamides, polyester-azomethines and polysilylesters) of ferrocene with siloxane as a co-monomere sequence (Scheme 5) has been recently reported [68,69].

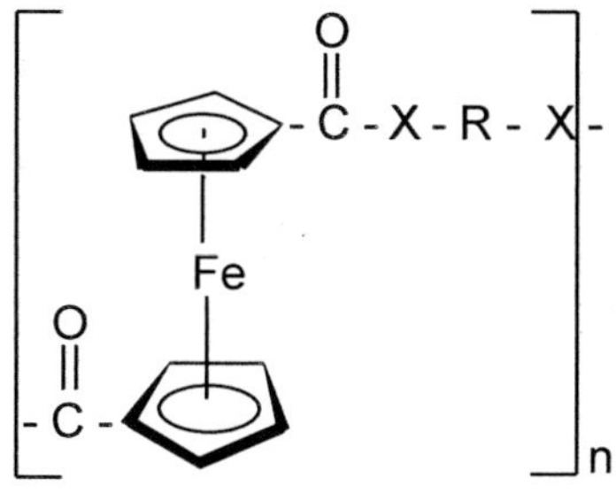

Sample	X	R	m
SAF1 [68]	NH	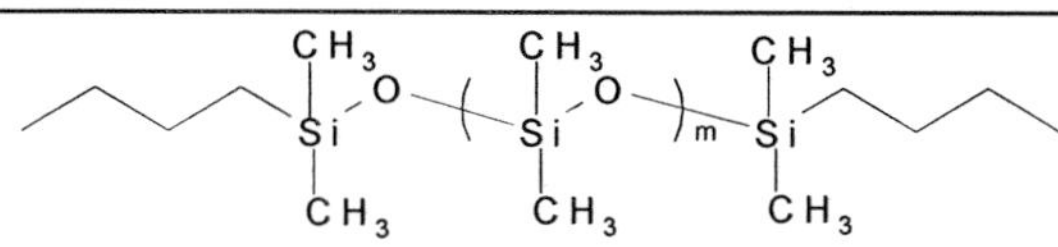	0
SAF2 [68]	NH	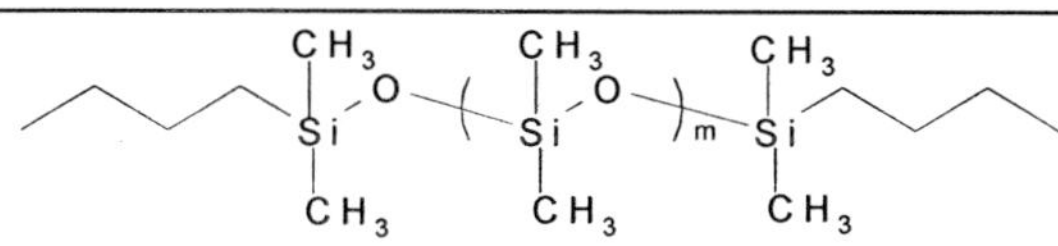	7
SEF1 [68]	O	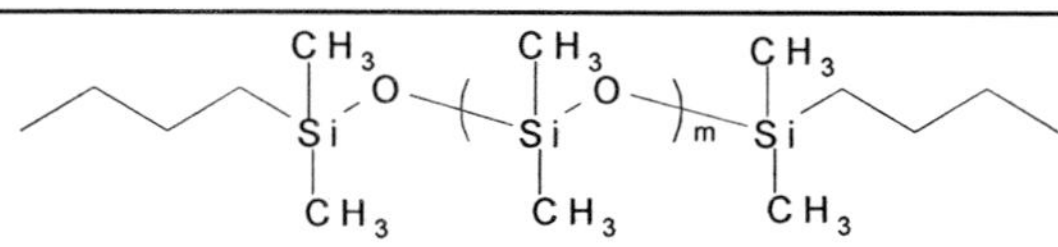	0
SEF2 [68]	O	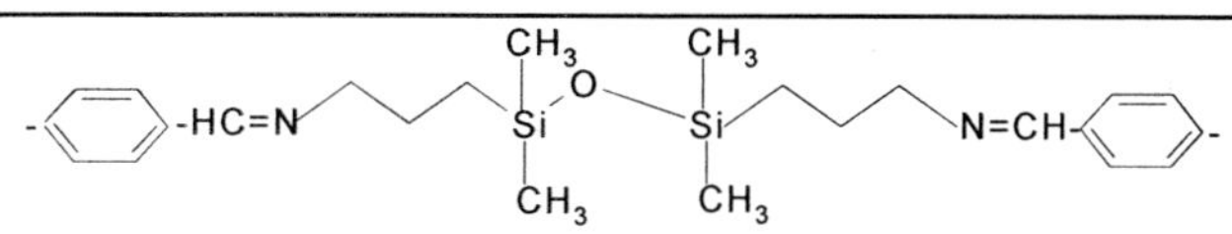	-
g-PSE [69]	O	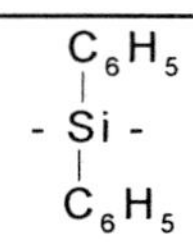	-
s-PSE [69]	O	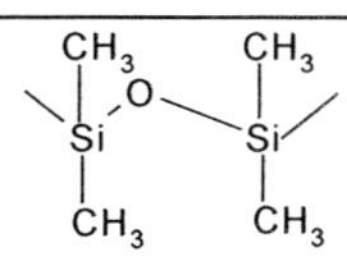	-

Scheme 5.

All polymers were soluble in common organic solvents. The ferrocene units confer high thermal stability and redox activity. The siloxane units improve solubility and decrease the transition temperatures. The presence of ester, amide or azomethine internal functions influences most of the copolymer`s behaviors (thermal, mesomorphic, photochemical and redox) [68]. The silyl-ester groups confer high degradability [69]. Thus, the linkage type between the two co-monomere units can be used as a parameter in designing compounds with desired properties. Therefore, the polycondensation procedure seems to be a good choice for the synthesis of such compounds. In addition, this procedure permits to use a large range of functionalized monomers. Model compounds can be obtained which to help in characterization of the polymers.

2.2. Polymers with Metal Atom in the Side Position

2.2.1. Metal-Containing Polymers Obtained by Polycondensation of the Bifunctionalized Chelate Monomers

2.2.1.1.Bifunctionalized Metal-Complexed Monomers

Various applications of metal-Schiff base complexes that can be used as catalyst, sensors, liquid crystalline materials and model systems for biological macromolecules have been extensively demonstrated. Zinc chelates coordinated with salicylaldehyde Schiff base have been attracted much attention due to their intense fluorescence, relatively high photoluminescence quantum efficiency and good thermal stability. However, the geometry rigidity of complexes results in insolubility of these materials in common solvents and difficulty in forming film. To solve this problem, the introduction of a flexible chain (long alkyl, glycolic or polystyrene chains) into complex has been reported recently [70-74]. The siloxanes were also used as flexibilizing sequences in such structures. Thus, siloxane-based ligands having hydroxyl and azomethine as chelating groups have been synthesized by condensation of 1,3-bis(3-aminopropyl)tetramethyldisiloxane or α,ω (3-aminopropyl)oligodimethylsiloxane with 2,4-dihydroxybenzaldehyde (Scheme 6) [75,76].

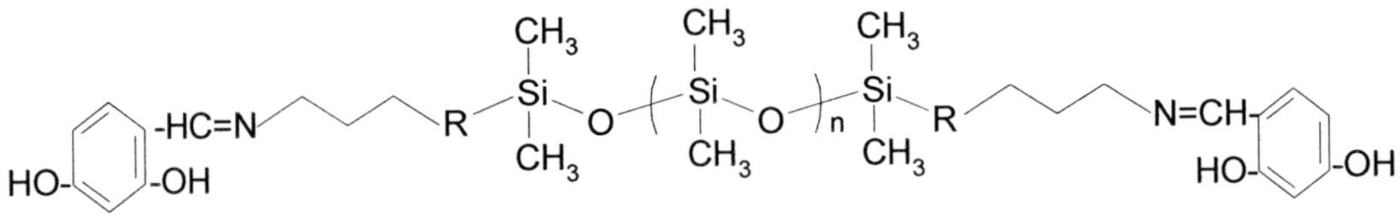

with m=0, 6.5 or 12.

Scheme 6 [75].

The resulted ligands were converted to the chelates of the divalent metals, copper, cobalt, zinc, nickel and cadmium, having the idealized structures presented in Scheme 7. The chelates were soluble in common organic solvents.

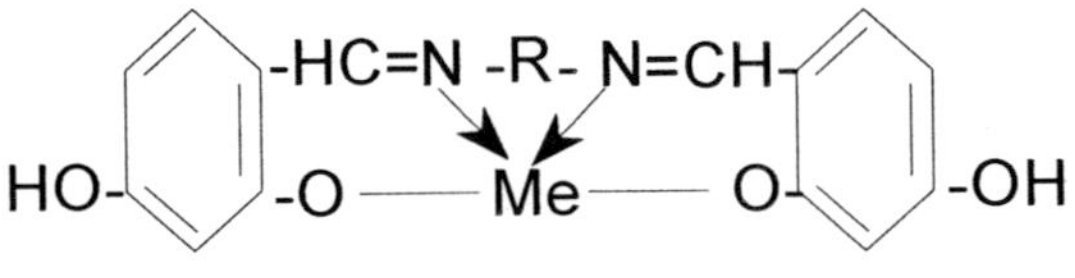

with R: $(CH_2)_3(CH_3)_2SiO[(CH_3)_2SiO]_nSi(CH_3)_2(CH_2)_3$

n = 0 or 8

Me: Cu(II), Ni(II), Co(II), Cd(II), Zn(II)

Scheme 7 [75].

Another series of siloxane-containing ligands (Scheme 8) has been synthesized by reaction of 1,3-bis(3-aminopropyl)tetramethyldisiloxane with a series of ketones (acetylacetone, 2,4-dihydroxybenzophenone and 2,4-dihydroxyacetophenone) [77].

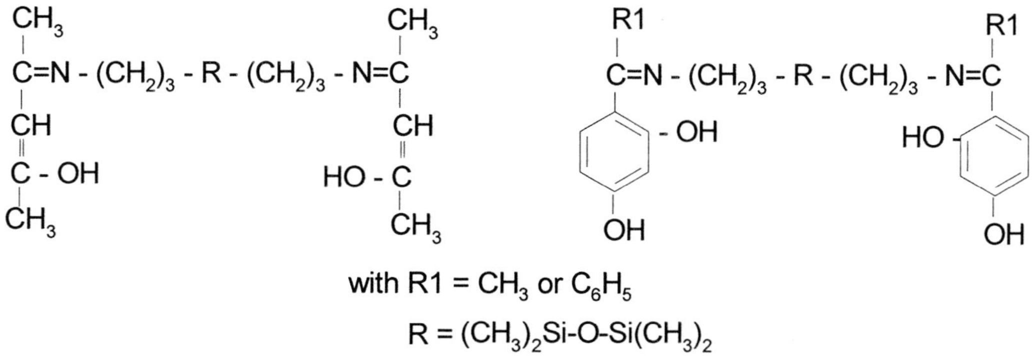

Scheme 8 [77].

These compounds were also used for coordination of some divalent metals (Cu, Ni, Co), resulting in soluble polymeric structures presented in Scheme 9.

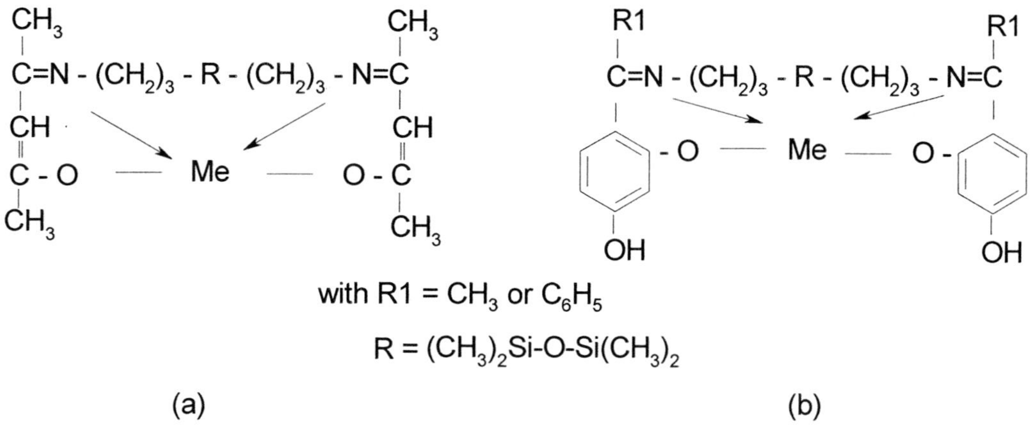

Scheme 9 [77].

Both ligands and their metal complexes were soluble in a wide range of organic solvents.

2.2.1.2. Polycondensation

Some of the obtained bis-phenolic chelates (i.e., those derived from 2,4-dihydroxybenzaldehyde, 4-dihydroxybenzophenone and 2,4-dihydroxyacetophenone) were covalently inserted as sequences in polymeric structures by their polycondensation with 1,3-bis(carboxypropyl)tetramethyldisiloxane as such [75] or bis(*p*-carboxyphenyl)diphenylsilane as diacid chloride [76,77], resulting in alternating polymeric structures (Scheme 10a,b). Polymers based on bis-phenolic chelates without siloxanes and a siloxane comonomer (1,3-bis-(chloromethyl)-1,1,3,3-tetramethyldisiloxane or α,ω-bis (chloromethyl) oligodimethylsiloxanes having different numbers of siloxane units in the chains) have also been synthesized [78]

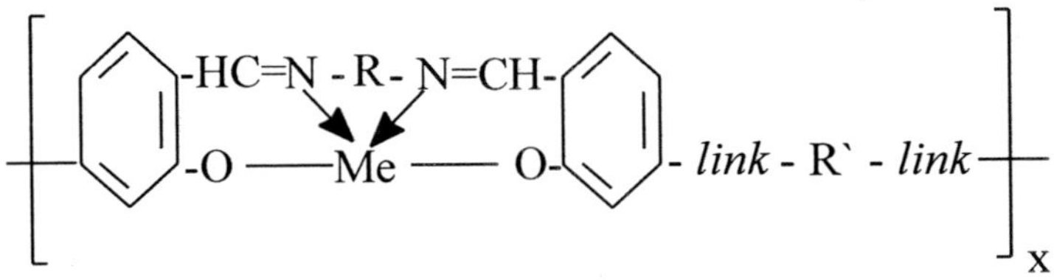

with: Me: Cu (II), Ni (II), Co (II), Cd (II), Zn (II)

R	R`	link	Ref.
$(CH_2)_3(CH_3)_2SiO[(CH_3)_2SiO]_n Si(CH_3)_2(CH_2)_3$, n = 0 or 8	$(CH_2)_3(CH_3)_2SiOSi(CH_3)_2(CH_2)_3$	OOC	[75]
$(CH_2)_3(CH_3)_2SiOSi(CH_3)_2(CH_2)_3$	$H_4C_6(C_6H_5)Si(C_6H_5)C_6H_4$	OOC	[76]
CH_2-CH_2	$CH_2(CH_3)_2SiO[(CH_3)_2SiO]_nSi(CH_3)_2CH_2$, n = 0, 6.5 or 13	O	[78]

(a)

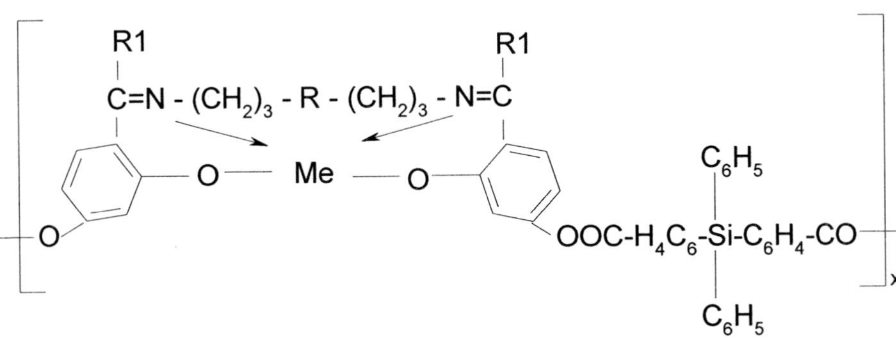

(b) [77]

with: R1: CH_3 or C6H$_5$
R: $(CH_2)_3(CH_3)_2SiOSi(CH_3)_2(CH_2)_3$
Me: Cu(II), Co(II)

Scheme 10.

The polymerization of the metal containing monomers has already been applied to the synthesis of chelate polymers [1]. These types of polymer-metal complexes are known for their well-defined coordination structures. Polymerization can occur by radical or ionic initiation. Interfacial polycondensation of bifunctional low molecular chelates dissolved in NaOH with bifunctional aromatic acid chlorides dissolved in CH_2Cl_2 to give polyesters, polyethers or polyamides was also reported [1,79]. Polyesters can be obtained by either direct or activated polycondensation. Direct polycondensation requires high temperature, which often leads to side reactions that limit the molecular weights of the polymers. N,N`-dicyclohexylcarbodiimide (DCC) was used as activating agent and complex 4-(dimethylamino)pyridinium 4-toluenesulfonate (DPTS) as a catalyst in direct polycondensation of a siloxane diacid, 1,3-bis(carboxypropyl)tetramethyldisiloxane or α,ω-bis(carboxypropyl)oligodimethylsiloxane, with bisphenol chelates above presented [75] resulting in new polymer-metal complexes.

Diacid chloride was also above used as activating agent in order to insert the chelate monomers in the polyester structures by working under mild conditions. Thus, metal bisphenol-chelates were reacted with bis(*p*-carboxyphenyl)diphenylsilane as diacid chloride [76].

The Schiff base of 2,4-dihydroxybenzaldehyde with ethylenediamine has been prepared and complexed with two different divalent metals: Cu (II) and Ni (II). The resulted bifunctionalized chelates were inserted in alternating polyetheric structures by their polycondensation with 1,3-bis-(chloromethyl)-1,1,3,3-tetramethyldisiloxane or α,ω-bis(chloromethyl)oligodimethylsiloxanes having different numbers of siloxane units in the chains. In order to activating them, the chelates were converted in salt forms by their previously treating with NaOH 0.1 N solution or *in situ* in presence of anhydrous K_2CO_3 [78].

The polycondensation of copper, nickel or cobalt resorcylaldehyde-o-phenylenediamine complexes with dimethyl- or diphenyl-dichlorosilanes [23] were earlier reported but the resulting polymers were insoluble [80].

Tetramthylsiloxanediolate is present as a bridge to externally join cubane type $(Li_4O_3Cl)_3$ lithium carbonato chloro hexamethylphosphoramide (HMPA) complex. A molecule consists of three distorted cubanes, which are formed by four lithium, three oxygen and one chlorine atoms, three bridging and three non-bridging HMPA molecules, three bridging $^-$O-Si(CH3)2-O-Si(CH3)2-O$^-$ dianions, one central carbon atom and one central chloride ion [81].

2.2.2. Metal-Containing Polymers Obtained via Macromolecular Ligands

A series of polyazomethines containing pyridine ring and oligosiloxane moieties has been synthesized as macromolecular tridentate ligands and complexed with cobalt (Scheme 11) in order to obtain highly oxygen permselective membrane [82-84].

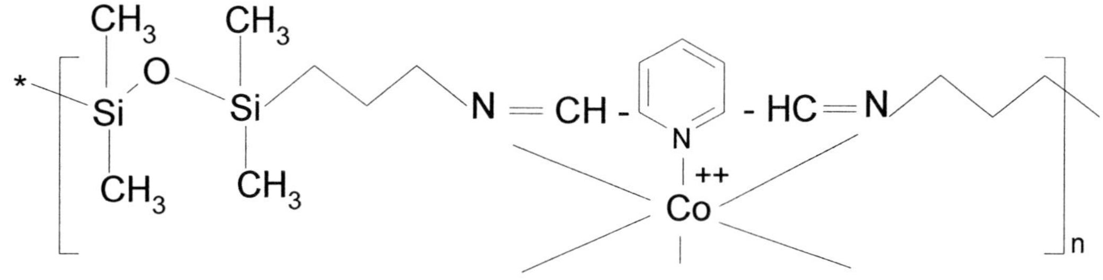

Scheme 11 [83].

Polydimethylsiloxane side (Scheme 12a) [85] or end (Scheme 12b) [85,86] modified by the N-pyrid-2-ylmethyl amide (MPY) group forms stoichiometric chelate complex with Zn salts [85].

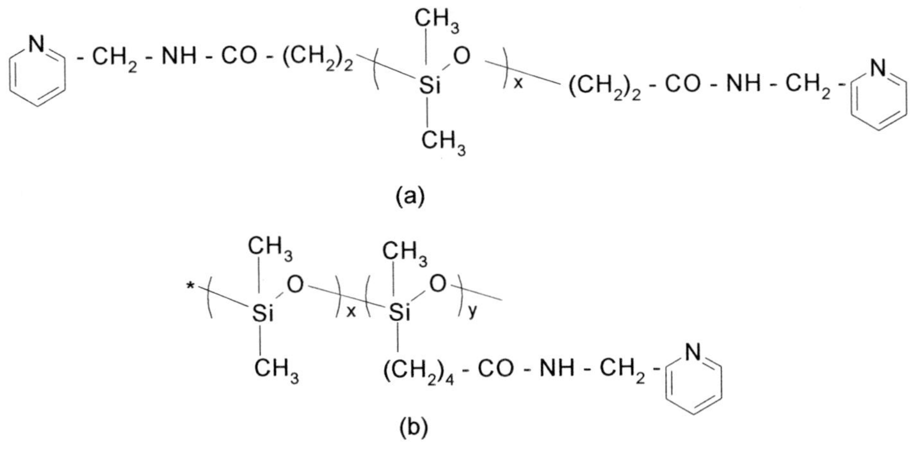

Scheme 12 [86].

1,3-bis(aminopyridinato) functionalized disiloxane was used as a ligand for lanthanide obtaining complexes. No polymeric structures are formed (Scheme 13). Complexes are useful as catalysts in ring-opening polymerization of lactones or ethylene polymerization [87].

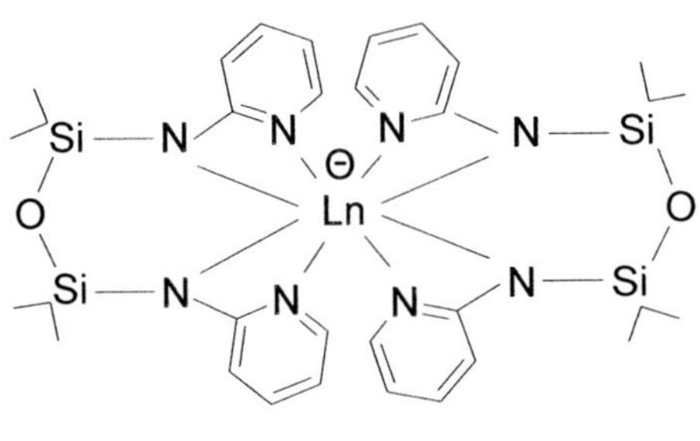

Scheme 13 [87].

Hypervalent pentacoordinated organosilicon complexes obtained from the reaction of disubstituted pyridine ligands with dichlorodiphenylsilane, dichloromethylphenylsilane and bis(dimethylamino)dimethylsilane are reported (Scheme 14) [88].

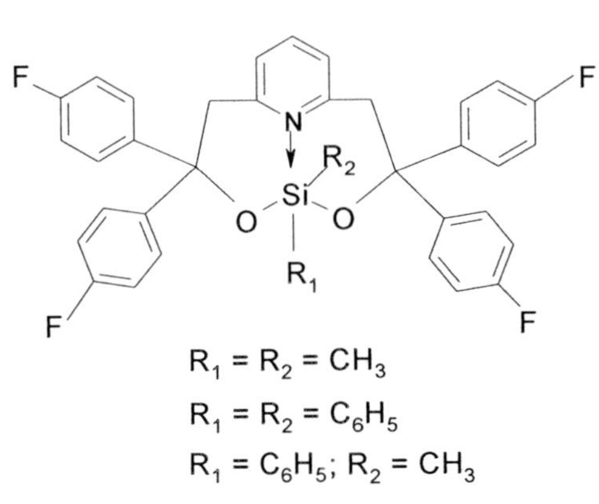

Scheme 14 [88].

It was reported that polyorganosiloxanes with pendant amino groups could give rise to catalytically active copper (II) complexes [89]. Polysiloxanes bearing $Ru(bpy)_3^{2+}$ pendant groups have been reported by Nagai et al. The emission of films on their base was quenched by O_2 in water more efficiently than in the gas phase. Based on these results, it has been concluded that polysiloxane chains have a specific affinity for O_2 molecules in water [12].

Many studies are dedicated to polyazomethines as chelating agents or as semiconductor materials [90-93]. The electrical properties of polyazomethines and their polymers depend in great measure on the kind of amine and aldehyde from which they were obtained [94]. The literature is poorly regarding siloxane azomethines synthesis [95], even though the siloxane presence can induce some interesting properties such as solubility, high gas permeability, etc.

5,5`-Methylene-bis-salicylaldehyde was reacted with siloxane diamines to form polyazomethines (Scheme 15) [96]. These were found readily to complexe with several metal ions: copper, cobalt, and nickel (Scheme 16). Two routes of preparation polychelates were used: first route consisting in the reaction of above prepared ligand with metal salt, and second route consisting in one step reaction by heating all components together.

Both soluble and insoluble fractions were isolated having almost the same compositions. The second route proved to be the more effective one permitting easy and rapid synthesis of the chelate polymers with high percent of soluble part. The siloxanes involving in such structures confer the improved solubility and ability to form films.

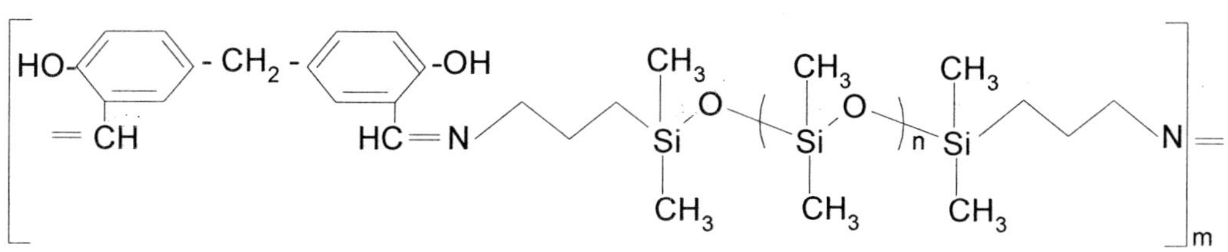

Scheme 15 [96].

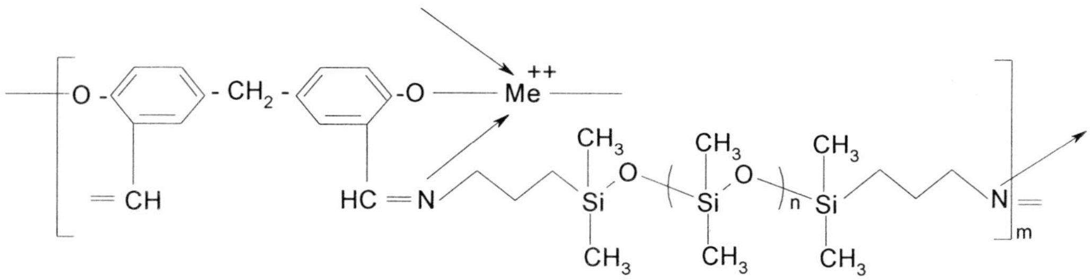

Scheme 16 [96].

Siloxane diamines were also reacted with 2,5-dihydroxy-1,4-benzoquinone resulting in structure presented in Scheme 17 [97]. These compounds behave as macromolecular ligands due to the presence of azomethine and hydroxyl groups along the chains that can form metallic complexes as in Scheme 18 [98].

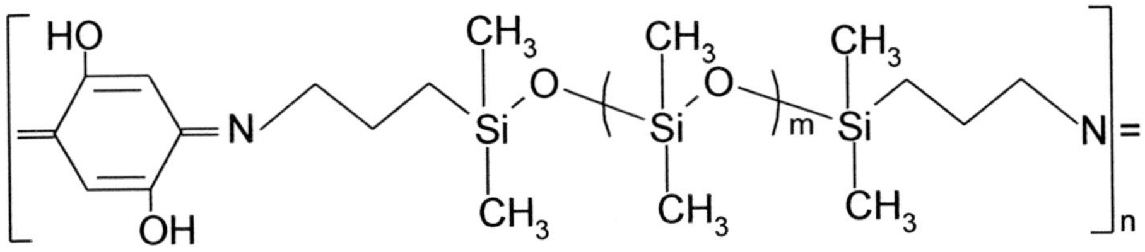

cu m=0, 6.5 sau 12.

Scheme 17 [97].

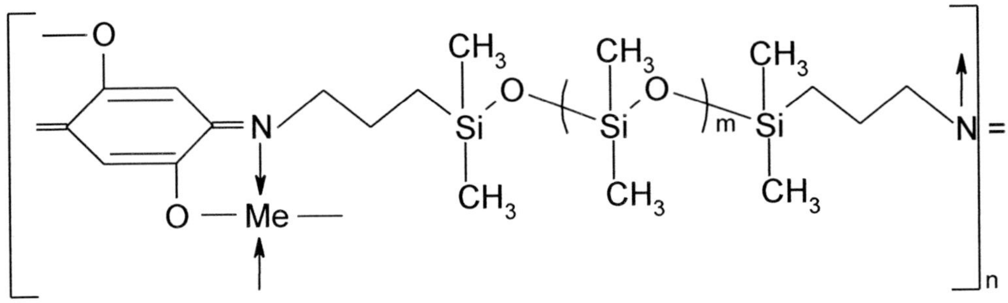

Scheme 18 [98].

In this case metal complexes were insoluble in all tested solvents.

1,3-Bis(N-imidazoyl)propyl-1,1,3,3-tetramethyldisiloxane prepared by modification of 1,3-bis(chloropropyl)-1,1,3,3-tetramethyldisiloxane and commercially 1,3-bis(3-aminopropyl)-1,1,3,3-tetramethyldisiloxane were reacted with metal thiocyanates obtaining complexes [M(1,3-bis(N-imidazoyl)propyl-1,1,3,3-tetramethyldisiloxane)$_2$(NCS)$_2$] (M=Mn, Ni). It has been proved that in the presence of imidazole donor groups, monomers containing 16-membered chelate rings are formed, while the using of amino groups results in the formation of chains of 24-membered rings of [Ni(1,3-bis(3-aminopropyl)-1,1,3,3-tetramethyldisiloxane)$_2$(NCS)$_2$] [99].

Polysiloxanes-based macromolecular ligands have also been prepared by attaching 1,4,7-triazacyclononane (TACN) groups as pendent groups to the siloxane chain. A α,ω-di—n-octyl-terminated poly(dimethyl)methyl-H-siloxane copolymers, prepared by copolymerization of octamethylcyclotetrasiloxane with 1,3,5,7-tetramethylcyclotetrasiloxane in a certain molar ratios, has been treated with 4-bromobutene in presence of platinum(0)-1,3-vinyl-1,1,3,3-tetramethyldisiloxane complex as a hydrosilylation catalyst. Then TACN derivative was bonded to the butyl chain attached to the polysiloxane backbone, resulting in structure presented in Scheme 19 [100].

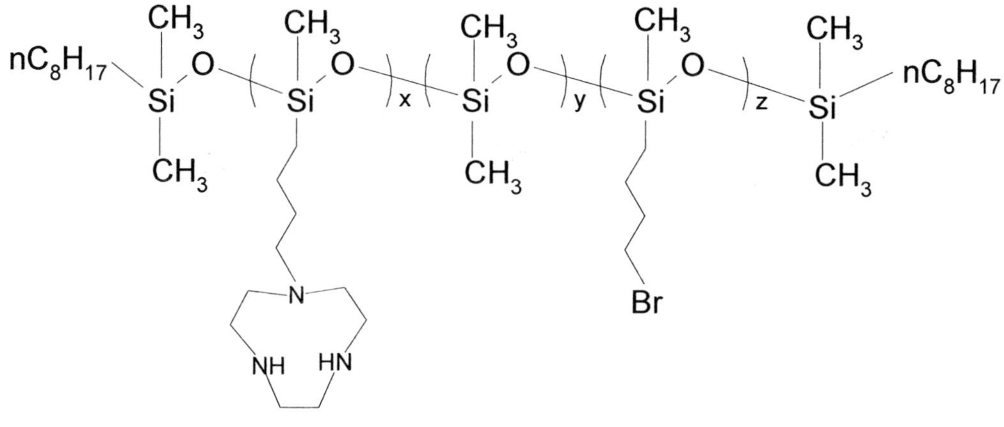

Scheme 19 [100].

Co(II) was then coordinated to the TACN, followed by cobalt oxidation or N-octyl TACN coordination and cobalt oxidation. In both resulted materials TACN forms thermodynamically and kinetically stable $Co(TACN)_2^{3+}$ complexes in which the six coordination sites of the Co(III) are occupied by nitrogens from the TACN. The polymers were coated on fused-silica capillary columns and spherical silica particles, which were used for capillary electrophoresis and capillary electrochromatography, respectively [100].

There is a high interest in the chemistry of transition metal complexes containing vinyl derivatives of organosilicon compounds as π-ligands due to both particular nature of the bond between vinylsilicon derivatives and transition metals, and to the possible application of this kind of complexes to catalysis of hydrosilylation processes. In addition, it has been proved that vinylsilicon π-bounded ligand to metal, possesses enhanced thermal stability relative to that of their silicon-free analogues. A suggested reason for this fact is the interaction between the silicon atom and the C=C bond. This allows delocalisation of electron density from the alkene p-orbital onto the d orbitals of the silicon atom, which enhanced the back-bonding component of the system [101]. An example of such complex is Karstedt catalyst, a platinum complex containing both bridging and chelating divinyl ligands $[\{Pt(\eta\text{-}CH_2{=}CHSiMe_2)_2O\}_2\{(\mu\text{-}CH_2{=}CHSiMe_2)_2\text{-}O\}]$. High reactivity of this complex has led to synthesis of other Pt(0) complexes containing both divinyltetramethyldisiloxane (LL) and maleic anhydride, styrene or various tertiary phosphines. There are also known vinylsiloxane-complexes of palladium and rhodium. The transition metal complexes with vinylsilicon ligands also include a number of organosilicon derivative-containing nickel complexes. Nickel equivalent of Karstedt catalyst $[\{Ni(\eta\text{-}CH_2{=}CHSiMe_2)_2O\}_2\{(\mu\text{-}CH_2{=}CHSiMe_2)_2\text{-}O\}]$ has been synthesized in the reaction of (*trans,trans,trans*-cyclododeca-1,5,9-triene)nickel(0) or bis(cycloocta-1,5-diene)nickel(0) with LL. There are known examples of phosphine-vinylsiloxane nickel(0) complexes $[Ni(LL)(PR_3)]$ [102]. Thus, 1,3-bis(vinyl)-1,1,3,3-tetramethyldisiloxane was used as chelating ligand for nickel, resulting in binuclear nickel(0) complex (Scheme 20a). A Ni complex, $[Ni(PPh_3)\{(\eta^4\text{-}CH_2{=}CHSi(Me)(\mu\text{-}O)\}_3]$, was synthesized using cyclotris[vinyl(methyl)siloxane], in which two vinyl groups are coordinated to the metal center but the third one is still free (Scheme 20b) [101].

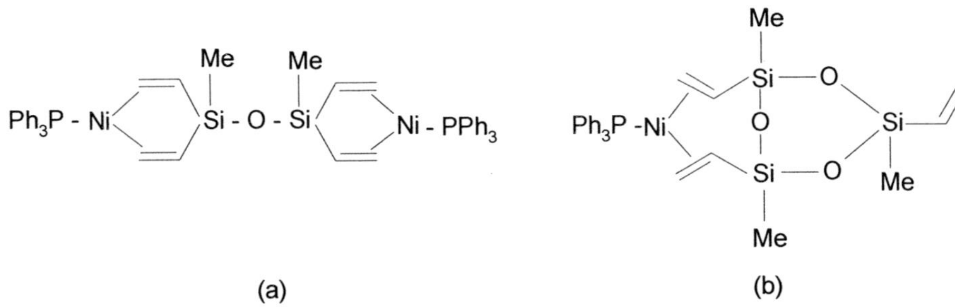

Scheme 20 [101].

Nickel complexes with cyclic p-vinylsilicon ligands have also been synthesized and characterized. Cyclotetrakis[vinyl(methyl)siloxane] and cyclotetrakis[vinyl(methyl)silazane] have been used to generate the nickel complexes $[\{Ni(PR_3)\}_2\{\eta\text{-}CH_2{=}CH(Me)Si(\eta\text{-}O))_4\}]$ and $[\{Ni(PR_3)\}_2\{\eta\text{-}CH_2{=}CH(Me)Si(\eta\text{-}NH))_4\}]$, respectively, where R=Ph, C6H4Me-4, C6H11-c. The first one is similar to the earlier described platinum complex [101].

Well-defined PDMS-b-PMVS diblock copolymers with controlled molecular weights and narrow molecular weight distributions were synthesized via anionic living polymerization of hexamethylcyclotetrasiloxane (D_3) with 1,3,5-trimethyl-1,3,5-trivinylcyclotrisiloxane (D_3^V) initiated by n-buthyllithium. The vinyl groups were modified by hydrosilylation attaching trimethoxysilethyl or triethoxysilethyl pendent groups. The diblock copolymers were used as dispersants for cobalt nanoparticles obtained by thermolysis of $Co_2(CO)_8$. It has been proved that the trialkoxysilethyl-functional blocks served as "anchor" segments for the superparamagnetic cobalt nanoparticles [103].

PDMS containing three or six terminal carboxylic acid groups at one end only (Scheme 21) were synthesized by chemical modification of trivinylsiloxy-terminated precursor with mercptoacetic acid or mercaptosuccinic acid. These were used as dispersion stabilizers for magnetite nanoparticles in PDMS carrier fluids. PDMS-magnetite nanoparticle complexes were formed by interfacial adsorption of the carboxylic acid functional PDMS stabilizers onto magnetite nanoparticles in slightly acidic media. The studies suggested that the binding process involved adsorption of PDMS carboxylate groups onto positively charged magnetite surfaces. The PDMS-magnetite complexes could be dispersed in PDMS carrier fluids, resulting in magnetic PDMS fluids that have potential to be used in biomedical applications [104].

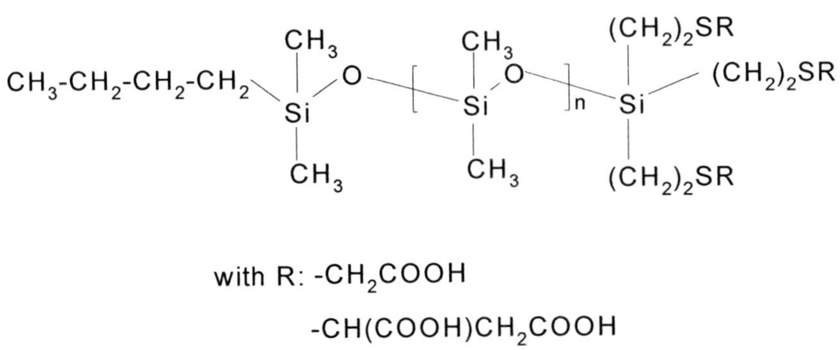

Scheme 21 [104].

Metallomesogenic or metal-containing liquid crystal consisting in a flexible polysiloxane backbone with side copper carboxylate complex mesogene and a flexible aliphatic spacer (Scheme 22) has been synthesized by hydrosilylation reaction between poly(methylhydrosiloxane) with 10-undecanoic acid followed by cooper complexation in presence of sodium stearate or sodium laurate. The applicability of the prepared compounds to the analysis of phtalate esters was evaluated [105].

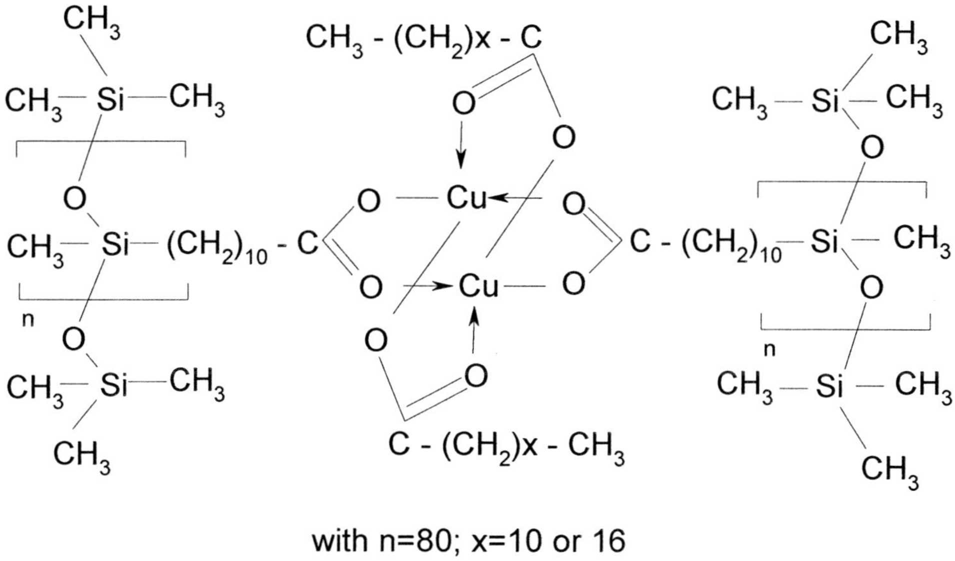

Scheme 22 [105].

Monometallic, dimetallic and polymetallic complexes of oligosiloxane ligands as models of metals on the silica surface were reported. The latter type is of particular interest since these compounds contain assemblies of four to six transition metal atoms in higher oxidation state (e.g. Ni^{II}). Structural studies reveal no direct metal-metal bonding in these complexes [102].

Crown ethers, cyclodextrines and calixarenes are the three generations of host molecules, which possess good complexation ability both with ions and neutral molecules. The application of their derivatives in gas-chromatography (GC) is limited due to the high melting point which leads to difficulties in coating them onto the internal wall of the capillary column, thus resulting in poor column efficiency [106]. A facile method to solve this problem is to attach the calixarene molecules directly onto a polymer substrate, such as polysiloxane. Thus calixacrown ether polysiloxanes, namely pendant hand-basket type calixarene for gas chromatography stationary phases were prepared by hydrosilylation of ω-undecenyl substituted calixacrown ether with dichloromethylsilane followed by condensation with silanol-terminated polydimethylsiloxane. The obtained materials are used as stationary phases of low polarity with good film-forming ability and excellent thermostability [106]. Such type of groups in siloxane porous materials proved also to exhibit preconcentration potential for divalent Cu, Pb, and Hg ions, especially for the trivalent auric ion [107].

Polytetramethyl-*p*-silphenylenesiloxane (I) and polymethylphenylsiloxane (II) were prepared and subsequently reacted with hexacarbonylchromium to generate their n^6-$Cr(CO)_3$ complexed analogs [108].

Soluble tricarbonylchromium complexes were made by reacting $Cr(CO)_6$ with a ladder polyphenylsilsesquioxane and a linear polydiphenylsiloxane. These new polymer $Cr(CO)_3$ complexes were evaluated as stereoselective hydrogenation catalysts [109].

A series of liquid crystalline polysiloxanes having attached groups of azo-type in side position were synthesized by hydrosilylation, starting from a poly(methyl-H-siloxane) and azo-type mesogenic compounds having an end allyl group. The resulted polymers were used to coordinate with palladium dichloride and potassium chloride, by which a series of palladium complexes were prepared (Scheme 23). The mesogenic properties of the liquid crystalline polymer ligands and their palladium complexes were varied gradually by changing the length of the alkoxy groups on the side chain. The polymers have a color emissive group and a highly flexible polysiloxane main chain and may potentially used as nonlinear optical materials [110].

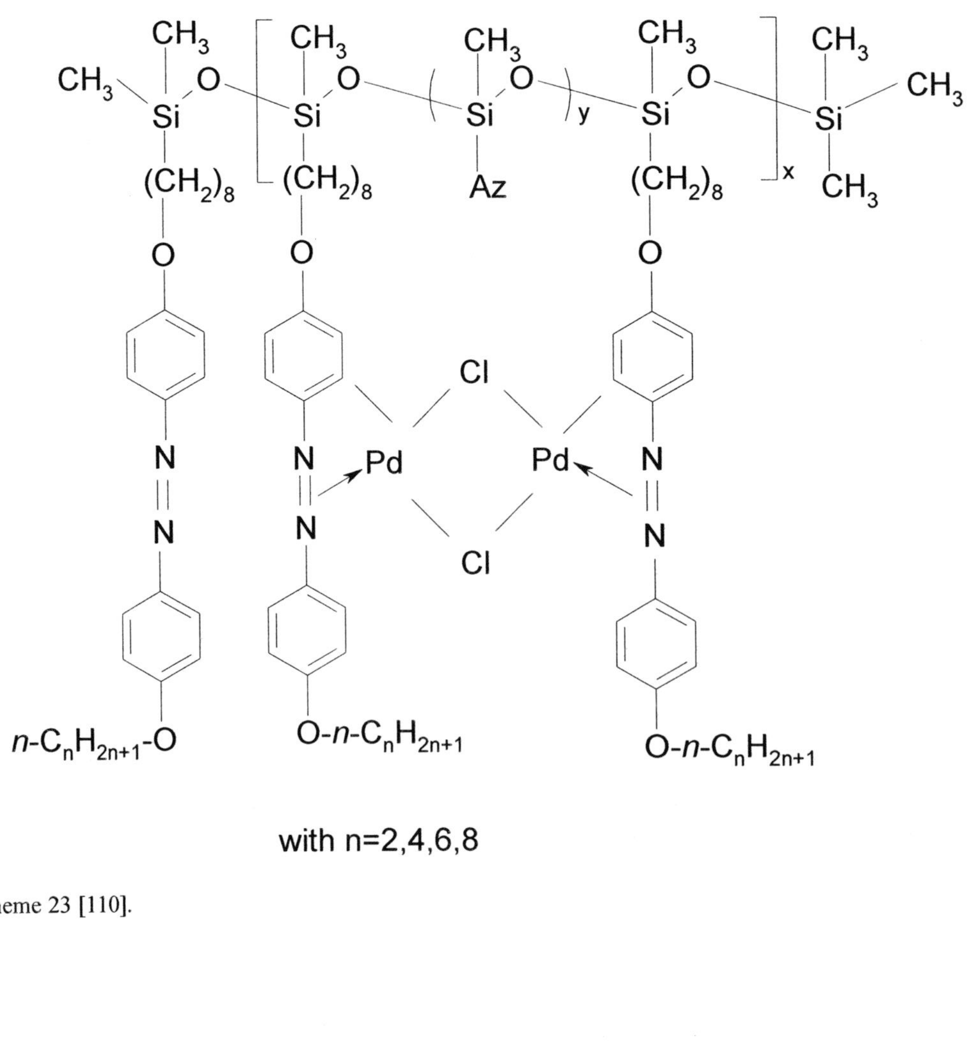

Scheme 23 [110].

3. SILOXANE NETWORKS COMPLEXING METALS

Crosslinked polysiloxanes as porous solid (i.e., silica) carrying a variety of ligand groups on the surface are of high interest due to their capacity to bind metals [18].

There are two major pathways to obtain silica with metal complexing ability:

- by chemical modification of the already prepared silica surface;
- by direct synthesis, using the sol-gel chemistry.

Immobilization of organic functional groups on a siliceous surface is used to produce a variety of modified silica. The silylation process is employed to generate the desired group. The silanol groups on the silica surface partially interact with a silane reagent that will form a covalent bond with the surface. The organofunctional groups grafted on silica can confer a series of properties to the surface, which considerably differ from the original matrix [111]. The ligand groups on the silica surface play an important role in co-ordination chemistry conferring multicoordinating capabilities. An extended list containing organic reagents/organic chelating groups examples immobilized on silica surface in order to its use for metal ions preconcentration are reported in literature [111].

Polysiloxane networks with immobilized pyrogallol ligands (Scheme 24) have been prepared by reacting aminopropyl-functionalized organosilica with first dibromopropane and then pyrogallol. This immobilized ligand system exhibits high potential for extraction of some metal ions (Fe^{3+}, Co^{2+}, Ni^{2+}, and Cu^{2+}) [112].

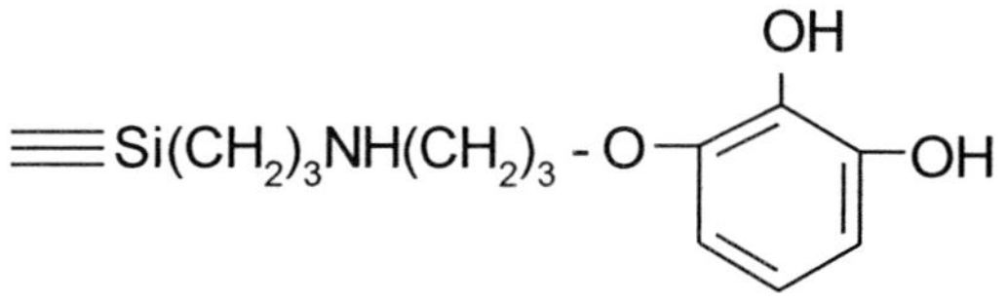

Scheme 24 [112].

The functionalization of the porous solid siloxane polymers with chelating groups based on glycine and iminodiacetic acid (Scheme 25) and a preliminary investigation of their metal uptake ability and using in chromatography has been reported [18].

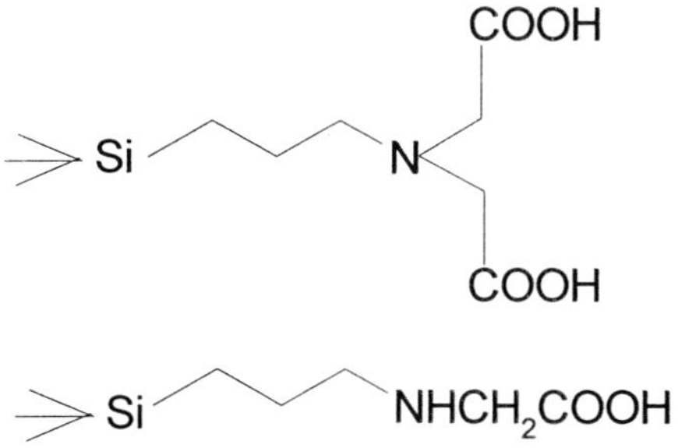

Scheme 25 [18].

Diethyliminidiacetate groups immobilized on the insoluble porous polysiloxane were modified with ethylenediamine [113] or diethylenetriamine [114] (Scheme 26) resulting in polysiloxane ligand systems that exhibits potential for the uptake of transition metal ions [101].

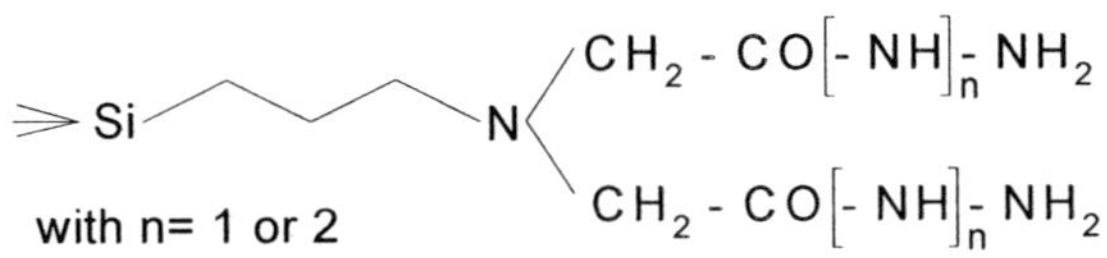

Scheme 26.

Vinyl-functionalized polysiloxanes having different degree of branching, grafted on silica were used as hybrid supports for immobilization of the rhodium and platinum complex catalysts. The catalyst activity was found to be dependent on the type of the parent complex, the degree of polymer branching and the vinyl ligands content [115].

Sol-gel chemistry also offers a convenient method to prepare silica modified with different chemical species such as complexing groups for metals [117]. Organics, organometallics [118], proteins and enzymes are just a few of the types of compounds that have been encapsulated in xerogels [118].

Sol-gels are framework polysiloxanes obtained by hydrolysis of alkoxysilane precursors. In general, tetralkoxysilanes are used as precursors. The addition of a trialkoxyorganofunctional silane in a desired percent permit to obtain a functionalized sol-gel [119].

Thus, by using tetraethylorthosilicate (TEOS) and n-[3(trimethoxysilyl)]-ethylenediamine, a silicon-organic hybrid matrix with pendant organic moiety was synthesized. The hybrid matrix was used as a substrate for the adsorption of chromium and iron from their chlorine aqueous solutions [117].

It has been reported that in situ coordination of metal ions to bifunctional ligands distal silicate ester groups permits the covalent incorporation of metal complexes as dopant species into silica xerogel matrices during conventional sol-gel syntheses. Subsequent calcinations and reduction of these silica xerogels doped with metal-ion complexes resulted in nanocomposites containing metal nanoclusters [120].

The polysiloxane-immobilized triamine ligand system (Scheme 27), has been prepared via the sol-gel process by hydrolytic polycondensation of $(EtO)_4Si$ and $(MeO)_3Si(CH_2)_3NH(CH_2)_2NH(CH_2)_2NH_2$. An alternative method consisted in the reaction of the iodopolysiloxane P-$(CH_2)_3I$ and diethylenetriamine in the presence of triethylamine. The immobilized triamine ligand forms metal chelate complexes when treated with aqueous metal(II) ion solutions (Co^{2+}, Ni^{2+}, and Cu^{2+}). The elemental analysis and FTIR results suggested that this ligand system undergoes a substantial leaching of diethylenetriamine ligand moieties from the siloxane framework [121].

$$P\text{-}(CH_2)_3NH(CH_2)_2NH(CH_2)_2NH_2$$

P - the siloxane network

Scheme 27 [121].

An organic precursor containing oxygen-based chelating groups, the acetylacetonatepropyltrimethoxysilane, was obtained and polycondensed with tetraethylortosilicate (TEOS) to obtain a hybrid polymer containing acetylacetonate groups. These groups are well known ligands in coordination chemistry and are suitable to chelate different transition metals, resulting in heterogeneous transition metal acetylacetonate catalysts [122].

Eu^{3+} or Tb^{3+}-doped organically modified silica xerogels were prepared by the reaction of (3-aminopropyl)triethoxysilane with 3-isocyanatepropyltriethoxysilane followed by the subsequent hydrolysis and condensation in the presence of Eu^{3+} or Tb^{3+} via sol-gel method. The formed xerogel shows a strong blue emission with the maximum excitation and emission wavelength at 351 and 420 nm, respectively [123]. Europium-complex was entrapped into tetramethoxysilane (TMOS)-based sol-gel and methyltrimethoxysilane (Me-TriMOS), phenyltrimethoxysilane (Ph-TriMOS), dimethyldimethoxysilane (DiMe-DiMOS) and diphenyldimethoxysilane (DiPh-DiMOS)-based ormosil (organically modified siloxanes) matrices [124].

Water-swellable chelated polymers were synthesized, starting from In(III) and Sn(II) and *N*-trimethoxysilylpropylethylenediamine with ethylenediaminetetraacetic acid (EDTA) as coordination ligands. *N*-[3-Trimethoxysilyl)propyl]ethylenediamine (TMSPEDA) is a coordination ligand because of its ethylenediamine moiety and aprecursor of the silicone polymers due to its trimethoxysilyl moiety. Due to the double crosslinking, by coordination and condensation polymerization, the polymer precursors became insoluble but swellable in water. This water-swellable polymer containing multi-inorganic ions was used for coating inorganic powder. A fine powder of metal oxide composites (In_2O_3-SiO_2, In_2O_3-SnO_2-SiO_2 or In_2O_3-SnO_2) on In_2O_3 powders is formed as a result of material combustion [34].

Two new chelating materials were synthesized by modifying the activated silica gel phase with Schiff bases of 4,4′-diaminodiphenylether (DDE)/4,4′-diaminodiphenylsulfone (DDS) and *o*-hydroxybenzaldehyde (*o*-HB) and used for the extraction of metal ions such as Zn^{2+}, Mn^{2+} and Cr^{3+} [125].

By adding an aldehyde (glutaraldehyde) to different sol-gel precursors containing amine groups (i.e., 3-(trimethoxysilyl)propylamine, N-[3(trimethoxysilyl)propyl]-ethylenediamine, [3-(trimethoxysilyl)-propyl]ethylenetriamine), hybrids containing Schiff bases are formed and used as copper adsorbant [126].

Porous polysiloxane resins have been prepared by the ring-opening polycondensation of epoxy groups of bisglycidyl calyx[4]arene with amino groups of polysiloxane microparticles that were prepared by the sol-gel technique. The resulted materials exhibited high thermostability and, relative to both the special cavity and the network structure of polysiloxane resin. The nitrogen atoms in resins also exert some auxiliary action [107].

CONCLUSION

Various complex polymer-metal structures containg siloxanes and metals in different position can be prepared. The siloxane presence in such structures confers improved solubility, film forming ability, flexibility, gas permeability, permitting in the same time the development of some properties due to the metal presence: redox, magnetic, electric, optic properties, liquid crystal behavior, increased thermal stability, etc. A new added value due to the co-existence of the siloxanes and metal moieties is also self-assembling ability. There are same parameters by which these properties can be designed: metal type, metal position towards the polysiloxane backbone, siloxane sequence length, chelating group nature. The silicone chemistry is a versatile one permitting large variations of these parameters. Therefore, it is expected a more intense research in this field in the future.

ACKNOWLEDGMENT

The author is grateful to Dr. M. Marcu for helpful discussions and ideas on this topic. This work has been occurred in the frame of Project CEEX-MATNANTECH No. 52/2006.

REFERENCES

[1] Kaliyappan, T.; Kannan, P. *Prog. Polym. Sci.* 2000, 25, 343-370.
[2] Kaminski, W.; Modrzejewska, Z. *Sep. Sci. Technol.* 1997, 32, 2659- 2668.
[3] Kolliar, A.; Morawelz, N. *J. Am. Chem. Soc.* 1955, 77, 3692.
[4] Martin, D. F.; Shamma, M.; Fernelius, W. *J. Am. Chem. Soc.* 1958, 80, 4891-4895.
[5] Korsak, V. V.; Krongaza, E. S.; Cladkov, A. M.; Scheipa, V. E.; Lypeva, L. K. *Visokomol. Soedin.* 1959, 1, 1764.
[6] Comba P, Hambley T W. Molecular Modeling of Inorganic Compounds, *VCH,* Weinheim, 1995.
[7] Korsak, V. V. Progrese in Chimia Polimerilor, Ed. *Stiintifica,* Bucuresti, 1968.
[8] Kickelbick, G. *Prog. Polym. Sci.* 2003, 28, 83-114.
[9] Voronkov, M. G.; Mileshkevich, V. P.; Yuzhelevski, Yu. A., *The Siloxane Bond - Physical Properties and Chemical Transformations,* New York, London, 1978, 47.
[10] Barry, A. J.; Beck, H. N. Siloxane Polymers. In: *Inorganic Polymers, Stone FGA, Graham WAG (Eds.),* Academic Press: New York, London, 1962; 272.
[11] Noll, W. *Chemistry and Technology of Silicones,* Academic Press: New York-London, 1968.
[12] Rehahn, M. *Acta Polymer.* 1998, 49, 201-224.
[13] Gupta, K. C.; Abdul Kadir, H. K.; Chand, S. *J. Macromol. Sci. Pure Appl. Chem.* 2002, A39, 1451-1474.
[14] Manners, I. Synthetic Metal-Containing Polymers, Willey-VCH Verlag GmbH and Co. KGaA, Weinheim, Germany, 2004, 1-37.
[15] Grubbs, R. B. *J. Polym. Sci. Polym. Chem.* 2005, 43, 4323-4336.
[16] Dobrawa, R.; Wurthner, F. *J. Polym. Sci. Polym. Chem.* 2005, 43, 4981-4995.
[17] Jal, P.K.; Patel, S.; Mishra, B.K. *Talanta.* 2004, 62, 1005–1028.
[18] El-Nasser, A. A.; Parish, R. V. *J. Chem. Soc. Dalton Trans.* 1999, 3463-3466.
[19] Casado, C. M.; Moran, M.; Losada, J.; Cuadrado, I. *Inorg. Chem.* 1995, 34, 1668-1680.
[20] Abd-El-Aziz, A. S. In: Macromolecules Containing Metal and Metal-Like Elements, Vol. 4: Group IVA *Polymers,* Abd-El-Aziz, A. S.; Carraher, C. E.; Pittman, U.C.; Zeldin, M.; Sheats J. E. (Eds.), John Wiley and Sons, 2005, 1-27.

[21] Goodgame, D. M. L.; Lickiss, P. D.; Rooke, S. J.; White, A. J. P.; Williams, D. J. *Inorg. Chim. Act.* 2001, 324, 218-231.
[22] Blagodatskikh, I. V.; Shchegolikhina, O. I.; Larina, T. A; Zhdanov, A. A.; Vasil'ev, V. G. *Vysokomol. Soedin. A* 1996, 38, 1876-1880.
[23] Ophir, R.; Shvo, Y. J. Molec. Catalysis 1999,140, 259-265 and references from herein.
[24] Murugavel, R.; Voight, A.; Walawalkar, M. G.; Roesky, H. W. *Chem. Rev.* 1996, 96, 2205-2236.
[25] Lazell, M.; Motevalli, M.; Shah, S. A. A.; Sullivan, A. C. *J. Chem. Soc. Dalton Trans.* 1997, 3363–3366.
[26] Wolczanski, P. T. *Polyhedron.* 1995, 14, 3335-3362.
[27] Siegel, G. A., Bartlett, R. A., Decker, D., Olmsted, M. M., Power, P. P., *Inorg. Chem.* 26, 1773, 1987.
[28] Kownacki, I.; Kubicki, M.; Marciniec, B. *Polyhedron.* 2001, 20, 3015-3018.
[29] Attar-Bashi, M. T.; Rickard, C. E. F.; Roper, W. R.; Wright, L. J.; Woodgate, S. D. *Organometallics.* 1998, 17, 504-506.
[30] Rattay, M.; Fenske, D.; Jutzi, P. *Organometallics.* 1998, 17, 2930-2932.
[31] Harrod, J. F.; Pelletier, E. *Organometallics.* 1984, 3, 1064-1069.
[32] Zhao, J.; Sakthivel, A.; Santos, A. M.; Kühn, F. E. *Inorg. Chim. Acta.* 2005, 358(14), 4201-4207.
[33] Scott McIndoe, J.; Nicholson, B. K. *J. Organomet. Chem.* 2002, 648, 237-245.
[34] Hong, L.; Ruckenstein, E. *J. App. Polym. Sci.* 1998, 67, 1891–1903.
[35] Sviderskii, V. A.; Voronkov, M. G.; Klimenko, S. V.; Klimenko, V. S.; Skakun, E. V., Russ. *J. Appl. Chem.* 2005, 78 (3), 375-378.
[36] Steiner, A.; Lawson, G. T.; Walfort, B.; Leusser, D.; Stalke, D. *J. Chem. Soc. Dalton Trans.* 2001, 219-221.
[37] Gunji, T.; Kubota, K.; Kishiki S.; Abe, Y. *Bull. Chem. Soc. Jap.* 2002, 75 (3), 537-543.
[38] Makarova, L. I.; Sergienko, N. V.; Trankina, E. S.; Zhdanov, A. A.; Ahmet"eva, E. A. Russ. *Chem. Bull.* 2003, 52 (1), 170-172.
[39] Brook, M. A.; Himmeldirk, R. S. Chelating silicone polymers, US Patent Issued on May 20, 2003, Application No. 579960 filed on 2000-05-26.
[40] Constable, E. C., Cargill Thompson, A. M. W.; Tocher, D. A. *Polym. Prepr.* 1993, 34 (1), 110.
[41] Vancso-Szmercsanyi, I.; Callo, A. *J. Polym. Sci. Polym. Chem.* 1982, 20, 639.
[42] Woehrle, D.; Schnurpfeil, G. In: *Comprehensive Polymer Science,* 2nd Suppl., Aggarwal SL, Russo S (Eds.), Elsevier, Oxford, 1996 p. 577.
[43] Cazacu, M.; Airinei, A.; Marcu, M, *Appl. Organomet. Chem.* 2002,16, 643-648.
[44] Cazacu, M; Vlad, A. Unpublished results
[45] Carraher Jr., C. E. *J. Inorg. Organomet. Polym. Mater.* 2005, 15(1), 121-145.
[46] Vassylyev, O. *Molecules.* 2005, 10, 585-586.
[47] Dombrowski, K. E.; Baldwin, W.; Sheats, J. E. *J. Organomet. Chem.* 1986, 302, 281-306.
[48] Sargent, J. R.; Weber, W. P. *Polymer.* 1999, 40, 3795-3802.
[49] Sargent, J. R.; Weber, W. P. *Polym. Prepr.* 1998, 39 (1), 274.
[50] Manners, I. *Pure Appl. Chem.* 1999, 71, 1471-1476.
[51] Manners, I. J. Polym. Sci. Part A Polym. Chem. 2002, 40, 179-191.

[52] Najafi-Mohajeri, N.; Nelson, G. L.; Benrashid, R. *J. Appl. Polym. Sci.* 2000, 76, 1847-1856.
[53] Ni, Y.; Rulkens, R.; Manners, I. *J. Am. Chem. Soc.* 1996, 118, 4102-4114.
[54] Power-Billard, K. N.; Manners, I. *Macromol. Rapid Comm.* 2002, 23, 607-611.
[55] Alonso, B.; Gonzalez, B.; Garcia, B.; Ramirez-Oliva, E.; Zamora, M.; Casado, C. M.; Cuadrado, I. *J. Organomet. Chem.* 2001, 637-639, 642-652.
[56] Losada, J.; Garcia Armada, M. P.; Cuadrado, I.; Alonso, B.; Gonzalez, B.; Casado, C. M.; Zhang, J. *J. Organomet. Chem.*, 2004, 689, 2799-2807.
[57] Boudjouk, P.; AL-Badri, Z. M. H.; Chauhan, B. P. S. *J. Organomet. Chem.* 2004, 689, 3468-3471.
[58] Garcia Armada, M. P.; Losada, J.; Cuadrado, I.; Alonso, B.; Gonzalez, B.; Casado, C. M. *Electroanal.* 2003, 15, 1109-1114.
[59] Massey, J.; Nicole Power, K.; Manners, I.; Winnik, M. A., *J. Am. Chem. Soc.* 1998, 120, 9533-9540
[60] Resendes, R.; Massey, J. A.; Temple, K.; Cao, L.; Power-Billard, K. N.; Winnik, M. A.; Manners, I. *Chem. Eur. J.* 2001,7, 2414-2424.
[61] Losada, J.; Garcia Armada, M. P.; Cuadrado, I.; Alonso, B.; Gonzalez, B.; Casado, C. M.; Zhang, J. *J. Organomet. Chem.* 2004, 689, 2799-2807.
[62] Pilar Garcia Armada, M.; Losada, J.; Cuadrado, I.; Alonso, B.; Gonzalez, B.; Ramirez-Oliva, E.; Casado, C. M. Sensor. *Actuat. B Chem.* 2003, 88, 190-197.
[63] Horacek, M.; Stepnicka, P.; Kubista, J.; Gyepes, R.; Cisarova, I.; Petrusova, L.; Mach, K. . 2002, 658, 235-241.
[64] Wang, X. S.; Winnik, M. A.; Manners, I. *Macromolecules.* 2002, 35, 9146-9159.
[65] Boudjouk, P.; AL-Badri, Z. M. H.; Chauhan, B. P. S., *J. Organomet. Chem.* 2004, 689, 3468-3471.
[66] Patterson, W. J.; McManus, S. P.; Pittman Jr., C. U. *J. Polym. Sci Polym. Chem. Ed.* 1974, 12, 837-850.
[67] Casado, C. M.; Moran, M.; Losada, J., Cuadrado, I. *Inorg. Chem.* 1995, 34, 1668-1680.
[68] Cazacu, M.; Vlad, A. ; Marcu, M.; Racles, C.; Airinei, A. ; Munteanu, G. *Macromolecules.* 2006, 39, 3786-3793.
[69] Cazacu, M.; Munteanu, G.; Racles, C.; Vlad, A.; Marcu, M., *J. Organomet. Chem.*, 2006, 691 (17), 3700-3707.
[70] Yang, Z.; Lu, J.-mei; Wang, L.-hua *Polym. Bull.* 2005, 53, 249-257.
[71] Wang, X. F.; Pan, Y. Z.; Liang, Y. Q. *Langmuir.* 2000, 16, 7538-7540.
[72] Luo, X. Z.; Wang, X. F.; Wu, S. X.; Liang, Y. Q. J. *Colloid Interface Sci.* 2003, 258, 432-434.
[73] Sukwattanasinitt, M.; Natalaksakul, A.; Potisatityuenyong, A.; Tuntulani, T.; Chailapkul, O.; Praphairakait, N. *Chem. Mater.* 2003, 15 (23), 4337-4339.
[74] Kaliyapan, T.; Rajagopan, S.; Kannan, P. *J. Appl. Polym. Sci.* 2004, 91, 494-500.
[75] Marcu, M.; Cazacu, M.;Vlad, A.; Racles, C. *Appl. Organomet. Chem.* 2003, 17, 693-700.
[76] Cazacu, M.; Marcu, M.; Vlad, A.; Rusu, Gh. I.; Avadanei, M. *J. Organomet. Chem.* 2004, 689 (19), 3005-3011.
[77] Vasiliu, M.; Cazacu, M. ; Marcu, M. ; Racles, C. ; Vlad A. *Appl. Organomet. Chem.* 2005,19, 614-620.
[78] Cazacu, M.; Marcu, M.; Vlad, A.; Vasiliu, M. *J. Macromol. Sci.* 2004, A41, 5, 565-575.

[79] Spiratos, M. ; Airinei, A. ; Ciobanu, A. *Rev. Roum. Chim.* 1980, 25, 1083-1089.
[80] Marcu, M.; Lazarescu, S. E.; Grigoriu, G. E. *Mater. Plast.* 1989, 26, 196-198.
[81] Hyvärinen, K.; Klinga, M.; Leskelä, M. *Polyhedron.* 1996, 15(13), 2171-2177.
[82] Shin, K.; Aoki, T.; Okawa, E. *Macromol. Rapid Comm.* 1995, 16, 599-606.
[83] Aoki, T.; Shin, K.; Oikawa, E. *Polym. J.* 1996, 28, 106-112.
[84] Aoki, T, *Prog. Polym. Sci.* 1999, 24, 951-993.
[85] Moriguchi, N. ; Tsugaru, T. ; Amiya, S. *J. Molec. Struct.* 2000, 523, 93-101.
[86] Moriguchi, N. ; Tsugaru, T. ; Amiya, S. *J. Molec. Struct.* 1999, 477, 181-189.
[87] Noss, H.; Oberthur, M.; Fischer, C.; Kretschmer, W. P.; Kempe, R. *Eur. J. Chem.* 1999, 2283-2288.
[88] Gomez, E.; Hernandez, Z.; Alvarez-Toledano, C.; Toscano, R. A.; Santes, V.; Sharma, P. *J. Organomet. Chem.* 2002, 648, 280-287.
[89] Nemoto, N.; Yotsuya, T.; Aoyagi, S.; Ueno, Y.; Ikeda, K.; Takamiya, N., *Polymer Comm.* 1990, 31, 65-67.
[90] Samal, S.; Das, R. R.; Sahoo, D.; Acharya, S.; Panda R. L.; Rout, R. C., *J. Appl. Polym. Sci.,* 1996, 62, 1437-1444.
[91] Samal, S.; Das, R. R.; Dey R. K.; Acharya, S. J. *Appl. Polym. Sci.* 2000, 77, 967-981.
[92] Samal, S.; Mohapatra N. K.; Acharya, S. *React. Funct. Polym.* 1999, 42, 37-52.
[93] Rusu, G. I., *Ann. Sci. Univ. Iassy Cond. Matt. Phys.* 1999-2000, 45/46, 229-239.
[94] Kaya I.; Koyuncu, S. *Polymer.* 2003, 44, 7299-7309.
[95] Belwal, S.; Singh, R. V. *Appl. Organomet. Chem.* 1998, 12, 39-46.
[96] Cazacu, M.; Marcu, M.; Vlad, A.; Toth, A.; Racles, C. *J. Polym. Sci. Polym. Chem.* 2003, 41(20), 3169-3179.
[97] Cazacu, M.; Vlad, A.; Munteanu, G.; Airinei, A.; Marcu, M., *Chem. Mater.*-submited, 2007
[98] Cazacu, M.; Vlad, A.; Racles, C., Unpublished results.
[99] Fereday, A.; Goodgame, D. M. L.; Lickiss, P. D.; Rooke, S. J.; White, A. J. P.; Williams, D. *J. Inorg. Chem. Commun.* 2002, 5, 805-807.
[100] Wu, Q.; Lee, M. L.; Harrison, R. G. *J. Chromatogr.* A 2002, 954, 247-258.
[101] Maciejewski, H.; Sydor, A.; Kubicki, M. *J. Organomet. Chem.* 2004, 689, 3075-3081, and references from herein.
[102] Gavioli, G.; Borsari, M.; Zucchi, C.; Palyi, G.; Psaro, R.; Ugo, R.; Shchegolikhina, O. I.; Zhadanov, A. A. *J. Organomet. Chem.* 1994, 467, 165-167.
[103] Vadala, S. D.; Rutnakornpituk, M. L.; Zalich, M. A.; St Piere, T. G.; Riffle, J. S. *Polymer.* 2004, 45, 7449-7461.
[104] Wilson, K. S.; Goff, J. D.; Riffle, J. S.; Harris, L. A.; Pierre, T. G. *Polym. Adv. Technol.* 2005, 16, 200-211.
[105] Liu, C.-Y.; Yang, S.-H.; Chau, M.-H.; Shiue, C.-C. J. *Chromatogr.* A 2001, 933, 117-128.
[106] Zhang, L. F.; Chen, L.; Lu, X. R.; Wu, C. Y.; Chen, Y. Y. *J. Chromatography.* A 1999, 840, 225-233.
[107] Li, C.; Gong, S. L.; Meng, L. Z.; Hu, L.; Huang, H.; He, Y. B. *J. Appl. Polym. Sci.* 2005, 95, 1310-1318.
[108] Pittman Jr., C. U.; Patterson , W. J.; McManus, S. P. *J. Polym. Sci. Polym. Chem. Ed.* 1975, 13(1), 39 – 47.

[109] Awl, R. A.; Frankel, E. N.; Friedrich, J. P.; Swanson, C. L., *J. Polym. Sci. Polym. Chem. Ed.* 1980, 18(8), 2663 – 2676.

[110] Zhang, S. Y.; Li, Z. F.; Cao, S. K.; Zhang, R. B. *Polym. Adv. Technol.* 1996, 7, 715-718.

[111] Jal, P.K.; Patel, S.; Mishra, B.K. *Talanta.* 2004, 62, 1005–1028.

[112] Saadeh, S. M. ; El-Ashgar, N. M. ; El-Nahhal, I. M. ; Chehimi, M. M. ; Maquet, J.; Babonneau, F. *Appl. Organomet. Chem.* 2005, 19, 759-767.

[113] El-Nahhal, I. M. ; El-Ashagar, N. M. ; Chehimi, M. M. ; Bargiela, P. ; Maquet, J. ; Babonneau, F. ; Livage, *J. Micropor. Mesopor. Mat.* 2003, 65, 299-310.

[114] El-Nahhal, I. M.; El-Shetary, B. A.; El-Khair A.; Mustafa, B.; El-Ashagar, N. M.; Livage, J.; Chehimi, M. M.; Roberts, A., *Solid State Sci.,* 2003, 5, 1395-1406.

[115] Michalska, Z. M.; Rogalski, L.; Rozga-Wijas, K.; Chojnowski, J.; Fortuniak, W.; Scibiorek, M. *J. Mol. Catal A Chem.* 2004, 208, 187-194.

[116] de Farias, R. F.; de Souza, A. G.; Nunes, L. M.; de A. Cardoso, V. *Coll. Surf. A Physicochem. Eng. Aspects.* 2002, 211 (2-3), 295-298.

[117] Lev, O.; Tsionsky, M., Rabinovich, L. Glezer, V.; Sampath, S.; Pankratov, I.; Gun, J. *Anal. Chem.* 1995, 67, 22A-30A.

[118] Rao, M. S.; Dubenko, I. S.; Roy, S.; Ali, N.; Dave, B. C. *J. Am. Chem. Soc.* 2001, 123, 1511-1512.

[119] Breitscheidel, B.; Zieder, J.; Schubert, U. *Chem. Mater.* 1993, 3, 559-566.

[120] El-Nahhal, I. M.; El-Shetary, B. A.; Salib, K. A. R.; El-Ashgar, N. M.; El-Hashash, A. M., Phosphorus. Sulfur, *Silicon Rel. Elem.* 2002, 177(3), 741-753.

[121] Brasil, M. C.; Benvenutti, E. V.; Gregorio, J. R.; Gerbase, A. E. *React. Funct. Polym.* 2005, 63, 135-141.

[122] Han, Y.; Lin, J. *J. Solid State Chem.* 2003, 171 (1-2), 396-400.

[123] Lobnik, A.;, Majcen, N.; Niederreiter, K.; Uray, G., *Sensor Actuat. B Chem.* 2001, 74, 200-206.

[124] Dey, R. K.; Jha, U.; Singh, A. C.; Samal, S.; Ray, A. R. *Anal. Sci.* 2006, 22 (8), 1105-1110.

[125] Airoldi, C.; Monteiro Jr., O. A. C. *Polym. Bull.* 2003, 50, 61-68.

INDEX

A

B

C

D

E

F

G

H

I

J

K

L

M

N

O

P

Q

R

S

T

U

V

W

X

Y

Z